Microbiology in
Civil Engineering

FEMS Symposium No. 59

Microbiology in Civil Engineering

Proceedings of the Federation of European
Microbiological Societies Symposium held at
Cranfield Institute of Technology, UK

3–5 September, 1990

Edited by

P. HOWSAM

Silsoe College
Cranfield Institute of Technology
UK

CRC Press
Taylor & Francis Group
Boca Raton   London   New York

CRC Press is an imprint of the
Taylor & Francis Group, an **informa** business

A TAYLOR & FRANCIS BOOK

CRC Press
Taylor & Francis Group
6000 Broken Sound Parkway NW, Suite 300
Boca Raton, FL 33487-2742

First issued in paperback 2019

ISBN-13: 978-0-419-16730-3 (hbk)
ISBN-13: 978-0-367-86341-8 (pbk)
ISBN-13: 978-0-442-31277-0 (USA)

British Library Cataloguing in Publication Data

Federation of European Microbiological Societies Symposium
  (*Cranfield, England: 1990*)
  Microbiology in civil engineering.
  1. Civil engineering. Microbiological aspects
  I. Title   II. Howsam, P.   III. Series
  624

Library of Congress Cataloging-in-Publication Data

Available

Visit the Taylor & Francis Web site at
http://www.taylorandfrancis.com

and the CRC Press Web site at
http://www.crcpress.com

Contents

Preface

In recent years understanding of processes such as biodeterioration and biofouling, in fields usually outside civil engineering, has increased tremendously. Yet these processes, involving the activity of bacteria and other microorganisms are of relevance to all civil engineers in the full range of the artificial systems and natural environments which they engineer. Examples of where microbial processes are now known to have a significant affect include: corrosion; deterioration of concrete and many other construction materials; impairment of flow in groundwater abstraction systems/in pipes and pumps/in pressure relief and drainage systems in embankment dams; clogging of irrigation and drainage systems; alteration of soil/rock properties, production of harmful gases.

For a long time many of these 'buried' problems have been thought of, purely in terms of physical and chemical processes. What is often happening, but going unnoticed, is some form of microbial activity, which initiates, enhances or is encouraged by, other processes. When it is considered that the average bacteria cell is only a few microns in size, then it is hardly surprising if their activity does often go unnoticed.

At present therefore, the problems are as much the lack of awareness and therefore lack of diagnosis and quantification, as are any adverse effects on structure or operation. Obvious cases, say of corrosion or clogging have of course been identified in the past, but the applied remedies will only have related to the perceived chemical/physical causes, thus often allowing reoccurrence of the problem. In the case of less obvious manifestations, the concern is that long term damage and/or reductions in efficiency are gradually occurring unnoticed over periods of time.

Civil engineering constantly involves the interface of artificial structures with a wide diversity of subsurface soil-water environments, where, and engineers need perhaps to be reminded of this, microorganisms are omnipresent. For example, that nice fresh earthy smell you notice when digging the garden or moving top soil about on site, is due to the presence of a particular bacteria group called *Actinomyces*. Microorganisms are able to adapt to wide ranges of pH, temperature, oxygen and nutrient (i.e. C, N, P. S) supply, even to incredible extremes. There is evidence to show for example that biofilms (microbial growths on surfaces) can exist, indeed thrive, in circumstances where turbulence and high velocities/pressures occur. Again as an example, the plaque that develops on your teeth and which needs vigorous brushing to remove properly, is an example of a biofilm.

This conference publication serves to highlight a wide variety of areas of civil engineering, where microbiological activity needs to be considered more thoroughly by engineers during the design, construction and operation/use phases of projects. While in many areas there is much still to be understood, increased awareness and consequently improved engineering could be achieved quite quickly if communication between civil engineers and microbiologists were to increase.

Peter Howsam
May 1990

Acknowledgements

This publication owes its existence to all the contributors.

The conference and therefore this publication would not have happened without the sponsorship and support of:

The Federation of European Microbiological Societies (FEMS)

and the support in different ways of the following organisations:

Construction Industry Research & Information Association (CIRIA)
European Federation of Corrosion (EFC)
Institution of Civil Engineers (ICE)
Institution of Corrosion Science & Technology (ICorrST)
International Biodeterioration Research Group (IBRG)
Mott MacDonald Group Ltd
National Physical Laboratory (National Corrosion Service)
Rofe Kennard & Lapworth
Scott Wilson Kirkpatrick & Partners
Society of Chemical Industries (Preservation Group) (SCI)
Southern Water

Finally members of the conference technical steering committee helped to make it all happen by providing much useful guidance and by performing numerous tasks with great care and enthusiasm.

The technical steering committee:

Roy Cullimore, Regina Water Research Institute, Canada
Robin Egerton, Consultant, (RKL), UK
Brian Hawkins, Bristol University, UK   (Committee Chairman)
Paul Hillman, Southampton University, UK
Fin Jardine, CIRIA, UK
Carolyn King, Silsoe College, UK   (Conference Secretary)
Ken Seal, Biotechnology Unit, Cranfield (CIT), UK
Ken Tiller, National Corrosion Service, (NPL), UK
Sean Tyrrel, Silsoe College, Cranfield (CIT), UK

My many thanks to all of them.

Peter Howsam
Silsoe College

PART ONE
INTRODUCTION

1 MICROBES IN CIVIL ENGINEERING ENVIRONMENTS: INTRODUCTION

D.R. CULLIMORE
Regina Water Research Institute, University of Regina, Canada

Abstract
Relatively few direct relationships have been established between the
activities of the indigenous micro-organisms and the structure and
processes designed and managed by civil engineers. Traditionally, the
main focus of this interface has been the generation of biologically
induced corrosive processes which has had serious economic costs
measured in the billions of dollars for the oil industry alone. In
the last decade, the ubiquitous and dynamic natures of the intrinsic
microbial activities are now being recognized as inherent factors in
events such as plugging, remediation of pollution events, hydro-
phobicity in soils and the bioaccumulation of potentially hazardous
chemicals. Some of these events can become manageable once the key
controlling factors become understood. A survey of the present
understanding of these interfaces is discussed.
Keywords:Corrosion, Plugging, Clogging, Water Wells, Iron Related
Bacteria, Bioremediation, Microbiological Process Management.

1 Introduction

It is intriguing to notice that this conference is devoted to the
impact that the activities of the smallest organisms can have on the
activities of what is probably the most sophisticated species, Homo
sapiens. In recent millennia, man has learned to construct and devise
many sophisticated systems which not only provide protection, safety
and transportation but also allow communication. In today's society,
civil engineering has created many advances which have.facilitated
both the intellectual and material growth of society. So rapid has
been these developments, particularly in the last century, that the
interaction between these systems and structures with other species
has received but casual secondary consideration. For the surface
dwelling plant and animal species, desirable strains are often
landscaped into or allowed to flourish within these structures of man
whilst others are excluded. Today, the recognition of the impact of
these developments on the variety of the animal and plant species
present on the planet is becoming acutely appreciated. Such concerns
raise the philosophical question that in the future protection of the
earth's environment as a place for the survival and proliferation of

the human species, the whole nature of the surface of the planet is going to become subjected to an engineered management structure. Such a challenge falls primarily to the professional engineers, and more specifically the civil engineers.

In recognizing the need to "engineer" the surface gaseous, liquid and solid strata of this planet, there will have to be generated a greater understanding of the natural (physical, chemical and biological) processes which allow these planet to function in a relatively stable and yet dynamic manner. To undertake such a management means that a clear appreciation has to be generated to understand the role of the biological components as contributors to the dynamics of this planet. As a surface dwelling species, it is relatively convenient for Homo sapiens to comprehend the role of the plants and animals since these are, by and large, either surface dwelling or flourish in the surface-waters around the planet. Perhaps not so recognized at this time are the role of the micro-organisms. These are insidious organisms of relatively small dimensions and simple abilities which are almost totally ubiquitous within the soils, surface- and ground-waters and even upon and within many of the living organisms which populate this planet. An example of the ubiquitous nature of micro-organisms can be found in the fact that 90% of the cells in the human body are, in fact, microbial, only 10% are the tissue cells which make up the human being as such. This is possibly symbolic of the role of micro-organisms throughout the surface structures of this planet. They are small, insidious, numerous and also biochemically very active.

This conference provides one of the first forums for truly examining but some of the aspects in which micro-organisms can interact with the structures and processes of engineering.

At this time, the major impact of micro-organisms in engineering falls within the realm of facilitating the disposal of waste products generated by human society and interference with some of the processes generated by engineering. At this conference, the maximum attention will be paid to the interface between micro-organisms and the efficiency and operation of particular engineering processes. The recovery and transmission of water and oil from natural reservoirs to the consumer both form excellent examples of this interaction. For water exploitation, a major problem which is becoming acutely recognized is the impact that the indigenous microbial flora can have upon the effective and reliable delivery of the water to the consumer. Major concerns arise from the generation of health risks, biofouling of water sources, treatment plants and transmission systems, and corrosion to the equipment and lines used for that purpose. In the oil industry, the primary concern is with the risk of corrosion within the transmission systems.

The reason for these events will become clear during this conference as examples of the growth of micro-organisms within specific processes and natural systems are discussed. In the last decade, the dedicated work of a relatively small number of applied and ecologically biased microbiologists has allowed a much clearer understanding of the mechanisms by which these events occur. Perhaps the major development has been the development of an understanding of the growth of different species of micro-organisms within common consortial structures bound by polymeric matrices to form biofilms which are attached to inert surface structures. These biofilms, more commonly known as slimes, create the catalyst which will then generate such events as corrosion (through the generation of hydrogen sulphide and organic acids), bioaccumulation of selected inorganic and organic chemicals travelling generally in an aqueous matrix over the biofilm. Biodegradation may also occur and, once the biofilm has enlarged sufficiently, a total plugging of a porous medium. These latter events hold a tremendous potential for the control of the many hazardous wastes which have inadvertently entered the surface- and groundwaters, soils and rock strata on this planet. By coming to understand the methods by which these microbial activities can be utilized, it will become possible to include these manipulations of microbial processes as an integral part of civil engineering practices. As more and more energy is devoted to obtain greater understanding of these significant microbial processes, undoubtedly, it will be discovered that the activities of micro-organisms on this planet can be utilized and engineered to provide a more wholesome environment for the survival of as many species as possible into the foreseeable future.

2 Definition of scope

Microbiology involves the development of an understanding concerning the many and various mechanisms involved in the maintenance of flourishing populations of the various predominantly single celled and relatively undifferentiated organisms on this planet. Civil engineering relates to the design, construction and maintenance of public works of various types which are, at least in part, funded from the public purse. At first glance, any link between microbiology and civil engineering would appear to be at best tenuous. This observation is perhaps magnified by the fact that civil engineers often create very large and impressive structures which would appear to be "cast in stone" to last until eternity. However, these structures and the processes integrated into the designs are vulnerable to environmental constraints with respect to climate, soil types, location, etc. One major factor now becoming recognised as being significant to the durability and ongoing efficiency of such engineered systems, relates to these microbially driven impacts.

Perhaps it is the insidious and ubiquitous nature of the microbial incursions into the functionality of engineered systems which has precipitated a failure to recognise the need to comprehend the dynamics of these microbial events as being significant. The range of events which has been recognised by microbiologists as having some potential significance to civil engineering is gradually growing. There now is developing a more acute need to understand not only the diversity of microbially initiated events, but also to the environmental factors which can be utilized to effectively manage the process to the benefit of the engineered systems. Before this can be achieved, it is therefore very necessary to generate a matured understanding of the microbiological principles involved.

3 Potential microbial impacts

There is a wide range of microbial events which can be recognised as having a potential or real importance to civil engineering. A summarized list of these include corrosion initiation, bioimpedance of hydraulic or gaseous flows (e.g., plugging, clogging), bioaccumulation (e.g., localised concentration of heavy metals, hydrocarbons and/or radionucleides within a groundwater system), biodegradation (e.g., catabolism of potentially harmful organic compounds), biogenesis of gas (including methane, hydrogen, carbon dioxide, nitrogen which can lead to such events as the fracturing of clays, displacement of water tables and differential movement of soil particles), water retention within biofilms (which, in turn, can influence the rates of desiccation and freezing of soils) as well as the more understood roles in the remediation of high oxygen demand organic effluents prior to acceptable disposal into the environment.

These events occur naturally within the biosphere as microbiologically driven functions within and upon the crust of the planet. As civil engineering responds to the ultimate challenge of designing, building and maintaining systems to control the global environment (the greatest public works project ever), a detailed understanding of the inherent microbially driven systems will need to be addressed. At this time, the focus of attention has been directed to the use of specific strains of micro-organisms which have been selected or manipulated to undertake very specific biochemical functions under controlled conditions. While the biotechnological industry has achieved some successes in producing useful products, it has failed to recognise the potential for manipulating the natural microflora to achieve a "real world" conclusion (such as the bioremediation of an eco-niche impacted upon by nuisance or potentially toxic chemicals). There would appear to be an urgent need to facilitate a bridge of understanding between applied microbial ecologists and civil engineers in order to allow more efficient structures and processes to be developed.

4 Functional forms of microbial growth

Micro-organisms are influenced by a range of physical and chemical constraints which are unique to each strain. Generalizations can therefore be made only for major groups of bacteria, but natural variations will always occur. Before these variables can be considered in depth, the form of microbial growth has to be considered. A popular conception of micro-organisms is that these are cells usually dispersed in water and many may be able to move (i.e., be motile). In actuality, microbial cells can commonly be found in three states. These are: planktonic (dispersed in the aqueous phase), sessile (attached within a biofilm to a solid usually immobile surface), and sessile particulate (incumbent within a common suspended particle shared with other cells).

In soils and waters, the vast majority of the microbial cells are recovered from the sessile phases rather than the planktonic. Sessile micro-organisms are normally found within a biofilm (Costerton and Lappin-Scott, 1989) created by the cells excreting extra-cellular polymeric substances (ECPS). These polymers act in a number of ways to protect the incumbent cells. This action includes the retention of bonded water, accumulation of both nutrients and potential toxicants, providing a structural integrity which may include gas vacuoles, separating the incumbent cells from each other and potential toxicants and, upon shearing from the biofilm, afford protection during the process of migration as suspended particulates prior to the reattachment and colonization of a pristine (eco-niche) surface.

5 Potential mechanisms for controlling microbial processes

In recognizing that there will be a need to exercise control over the rate of a selected microbial activity within a defined environment, it will become essential to build into that control all of the major factors of influence. These can primarily be subdivided into four major groups which are: (1) physical, (2) chemical, (3) biological, and (4) structural. Each group not only very significantly reacts with the other groups but also includes a series of significant subgroups each of which can also generate a dominant function on occasions. For the individual microbial strain or consortium of strains, there are a set of environmental conditions relating to these four groups that have allowed this particular definable eco-niche to become established. Such microbial entities are, however, transient in both the qualitative and quantitative aspects due to the dynamic and competitive nature of the incumbents. In chronological terms, there is a very limited understanding of the maturation rate of these natural consortia. For example, what is the length of time that an active microbial biomass can remain integrated within an eco-niche positioned within the totally occlusive zone around a plugged water well? This has not been addressed. Many consortia of fungi and algae form very specific forms of growth as surface dwelling lichens growing at rates as slow as 0.1 mm per year often in an environment (i.e., the surface of a rock) in which nutrients are scarce and water rarely available. Here, the microbial activity is at a level where maturation can be measured in decades. This illustrates a condition in which a micro-organism has adapted to an apparent hostile environment and formed an eco-niche within which growth of a consortial nature can flourish for centuries.

6 Physical factors for microbial manipulations

Of the physical factors known to impact on microbial processes, the most commonly dominant factors are temperature, pH, redox and water potential. Functionally, microbial processes appear to require the presence of a temperature and pressure regime which would allow water to occur in the liquid form. Low temperatures cause freezing in which ice crystals form a matrix around any particulates containing micro-organisms and polymerics. Water may then be expressed from the liquid particulate to the solid ice shrinking the particulate volume until a balance is achieved between the bound liquid water and the ice. Such low temperatures inhibit biochemical activities causing a state of suspended animation to ensue. At higher temperatures, water can remain in a liquid phase under increasing pressures up to at least $374^{o}C$ rendering it a theoretical possibility that microbial activities could occur in these extreme environments.

Normal surface dwelling micro-organisms, however, generally function over a relatively limited temperature range of 10° to $40^{o}C$ with the optimal activities occurring at a point commonly skewed from the mid-point of the operational range. These optima are different for each strain of bacteria to some extent and therefore temperature changes can cause shifts in the consortial structure of micro-organisms.

Optimal and functional ranges of activity also exist when the micro-organisms interact with the environmental pH regime. Most microbes function most effectively at neutral or slightly alkaline pH values with optima commonly occurring at between 7.4 and 8.6. Acidoduric organisms are able to tolerate much lower pH values as low as 1.5 to 2.0 pH units but are generally found in specialised habitats. Where biofilms have been generated within an environment, the pH may be buffered by the polymeric matrices to allow the incumbent micro-organisms to survive and function within an acceptable pH range which is maintained by the consortial activities.

Redox has a major controlling impact on micro-organisms through the oxidation-reduction state of the environment it reflects. Generally, an oxidation state (+Eh values) will be supportive of aerobic microbial activities while a reduction state (-Eh) will encourage anaerobic functions which, in general, may produce a slower rate of biomass generation, a downward shift in pH where organic acids are produced, and greater production potential for gas generation (e.g., methane, hydrogen sulphide, nitrogen, hydrogen, carbon dioxide). Aerobic microbial activities forming sessile growths are often noted to occur most extensively over the transitional redox fringe from 0 to +150 Eh.

7 Chemical factors for microbial manipulations

Chemical factors influencing microbial growth can be simply differentiated into two major groups based upon whether the chemical can be stimulative or inhibitory to the activities within the biomass. The stimulative chemicals would in general perform functional nutritional roles over an optimal concentration range when the ratio of the inherent nutritional elements (e.g., carbon, nitrogen, phosphorus) are in an acceptable range to facilitate biosynthetic functions. These concentrations and ratios vary considerably between microbial strains and form one of the key selective factors causing shifts in activity levels within a biomass. In the management of a biological activity, it is essential to comprehend the range and inter-relationships of any applied nutrients in order to maximise the biological event that is desired.

Inhibitory chemicals will generate some negative impact on the activities of the biomass. These can take the form of a general or selective toxicity. For example, heavy metals tend to have a general form of toxicity while an antibiotic generated by one member of a consortium may be selectively toxic to some transient or undesirable microbial vectors. The biofilm itself may tend to act as an accumulator for many of these potentially toxic elements which may be concentrated within the polymeric matrices of the glycocalyx. When concentrated at these sites, these toxicants form a barrier to predation by non-consortial vectors. At the same time, these accumulates are presumable divorced from direct contact with the incumbent microorganisms.

8 Biological factors in microbiological manipulations

In recent science, there has been generated a fundamental shift in conceptual mechanisms by which a biological vector can be used to directly control the activities of a targeted group of organisms. Traditionally, the concept of control focused on the deliberate inoculation of a predator (to feed upon) or pathogen (to infect with disease) into the eco-niche. Many of these attempts have met with failure or lack of statistical validation from the limited successes due to the natural defence mechanisms (e.g., extra-cellular polymeric substances) and variable composition of the consortium itself.

In the last decade, the impressive advances in the manipulation of the genetic materials particularly in micro-organisms has generated a new concept. This concept involves the inoculation of a genetically modified organism which has been demonstrated, in the laboratory setting, to have superior abilities (usually to degrade a specific nuisance chemical) than comparable microbial isolates taken from the environment. The promise from these biotechnologies is that the superior function can be transplanted into the natural environment and achieve similar equally superior results to that observed in the laboratory. Such concepts still have to address the ability of the inoculated genetically modified organism to effectively compete within the individual eco-niches (some of which may be as small as 50 microns in diameter) with the incumbent microflora and still perform the desired role effectively.

9 Structural factors influencing microbial manipulations

Another major factor influencing any attempt at microbial manipulation within a natural environment is the impedances created by the physical structures between the manipulator and the targeted eco-niche. These structures also influence the environmental characteristics of that habitat. Such manipulations may be relatively easy to achieve where the target activity is either at a visible solid or liquid surface/air interface or within a surface water (preferably non-flowing). Subsurface interfaces involve a range of impedances not only to the manipulator but also to the incumbent organisms.

In order to manipulate a subsurface environment, there are two potential strategies. The first is physical intercedence where a pathway is created to the targeted site by drilling, excavation, fracturing or solvent-removal processes. Here, there will be some level of contamination of the target zone with materials and organisms displaced by the intercedence process. As an alternate method, the second strategy involves a diffusive intercedence where the manipulation is performed remotely and gradually, by diffusion or conductance, moves towards the targeted zone. Such techniques may involve the application of gases (e.g., air, methane), solutions (containing optimal configurations of the desired chemicals to stimulate the event through modifying the environment), suspensions of micro-organisms in either vegetative, sporulated or suspended animation states (such as the ultramicro bacteria), thermal gradients (through the application of heat or refrigeration) or barriers of some type of permeability that will cause re-direction or impedance of flows in such a way as to induce the desired effect.

Diffusive intercedence can offer simpler and less expensive protocols but may be more vulnerable to the effects that any indigenous microflora may have upon the process. These effects could range from the direct assimilation of the nutrients, modification to the physical movement of the materials being entranced, to the releases of metabolic products which may amend the desired effect.

10 Conclusions

Engineering structures and processes in an environment free from micro-organisms may be relatively easy to design and simpler to manage due to the absence of these vectors. On this planet, the microbial kingdom forms a diverse group of species which occupy a wide variety of environments. When engineering projects are undertaken, these indigenous organisms respond in ways that are often subtle and go undetected. The demands on civil engineering are increasing to produce structures and processes which are more durable, better controlled and limit the potential impacts on the general environment. These changes will involve a better understanding of the effects of the global microflora on engineered systems. This process is just beginning.

11 Acknowledgements

The author wishes to thank the Natural Sciences and Engineering Research Council of Canada (grant in aid #OGP0005073), the National Research Council of Canada Industrial Research Assistance Programs, the Saskatchewan Research Council, the U.S. Corps of Engineers, the Province of New Brunswick, Ontario Ministry of the Environment, and the towns of Bulyea (Saskatchewan), Rothesay and Newcastle (New Brunswick) for financial assistance and placing facilities at the disposal of researchers associated with the various projects. Also, the help, advice and criticisms of George Alford, Peter Howsam, Neil

Mansuy, Stu Smith, Abimbola Abiola, Jeff Reihl, Marina Mnushkin, Roy
Leech, Jim Gehrels and the members of the 1986 IPSCO Think Tank on
Biofilms and Biofouling is acknowledged. In addition, the preparation
of this manuscript could not have been completed without the time and
dedication of Natalie Ostryzniuk.

12 References

Costerton, J.W. and Lappin-Smith, H.M. (1989) Behavior of bacteria in
   biofilms. American Society of Microbiology News, 55(12), 650-654.

Further reading

Atlas, R.M. and Bartha, R. (1981) Microbial Ecology: Fundamentals and
   Applications. Addison-Wesley Publishing Co., Menlo Park, CA.
Klug, M.J. and Reddy, C.A. (1984) Current Perspectives in Microbial
   Ecology. American Society for Microbiology, Washington, D.C.
Lynch, J.M. and Hobbie, J.E. (1979) Micro-organisms in Action:
   Concepts and Application in Microbial Ecology. Blackwell Scientific
   Publications, Oxford, U.K.
Prescott, L.M., Harley, J.P. and Klein. D.A. (1990) Microbiology.
   Wm.C. Brown Publishers, Dubuque, Indiana.

PART TWO

OVERVIEWS OF THE MAIN MICROBIOLOGICAL PROCESSES RELEVANT TO CIVIL ENGINEERING

2 MICROBES IN CIVIL ENGINEERING ENVIRONMENTS: BIOFILMS AND BIOFOULING

D.R. CULLIMORE
Regina Water Research Institute, University of Regina, Canada

Abstract
The terms biofilms and biofouling are defined as relating to the cause
and effect respectively of any physical intercedence into a natural or
engineered process by a biologically generated entity. The structure
of the biofilm is defined in terms of the incumbent qualitative and
quantitative aspects with particular emphasis on the phases of maturation through to a complete occlusion (total plug). Environmental
factors influencing the rate of growth of a biofilm with the C:N:P
ratio, temperature, redox potential, and pH being considered as
particularly dominant factors. Subsequent dispersive mechanisms are
also addressed which would allow the consortial organisms to migrate
as either suspended particulates (due to shearing) or non-attachable
ultramicro bacteria (generated as a response to severe environmental
stress). The net effect of biofouling as considered in terms of
causing various plugging, corrosion, bioaccumulation, biodegradation,
gas generation, and changes in water retention effects in soils,
ground- and surface- waters, and in engineered structures.
Keywords: Biofouling, Biofilms, Iron Related Bacteria, Sulfur Reducing
Bacteria, Ultramicro Bacteria, Plugging, Bioaccumulation, Corrosion.

1 Introduction

Marcus Aurelius Antoninus in his meditations written in the second
Century A.D. (volume V,23) wrote "For substance is like a river in a
continual flow, and the activities of things are in constant change,
and the causes work in infinite varieties; and there is hardly anything which stands still" and (volume IX,36) "The rottenness of matter
which is the foundation of everything! water, dust bones, filth, or
again, marble rocks, the callosities of the earth; and gold and
silver, the sediments; and garments, only bits of hair; and purple
dye, blood; and everything else is of the same kind. And that which is
of the nature of breath, is also another thing of the same kind,
changing from this to that" referenced many of the biological events
which, at various rates, recycle the elements found within and upon
the surface crust of this planet. Much of the microbiological
activities associated with these events occur within complex consortial communities formed within surface dwelling biofilms. Popularly

15

referred to as "slimes", these communities often stratify, act as accumulators of both nutritional and toxic chemicals, and cause plugging and corrosion. Darwin writing in 1845 on "The Voyage of the Beagle" referenced the gold mine at Jajuel, Chile and reports on the tailings piles after the gold has been separated as being heaped after which "a great deal of chemical action then commences, salts of various kinds effloresce on the surface, and the mass becomes hard (after leaving for two years and washing) ..it yields gold....repeated even six or seven times...there can be no doubt that the chemical action...liberates fresh gold". Here, Darwin was reporting a mine ore leaching process which was, for many years, thought to be a chemical process but is now recognised to be biological and to involve the formation of biofilms.

Within these growths, there are incumbent strains of micro-organisms which compete with each other for dominance as the biofilm matures and causes such secondary effects as plugging, corrosion and bioaccumulation. These factors along with others are discussed in this paper.

2 Interspecies competition

Within a biofilm there is an ongoing competition between the incumbent strains which may result in three effects. Firstly, the range of strains which may be present can become reduced or stratified. Qualitative reductions will occur where one or more of the strains becomes more efficient than the competing organisms and gains dominance. Young natural biofilms can contain as many as 25 or 30 strains which will, during the process of maturation, become reduced in numbers to five strains or less. Another response commonly seen in a maturing biofilm is the stratification of the biofilm in which the primary layering differentiates an exposed aerobic (oxygen present) and protected inner anaerobic (oxygen absent) layers. It is these inner MICCE2 surface interfacing layer that commonly will generate hydrogen sulfide and/or organic acids and initiate the processes of corrosion (Iverson, 1974).

A second major effect relates to the incumbency of cells within the biofilm which may be dense or dispersed. This incumbency density may be measured by undertaking an extinction dilution spreadplate enumeration of the viable cells as colony forming units per ml (cfu/ml) and the volume of the particulates after dispersion by laser driven particle counting as mg/L total suspended solids (TSS). From this data, it becomes possible to project the density of the cells within the biofilm as colony forming units per microlitre (cfu/uL). Stressed biofilms impacted upon, for example, by high bioaccumulates of gasoline (e.g., 16mg/ml biofilm) have been recorded as having incumbency rates of as low as 20 cfu/uL. When organic nutrients are concurrently applied, the incumbency rates recorded can increase by as much as one hundred fold.

Competition between the incumbent cells in a maturing biofilm in response to environmental conditions can lead to a third effect which is related to stress. Responses can occur at the individual cell

level or to the biofilm itself. Where stress impacts upon an individual cell as a result of trauma, it can cause a response. These
trauma can be created by a shift in one or more environmental factors
which no longer allows growth. Such events can include a shortage of
nutrients or oxygen, a build up in toxic end-products of metabolites
or antibiotics, and losses in structural integrity in the biofilm.

In general, cells under extreme stress will void non-essential
water and organic structures, shrink in volume with a tenfold
reduction in cell diameter, generate an electrically neutral cell wall
(to reduce attachability), undergo binary fission and enter a phase of
suspended animation. These minute non-attachable cells can remain
viable for very long periods of time as ultramicro bacteria. In this
state, bacteria can move with hydraulic flows until the environmental
conditions again become favorable. Once such facilitating conditions
recur, the cells again expand to a typical vegetative state and
colonize any suitable eco-niches. A newly developed water well can
provide such a suitable eco-niche for the transient aerobic ultramicro bacteria passing through the ground water.

3   Biofilm maturation

Biofilms may respond to stress as a normal part of the maturation
cycle. From studies undertaken using biofouling laboratory model
water wells (1 L capacity), a sequence of events has been recorded
following a basic pattern for the development of natural biofilms. In
this maturation, there are a number of distinct events. These can be
catagorised into a number of sequential phases: (1) rapid biofilm
volume expansion into the interstitial spaces with parallel losses in
flow; (2) biofilm resistance to flow next declines rapidly causing
facilitated flows which can exceed those recorded under pristine
(unfouled) conditions; (3) biofilm volume compresses by up to 95%,
facilitated flow continues; (4) biofilm sloughing causes increases in
resistance to hydraulic flow in a repeated cycling with a primary
minor biofilm volume expansion followed by secondary increased resistance to flow ending in a tertiary stable period; and (5) interconnection of individual biofilms now generate semi-permeable biological
barriers within which free interstitial water and gases may become
locked to form an impermeable barrier (plug). The speed with which
these phases are generated is a reflection of the environmental
conditions applied. For example, phases 1 to 3 can be accomplished
within ten days when generating a black plug layer using sulfur coated
urea fertilizer in sand columns (Lindenbach and Cullimore, 1989). The
pulses which occur in phase 4 as the biofilm expands slightly,
increases resistance to flow and then stabilizes has been recorded in
a model biofouled well to take 25 to 30 days per cycle while in
producing wells, cycles of as short as ten to twelve days have been
recorded.

During the phase 4 increases in resistance to flow, it can be
projected that the polymeric matrices forming the biofilm may now have
extended into the freely flowing water to cause radical increases in
resistance. At the same time as the resistance is increasing, some of

17

the polymeric material along with the incumbent bacteria will be
sheared from the biofilm by the hydraulic forces imposed. Such
material now becomes suspended particulates and forms survival
vehicles through which the incumbent organisms may now move to
colonize fresh eco-niches. In studies using laser driven particle
counting, these suspended particles commonly have diameters in the
range of 16 to 64 microns. In some cases the range in particle size
can be narrow (+/- 4 microns) around the mean particle size.

4 Bacteriology of plugging

Biological impedance of hydraulic flows, whether in a water well, heat
exchanger or cooling tower, involves the sequences of maturation
within the biofilm prior to the plugging event. The types of micro-
organisms involved in such events vary considerably with the environmental conditions present. In water wells, the iron bacterial group
are commonly associated with plugging (i.e., significant loss in
specific capacity due to a biofouling event involving biofilm
formation). It should be noted that an alternate term to plugging is
clogging. This later term is taken to mean any event (including
chemical and/or physical factors) which cause a total obstruction of
flow from a water well.
        Iron bacteria can also be defined as iron related bacteria (IRB) in
which iron and/or manganese forms natural accumulates (Ghiorse, 1984)
either within the cell, or surrounding extracellular polymeric
matrices in such a way that the product oxides and hydroxides render
visible orange to red to brown colors to the conglomerated growths
(e.g., slimes). These bacteria can be subdivided into three major
groups based upon the nature of the accumulates. In the first group,
the oxides and hydroxides are accumulated within a spiraling ribbon of
ECPS as it is excreted from the lateral wall of individual bacterial
cells. Such ribbons can exceed the length of the cell ten to fifty
fold and includes ten to fifteen harmonic cycles. Where these ribbons
are observed suspended in the product water from a water well upon
microscopic examination, the dominant IRB in any associated biofouling
is thought to be <u>Gallionella</u> (Hallbeck and Pederson, 1987). Little is
known of the life cycle for this bacterium and it has yet to be
cultured in a pure culture. It is, however, clear that the bacterium
producing these ribbon-like excrescences are Gram-negative rod-shaped
or vibrioid cells. Rarely is the cell seen still attached to the
ribbon which raises questions relating to why only the ribbon is found
and where can a bacterial cell grow within a heavily biofouled
environment and yet still produce such a ribbon. One hypothesis which
would address these questions is that these cells have grown on the
surfaces of the biofilm from which the ribbon-like stalks (false
prosthecae) extends into the zone of free liquid flow. The ribbon-
like nature of the prosthecae would cause some water to spiral down
towards the cell in an Archimedian screw-like action bringing oxygen
and nutrients into the proximity of the cells beneath. This would
give the cells of <u>Gallionella</u> an advantage over the other cells incumbent in the biofilm itself. As the prostheca expands into the water,

so the shear forces would increase until eventually the ribbon would break off to become a suspended particle in the water. As such, the fragmented ribbon enters the water phase where it can easily be recognised microscopically. Once the prostheca has sheared, the cell can now excrete a replacement stalk and continue to benefit from the ecological advantage this gives. The iron and/or manganese oxides and hydroxides would give a greater structural integrity to the prostheca. It is also generally believed that <u>Gallionella</u> is able to gain energy through the oxidation of these metallic ions from their reduced states.

The second group of iron bacteria also produce special structures within or upon which the iron and/or manganese oxides and hydroxides are accumulated. Here, the structures are tube-like (sheaths) with the bacteria able to function within the hollow central core. In some cases, the bacteria are able to migrate out of the sheath. Frequently when scanning electron microscopy is conducted upon biofouled material, abandoned sheaths are commonly seen embedded in the surfaces of the biofilm. This group is called the sheathed iron related bacteria and include the genera <u>Sphearotilus, Leptothrix</u> and <u>Crenothrix</u>. These bacteria tend to occur in waters with a relatively low organic carbon loading (<2 ppm total organic carbon).

The above groups of IRB are easily recognised by very distinctive morphological characteristics while the third group proves to be much more challenging since these organisms are commonly forming consortia including different strains of bacteria which may or may not be dispersed within the biofilm. These organisms can become stratified with the aerobic phase tending to be dominated by pseudomonads belonging to Section 4, family 1 of the Bergey's Manual while the anaerobic strata may be dominated by Sections 7 (sulfate reducing bacteria), 6 (anaerobic Gram-negative rods) and 5, families 1 and 2 (the enteric and vibrioid bacteria). The precise dominance sequence is a reflection of the nutrient and oxygen loadings of the causal water being delivered to the specific eco-niches. The amorphous nature of the sheared suspended particulates from such a biofouling renders microscopic examination frustrating but the more traditional bacteriological examinations on selective culture media such as R2A and Winogradsky Regina media or simple biological activity and reaction tests (BART^{TM}, Droycon Bioconcepts Inc., Regina, Canada) can be used to aid in the identification of the causative organisms.

Water samples being taken of pumped supplies from a well may not necessarily give an accurate indication of the microbial events occurring since the water could also contain the intrinsic planktonic population along with whatever sessile organisms that may be present in such sheared particulates as are suspended in the water. These occurrences are frequently random but a greater chance of observing these events can be achieved by entrapping the particulates in a moncell filtration system (Howsam and Tyrrel, 1989) or applying a pristine surface to the well water column onto which such organisms may now attach and form observable growths.

In practice, an understanding of the nature of a biofilm can be utilised in support of particular intercedent events such as the bioremediation of a gasoline plume, removal of potential toxic heavy metals or dangerous aromatic hydrocarbons. These would be alternative events to the more appreciated roles performed when biofouling causes plugging and corrosion.

5 Environmental factors influencing biological activities

Many environmental factors will interact with the growth dynamics of a maturing biofilm. These range through physical, chemical and biological factors (Atlas and Bartha, 1981). Of the physical factors, the three pre-eminent constraints relate to temperature, pH and the redox potential where the site of activity is in a saturated medium (Cullimore, 1987).

Temperature influences microbial growth in a number of ways ranging from the extremes where low temperatures cause an inhibition of the cellular metabolic processes, to high temperatures which impact particular on the protein constituents of the cell through thermal denaturation to cause cellular dysfunctions often leading to the death of the cell. Between these two extremes lies a temperature range within which the cells can be metabolically active. Such a range will include a narrower band within which growth and reproduction can occur with maximal responses being observed within a relatively narrow temperature range of commonly $<10^{\circ}C$. For temperate region ground waters in shallower aquifers of up to 300 meters depth, the normal temperatures experienced without geothermal influences would range from 1° to $25^{\circ}C$. This range is commonly associated with bacterial activities dominated by the lower mesotrophs ($>15^{\circ}$ to $<45^{\circ}C$ with optima at $<35^{\circ}C$), eurythermal psychrotrophs (grows both above and below $15^{\circ}C$, generally with the optimal temperature above that temperature) and stenothermal psychrotrophs (grows at $<15^{\circ}C$ only). Temperature will therefore influence the makeup of an incumbent consortium within a biofilm. For example, a water with a temperature of $10^{\circ}C$ (+/- $2^{\circ}C$) may be expected to include eurythermal and stenothermal psychrotrophs but no low mesotrophs. Shifting groundwater temperatures may be expected to cause changes in the dominant microbial strains within the biofilm.

The biofilms polymeric structures tends to impart a buffering activity which reduces the influence of pH shifts upon the activities of the incumbent organisms. In general, the maximum species divergence in incumbent flora when the pH of the passaging water is at between 6.5 and 9.0. Where the pH falls periodically to below this pH range, the biofilms have some capacity to buffer these effects. Additionally, the pH within a biofilm may become stratified with higher neutral or slightly alkaline conditions occurring in the upper aerobic zones while the anaerobic zones may be more acidic and support, and as a result, a narrower spectrum of microbial activities. In severely acidic conditions (e.g., pH <2.5) the spectrum may become so narrow as to include only only one or two different genera of micro-organisms. For example, many of the Thiobacillus species are able to function at very low pH values during the oxidation of sulfur and reduced sulfur compounds with the exclusion of other species (Kuenen, Boonstra, Schroder and Veldkamp, 1977).

In ecological investigations at sites of magnified biofouling, it has often been noted that these sites are in transitional zones between reductive and oxidative regimes where the redox values shift from a negative to a positive Eh value (e.g., from -50 to +150 Eh). Sequential dominance of IRB bacteria have been observed occurring along these redox gradients. Such sequential events have been manipulated through shifting and enlarging the redox transitional zone within an aquifer by recharging with oxygenated water to force the biofouling outwards (and away from the producing water well). This extension of the area occupied by the biofouling away from the well causes any bioaccumulation to take place more distantly from the well. The postdiluvial water may consequently be of a higher quality since there would have been a slower passaging of the causal water through the biofouled zone which is now further away from the turbulent zones of influence created by the pumping action within the well when activated.

Chemical factors can also influence microbial activities associated with biofouling events. These factors can be summarized into nutritionally supportive or inhibitory groups. The nuritionally supportive group would include compounds able to be utilized by the microbial cells for catabolic and synthetic functions essential to the continued survival, growth and dissemination of the species. Inhibitory compounds directly interfere with any of the supportive functions of a given species in such a way as to minimally retard specific competitive functions or maximally cause the death of the species group. In the processes involved in natural competition between micro-organisms, the supportive chemicals for one strain may, in fact, be inhibitory to another species. It may therefore be commonly expected that, where an inhibitory compound is introduced into an environment, there will be a radical shift in the component species within the impacted flora culminating in a tolerance or direct utilisation of the compound by the survivors. Additionally, the polymeric matrices of the biofilm will form a barrier through which such inhibitors would have to pass prior to imparting a direct impact on the incumbent microbial cells. Such matrices can often form a repository for toxic bioaccumulates away from these incumbent cells.

6 Nutritional factors influencing microbial activities

Nutritionally supportive chemicals are essential to the ongoing survival and growth of any given species. The nutrient elements essential to such a process can be determined by an examination of the ratio of these elements in the microbial cell. A typical ratio for these elements, for example, is C:O:N:H:P:S:K:Na:Ca:Mg:Cl:Fe which can be expressed gravimetrically as percentages of the dried weight for Escherichia coli as 50:20:14:8:3:1:1:1:0.5:0.5:0.5:0.2 respectively (Stanier, Doudoroff and Adelberg, 1970). Critical to the growth of micro-organisms is the C:N:P ratio since these are the three macronutrients which are most frequently observed as critical controlling factors for microbial growth. If there was a 100% efficiency in the utilisation of these elements, the ratio in the nutrient feed could be expected to be optimised at a ratio of 3.6:1:0.21 respectively. In reality, for the heterotrophic micro-organisms, the organic carbon consumed far exceeds the maintenance requirement due to the heavy catabolic demands to create energy as efficiently as possible with the venting of the totally oxidised carbon as carbon dioxide. The net effect of this shifts the C:N ratio to >40:1 in order to compensate for these carbon losses due to catabolism. An additional diversion is the synthesis of ECPS outside of the cell. The optimal ratios for the efficient growth of heterotrophs has not been established for C:N ratio due to these natural variations in the catabolic and synthetic functions. However, for the N:P ratio the bulk of these elements are retained within the cells so that it has generally been considered optimal within the range of 4 to 8:1 with excessive levels of phosphorus (i.e., ratio of <4:1) potentially causing build ups in the cellular reserves of stored polyphosphates and dinitrogen fixation.

In generating a comprehension of the impact of organic carbon on the growth of sessile or planktonic micro-organisms, concentrations of 1 ppm (mg/L) of organic carbon would appear to be insignificant. However, where an saturated environment contained one million viable units of bacteria per ml of total volume and each cell had a dried weight of 2×10^{-13}g and a 50% gravimetric carbon composition, the gross net weight of these viable units would be 2×10^{-7}g with 1×10^{-7}g occurring as carbon. If these organisms were in an environment with a total organic carbon of 1×10^{-6}g /ml of total volume, the ratio of cellular:free organic carbon would be 1:9. If one thousand bacteria were present per ml as viable units, then the cellular:free organic carbon ratio would shift to 1:9,000. It can therefore be projected that the critical dissolved organic carbon which can influence microbial activities in water should be considered to lie with the microg/L (ppb) range rather than the traditionally accepted mg/L (ppm) range. In developing a system to generate the bioaccumulation and/or biodegradation of specific nuisance chemical compounds in the environment, it becomes important to set the C:N:P ratios in a format beneficial to the required activity. If the target compound is organic, then it may contribute to the carbon ratio as a result of degradation and/or assimilation. A typical ratio to be established would range from 100 to 500:1:0.25 depending upon the amount of carbon that may become incorporated into the ECPS, the rate of biological activity on the targeted compound and the availability of alternate carbon substrates.

7 Conclusion

Civil engineers have learned over the past four millennia to design, construct and maintain public works on various scales of significance, but now the art of manipulating the microbial kingdom to become manageable factors within engineering needs to be developed. Through these understandings, the design and construction of new works may be achieved with an improved ongoing management system which will minimise the interferences and maximise the beneficial processes imparted by the microbial activities.

8 Acknowledgements

The author wishes to acknowledge the financial support of the Natural Sciences and Engineering Research Council of Canada (grant-in-aid #OGP0005073) and the Petroleum Association for a Clean Canadian Environment (joint funding of project with Ortech International of Ontario), Kevin Bellamy of Ortech International, the technical assistance of Abimbola Abiola, Jeff Reihl and Marina Mnushkin, and Natalie Ostryzniuk, for the the preparation of this manuscript.

9 References

Atlas, R.M. and Bartha, R. (1981) Microbial Ecology: Fundamentals and Applications. Addison-Wesley Publishing Co. Reading, MA.

Costerton, J.W. and Lappin-Smith, H.M. (1989) Behavior of bacteria in biofilms. American Society of Microbiology News, 55(12), 650-654.

Cullimore, D.R. (1987) Physico-chemical factors influencing the biofouling of groundwater, in International Symposium on Biofouled Aquifers:Prevention and Restoration (ed D.R. Cullimore), American Water Resources Association, Bethesda, pp. 23-36.

Ghiorse, W.C. (1984) Biology of iron- and manganese-depositing bacteria. Ann. Rev. Microbiol. 38, 515-550.

Hallbeck, E. and Pederson, K. (1987) The Biology of Gallionella, in International Symposium on Biofouled Aquifers: Prevention and Restoration (ed D.R. Cullimore), American Water Resources Association, Bethesda, pp. 87-95.

Howsam, P. and Tyrrel, S. (1989) Diagnosis and monitoring of biofouling in enclosed flow systems--experience in groundwater systems. Biofouling, 1, pp. 343-351.

Iverson, W.P. (1974) Microbial corrosion of iron, in Microbial Iron Metabolism (ed J. Neilands), Academic Press, Inc., New York, pp. 476-517.

Kuenen, J.G., Boonstra, H.G., Schroder, H.G.J., and Veldkamp, H. (1977) Competition for inorganic substrates among chemoorganotrophic and chemolithotrophic bacteria. Microb. Ecol., 3, 119-130.

Lindenbach, S.K. and Cullimore, D.R. (1989) Preliminary in vitro observations on the bacteriology of the black plug layer phenomenon associated with the biofouling of golf greens. J. Applied Bact., 67, 11-17.

Stanier, R.Y., Doudoroff, M. and Adelberg, E.A. (1970) The Microbial World. Prentice Hall International Inc., London.

3 BIOCORROSION IN CIVIL ENGINEERING

A.K. TILLER
National Corrosion Service, NPL Teddington, UK

Abstract
Biocorrosion (or microbially induced corrosion) is a major problem affecting most engineering alloys. It is ultimately associated with biofilms and the metal substrate. The composition and structure of the biofilm and its chemistry can lead to both aerobic and anerobic corrosion. The extent and nature of the corrosion sustained depends on the metal and the ecology of the biofilm. The paper reviews aspects of the problem which are considered important to engineers such as the micro-organisms involved, corrosion mechanisms, morphological data which will aid identification of the phenomena and the development of test procedures which allow rapid diagnosis. Examples of case histories are also presented together with comments on the various methods of prevention and control which are available.
Keywords: Biocorrosion, Microbially induced corrosion, Biofilms, Aerobic and anaerobic bacteria, Corrosion and prevention, Case histories.

1 Introduction

There are several important aspects of biocorrosion (or microbially induced corrosion), which engineers should appreciate. The first is that the phenomenon does not involve any new form of corrosion process; secondly it is inextricably associated with the presence of biofilms and bacterial extracellular polymers; thirdly, most engineering alloys are susceptible to this form of degradation. Tiller, A.K. (1988) Videla, H. (1987) Hamilton, W.A. (1985). It is also important to emphasise that when the biocorrosion problems occur, they are seldom straightforward and usually incur costly down time of operating plant. Since micro-organisms are widespread in nature, most natural and man-made environments are sufficiently contaminated to allow bacterial activity to proceed to a greater or lesser extent. Hence a wide range of operational problems arising from bacterial activity can be experienced and typical examples include the plugging of filter screens, down-hole corrosion of suction pumps, the internal and external corrosion of pipelines, deterioration (spalling) of concrete souring of boreholes and reservoirs and in some cases, contamination of the reservoir rock pores.

From the wealth of data which is now available and which has been produced mainly over the past decade, it is recognised that biocorrosion not only stimulates general, pitting, crevice and stress corrosion but is capable of enhancing corrosion fatigue and hydrogen embrittlement and cracking. Edyvean, R.G.J., Thomas, C.J. and Austen, I.M. (1985) Walch, M. and Mitchell, R. (1985). However, although it can be assumed that biocorrosion will occur in any environment in which micro-organisms can thrive, the extent of the activity of any specific species may be limited and hence conditions favourable for one type of micro-organism may be quite inimical to another. Although biocorrosion is a challenge to the materials and design engineer, many of the problems which arise can be prevented and or controlled by various new and established anti-corrosion procedures. Where engineers are aware of the problem, it is not uncommon to find well-planned maintenance and monitoring regimes.

The purpose of this paper, however, is not to discuss all the issues associated with the problem of biocorrosion in depth, but to provide a perspective or over-view of the problem, its impact on engineering alloys and materials and to create an awareness of the various techniques available for assessing and predicting the problem including prevention and control.

2 The nature and mechanism of biocorrosion problems

It is important to emphasise that microbial involvement in the field of corrosion does not involve any new form of corrosion process. Thus, the underlying processes are electrochemical in nature but the function of the micro-organisms is complex and in some cases bacterial activity and metabolism may generate inorganic intermediaries which eventually become the corrosive agent. Gragnolino and Tuovinen (1984) Newman et al (1986).

When a metal is exposed to an aerobic environment, ie. immersed in water or in contact with damp soil, the initial reaction is the dissolution of metal as metallic cations which leave behind an excess of electrons:

$$M \rightleftharpoons M^{2+} + 2e \quad \text{anodic reaction} \qquad (1)$$

These electrons are consumed at nearby cathodic sites by the balancing reaction, which in near neutral solutions is usually the reduction of oxygen to hydroxyl ions:

$$O_2 + 2 H_2O + 4e \rightarrow 4 OH^- \quad \text{cathodic reaction} \qquad (2)$$

The overall reaction is the formation and subsequent precipitation of insoluble products formed by the reaction of ferrous metallic ions with hydroxyl ion (and frequently oxygen in solution):

$$3 Me^{2+} + 6 OH^- \rightarrow 3 Me(OH)_2 \quad \text{corrosion product} \qquad (3)$$

In the simplest case the reactions occur uniformly over the surface of the metal. The anodic sites are determined either by the heterogeneity of the surface or by random failure of a tenacious pre-existing oxide film. The overall rate of corrosion in these circumstances depends on a number of factors. The important criterion, however is that the anodic and cathodic reactions must proceed in balance to preserve overall electroneutrality. In the absence of oxygen, the usual cathodic reactions for corrosion processes is the reduction of hydrogen or water:

$$2 \ H^+ + 2e \rightarrow H_2 \tag{4}$$

$$2 \ H_2O + 2e \rightarrow H_2 + 2OH^- \tag{5}$$

This reaction is limited in neutral solutions and therefore the corrosion process is stifled. At lower pH values hydrogen evolution usually predominates both in the presence and absence of oxygen.

In addition to these basic reactions, there are other factors which have a bearing on the overall process, in particular the micro-geometry of the metal surface and variation in the oxygen concentration of the environment creating differential aeration cells. The cathodic sites in this case are the more oxygenated regions. Similarly, if the metal is in contact with the soil or solution in which concentrations of aggressive anions vary from one site to another a concentration cell is produced. These phenomena usually encourage localised pitting corrosion which is exacerbated by the action of certain microbiological species.

Attention has already been drawn to the fact that micro-organisms are widespread in nature and it is because of this they are able to colonise surfaces and, by genetic mutation, acquire the ability to adapt to environmental changes. For this reason, they represent a dynamic system which is able to change with time.

Colonisation of metal surfaces results in the formation of biofilms and the adherence of these to the substrate is brought about by the release of extracellular polymers. White, D.C., Nivens, D.E. et al (1985) Ford, T.E., Maki, J.S. and Mitchell, R. (1988). Biofilms up to 100 μm thick are not unusual and in nearly all cases contain entrapped bacteria. Such films eventually develop into complex biological ecosystems, comprising active bacteria, their metabolites and the chemical changes generated by the system. Numerous corrosion processes and mechanisms develop as a result of this complexity.

2.1 Aerobic corrosion

Bacterial colonisation and biofilm formation which occurs under aerated conditions encourages the formation of differential aeration and concentration cells due to the uptake of oxygen by the microbial colony or biofilm present. The oxygen concentration under these colonies becomes reduced and localised pitting takes place. Where the iron oxidising bacteria, such as Gallionella, are involved corrosion tubercles develop and these encourage the co-accumulation of other aggressive anions such as chlorides. Since these organisms have the ability to oxidise soluble iron and manganese, deposition of ferric and manganic chlorides is common. Tatnall, R.E. (1984) Tiller, A.K.

(1988). Ferrous metals under these conditions sustain pitting damage whilst stainless steels often develop volcano shaped tubercles with sub-surface bottle shaped cavities beneath them. Kobrin, G. (1976) Stoecker (1981) Tiller, A.K. (1983). Tuberculations behave similarly to slime films and provide a suitable habitat for anaerobic bacteria such as the sulphate reducers. These stimulate corrosion by a number of methods but under these conditions it is probably due to the release of the metabolite sulphide.

Hence, all metals which possess a stable oxide film which provides corrosion resistance, will be susceptible to this form of corrosion as and when the film is damaged or the oxygen screened from the metal by the presence of biofilms. Metals which fall into this category are the aluminium alloys stainless steels, high alloyed steels, copper and its alloys and possibly titanium.

Corrosion can also be stimulated by the chemistry which develops within the biofilm and the extracellular polymers generated within them. Usually these are of a polyssacharide nature with a chemistry which allows them to act as a chelating agent for metal ions. The ability of an exopolymer to chelate specific anions from the bulk solution can influence its adhesion to the metal surface. The development of a concentration of metal ions within the biofilm or attached to the metal surface produces galvanic cells thus creating an imbalance in the electrochemistry of the system such that increased corrosion takes place.

2.2 Corrosion by acidic metabolites

The release of aggressive metabolites such as inorganic or organic acids is another mechanism by which biocorrision can proceed. During the metabolism of acidophilic bacteria such as the Thiobacillus thiooxidans or certain fungal growths, a range of inorganic and organic acids can be liberated. Booth, G.H. (1971) Purkiss, B.E. (1971) Calderon O.H. et al (1950). Corrosion under these conditions proceeds with the liberation of hydrogen as the main cathodic reaction. Typical reactions are given below for the oxidation of sulphur by sulphur oxidising bacteria:

$$H_2S + 2O_2 \rightarrow H_2SO_4 \tag{6}$$

$$2S^O + 3O_2 + 2H_2O \rightarrow 2H_2SO_4 \tag{7}$$

$$5Na_2S_2O_3 + 4O_2 + H_2O \rightarrow 5Na_2SO_4 + H_2SO_4 + 4S^O \tag{8}$$

Some bacteria generate other types of metabolites such as carbon dioxide and ammonia. In the case of ammonia this will encourage stress corrosion cracking of copper and its alloys. Bengough, H.G.D., May R. (1924).

2.3 Anaerobic corrosion

Perhaps the most important of the corrosion mechanisms associated with bacteria is that which is able to take place under anaerobic conditions. It has already been mentioned that the inner-layers of microbial films are frequently anaerobic and this is particularly the case when the system is in contact with soil and water under reducing conditions. Anaerobic soils and the anoxic regions of biofilms are the habitat of the sulphate reducing bacteria. These are obligate anaerobes which can only function in the absence of oxygen. They have a unique physiological and enzyme make-up which gives them the capability of being able to ultilise hydrogen for growth purposes with sulphate or other reduced sulphur compounds as the terminal electron acceptor. Hence, their metabolic product is sulphide. Besides these organisms, there are many obligate and facultative anaerobes that can utilise hydrogen and some are able to produce sulphide from sulphite and thiosulphate. Pankhania, I.P. (1988).

The mechanisms of corrosion involving these bacteria and in particular the sulphate reducers is very much more complex and is still far from being completely understood. Pankhania, I.P. (1988) Tiller, A.K. (1988) Videla, H. (1987) Iverson, W. (1987). At present five hypotheses exist to explain their involvement in the corrosion process. These are (a) the depolarisation of the cathode and neutralisation of molecular hydrogen; (b) corrosion by sulphide ions; (c) galvanic corrosion due to the formation of iron sulphide films; (d) corrosion due to the formation of elemental sulphur; and (e) the production of a corrosive volatile phosphorous compound.

The theory of cathodic depolarisation is based on the understanding that these organisms have the ability to utilise the cathodic hydrogen for dissimilatory sulphate reduction thus stimulating the anodic reaction. At the same time their metabolic product sulphide will react with metal ions in solution to form iron sulphide which in itself is a corrosive agent and depending on the chemistry of the sulphide, will encourage cathodic depolarisation. King, R.A. and Miller J.D.A. (1971) Miller J.D.A. (1971) Booth et al (1968). The main features of the classical theory are best illustrated in terms of chemical reactions:

$$4Fe \rightarrow 4Fe^{2+} + 8e \qquad \text{anodic reaction} \qquad (9)$$

$$8H_2O \rightarrow 8H^+ + 8OH^- \qquad \text{dissolution of water} \qquad (10)$$

$$8H^+ + 8e \rightarrow 8H \qquad \text{cathodic reaction} \qquad (11)$$

$$SO_4^{2-} + 8H^+ \rightarrow S^{2-} + 4H_2O \qquad \begin{array}{l}\text{cathodic depolarisation}\\ \text{by bacteria}\end{array} \qquad (12)$$

$$Fe^{2+} + S^{2-} \rightarrow FeS \qquad \text{corrosion product} \qquad (13)$$

$$3Fe^{2+} + 6OH^- \rightarrow 3Fe(OH)_2 \qquad \text{corrosion product} \qquad (14)$$

The alternative mechanism involving the iron sulphide as the cathodic depolarising agent has already been mentioned. Both of these theories are illustrated in Figure 1.

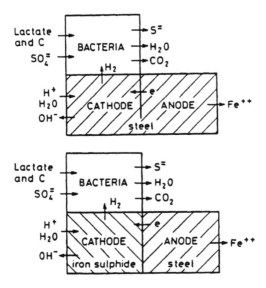

Fig. 1 Schematic representation of the classical mechanism (top) and
alternative mechanism (bottom) of microbial corrosion under
anaerobic conditions (after Miller and King)

One of the important features of cathodic depolarisation is the nature
of the sulphide film. The chemistry of the formation of iron sulphide
minerals as shown in Figure 2 explains how the primary corrosion
products mackinawite and siderite can under appropriate conditions be
transformed to other materials which are less protective and more
corrosive. Rickard, D.T. (1969) Mara, D.D., Williams, D.S.A (1972).

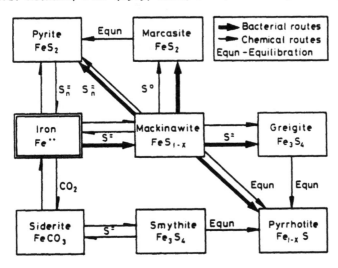

Fig. 2 Interrelationship between the sulphides of iron (after Rickard)

2.4 Anaerobic/aerobic conditions

In practice, both anaerobic and aerobic cycles occur and when these conditions arise, elemental sulphur is produced. Sulphur can promote corrosion by a concentration cell mechanism similar to that described for differential aeration cells. Schaschle, E. (1980). Usually there is an anodic area beneath a porous material which shields the metal from the dissolved sulphur, the adjacent area where dissolved sulphur is readily being cathodic. The mechanism is illustrated below:

$$Fe^O \; . \; \rightarrow Fe^{2+} + 2e \qquad\qquad \text{anodic reaction} \qquad\qquad (15)$$

$$S^O + H_2O + 2e \rightarrow HS^+ + OH^- \qquad\qquad \text{cathodic reaction} \qquad\qquad (16)$$

$$. \; Fe^{2+} + HS^- \rightarrow FeS + H^+ \qquad\qquad \text{overall reaction} \qquad\qquad (17)$$

The rate of corrosion which can arise from sulphate reducing bacterial activity can be very high particularly when it is controlled by the action of sulphur or its compounds. Intergranular stress corrosion cracking of sensitised austenitic stainless steels has been reported under these conditions. Gragnolino, G., MacDonald, D.D. (1982).

The intrinsic corrosive nature of the sulphide film can be very high. For example, for steel, the presence of H_2S alone encourages corrosion at a rate of 12.8 mm/year, whereas in the presence of mackinawite, a protective sulphide, it is reduced to 5.3 mm/year. Greigite, however, is much more corrosive than either of these compounds and is reported to stimulate a rate of corrosion of 120 mm/year. In the case of elemental sulphur, the rate may be anything up to 8 times that for greigite.

In addition to the above mechanisms these organisms and their biofilms can also encourage other forms of corrosion such as hydrogen embrittlement and hydrogen-induced cracking of high strength steels and stainless steels. Walch, M., Mitchell, R. (1987). The mechanisms of corrosion cracking in this case is associated with poisoning of the surface by hydrogen sulphide. This retards the formation of molecular hydrogen thus encouraging the (atomic) hydrogen to diffuse into the metal. Hydrogen embrittlement and cracking is an area which is currently being examined extensively and particularly in those situations where biofilms predominate. Mitchell, R., Ford, T.E. (1989). Where bacteria produce hydrogen per se within the biofilms this can also be absorbed by the metal. Biogenic sulphide is also a factor in corrosion fatigue and is reported to stimulate the rate of crack growth. Mitchell, R., Ford, T.E. (1989).

Although interaction of bacteria and biofilms with metals can give rise to severe corrosion problems, techniques to control or avoid them are readily available.

3 Assessment and prediction of biocorrosion

The assessment prediction and control of microbial corrosion require both laboratory studies and field investigations. The experimental study of the problem both in the laboratory and in the field has progressed markedly over the past decade. Tests using sophisticated electrochemical cells and techniques such as AC impedance and electrical noise, have been used to study hydrogen permeation, the corrosion of buried cast and ductile iron pipes and the microbial corrosion of rebars in reinforced concrete. Dexter, S.C. (1983). Also several new techniques such as ATP photometry, epifluorescence microscopy and various types of test kits for field and laboratory studies have been introduced by microbiologists. Biodeterioration 6 (1984) Biodeterioration 7 (1988) Sequeira, C.A.C. and Tiller, A.K. (1988). These are in addition to existing microbiological and chemical assessment procedures. In the case of field studies, the use of corrosion monitoring techniques such as probes for polarisation resistance and electrical resistance has increased. Bioprobes installed in pipelines and other flowing systems provide a facility for studying the development of biofilms on metal surfaces. This is an area which has expanded considerably over the last few years and it is recommended that engineers should acquaint themselves with these techniques.

Techniques for assessing the corrositivity of soils have been available for some time. The main factors responsible for making the soil aggressive are anaerobic bacteria, the soil chemistry (particularly acidity and salt content), the redox potential, resistivity, differential aeration and the formation of concentration cells. These are not all independent parameters but often have to be considered in order to allow the corrosiveness of the environment to be assessed. Table 1 below is an example of a simple scheme which has proved to be reasonably successful in practice and which will allow the pipeline engineer to acquire data sufficient to implement good corrosion engineering practices.

Table 1. Scheme for Assessment of Corrosiveness of Soils

	Aggressive	Non-Aggressive
Soil resistivity (Ω cm)	< 2000	> 2000
Redox potential at pH 7.0 [Vonhe)]	< 0.40 V < 0.43 V (if clay)	> 0.40 V > 0.43 (if clay)
Water content (Wt %)	> 20%	< 20%

In the case of natural waters, there are a number of well established criteria and the problem of predicting whether microbial factors will make a significant contribution to the rate or distribution of corrosion, will depend on the assessment of the probability of conditions developing in which the microbial activity can proliferate. Obviously analytical data are required including the microbiological characteristics of the system. It must be emphasised that a high oxygen value does not mean that bacteria such as the sulphate reducers are absent in debris or in situations where there is stagnant water. The reverse also applies to the aerobic bacteria such as the iron oxidisers. Microbiological monitoring is important in this instance but the data obtained will depend primarily upon (a) the ability of the media and procedures used to detect the bacteria efficiently and quickly; (b) an ability to relate the results obtained to an actual risk of corrosion or other problems. Hence it is important to monitor both the planktonic and sessile bacteria, that is those organisms adhering to the surface of the metal. By using the wide range of techniques now available including available procedures discussed above it is possible to predict and assess the corrosive nature of many environments.

4 Prevention and control

No universal approach to the prevention and control of microbially induced corrosion is available. Several options have to be considered before any particular procedure is implemented. However, consideration should be given at the design stage ensuring that aspects of the design and operating procedures will not encourage corrosive attack. For example, the presence of crevice regions should be avoided since these act as sites for colonisation by the bacteria and deposition of residual detritus contained in the system would subsequently act as a micro-habitat for bacterial growth. Crevice regions which should be avoided are those which are formed by poorly laid-down welds which have raised geometry, porosity and blow-holes. Welds where possible, should be ground and polished. Bolted overlap joints and the use of fibrous gaskets in flanged joints also represent poor design features. Biodeterioration of the fibrous gaskets can occur and depending upon the nature of the fibrous material used, bacteria can be drawn by capillary action into the joint. In addition to the corrosion stimulated by the oresence of the micro-organisms, build up or co-accumulation of other aggressive anions such as chloride and sulphate can occur.

4.1 Effect of flow
Hydrodynamic conditions and flow regimes are also important. Stagnant conditions in pipe-lines, water storage tanks and pumps should be avoided if possible. Low flow regimes and stagnant conditions encourage colonisation of the surfaces by planktonic bacteria; biofilms develop and subsequent corrosion of the material takes place. Increased flow velocity will markedly reduce the potential for this to occur and where biofilms already exist, can help to reduce the film thickness, However, changes in the flow velocity will not necessarily remove the biofilm completely.

4.2 Operational factors

Operational considerations also play an important part in influencing the risk of microbially induced corrosion. Successful operation of any plant likely to be susceptible to this type of attack will depend on a number of factors such as initial conditioning, subsequent operation and maintenance. The frequency of periodic inspections is an important factor in control of microbial corrosion. Accessibility of equipment and ease of maintenance should be ensured at the design stage to ensure that inspection and maintenance can be readily carried out. It cannot be over-emphasised that microbial infections and reinfections occur frequently but can be predicted.

4.3 Material selection

As far as material selection is concerned the most important factor is to ensure that the material selected has been fully evaluated with respect to resistance to biocorrosion.

4.4 Protective systems

For pipeline systems, it is possible to use a non-aggressive backfill such as chalk, provided that the trench into which the pipe is being laid is reasonably well-drained. Chalk primarily provides an alkaline environment, producing a sufficient pH change to inhibit any activity or growth of most soil bacteria. A range of sands is also available and these are useful in providing a physical barrier of uniform compaction. However, little protection is provided against subsequent infection by water-born organisms or contamination by metabolic products arising from the activity of bacteria in adjacent regions. Usually pipelines are protected by the use of protective coatings or tapes. Detailed discussion of the various systems available is outside the scope of this paper and the reader is advised to consult the specialist publications which are available from manufacturers and suppliers. Modern protective surface coatings will provide a very high standard of corrosion protection but they are prone to mechanical damage and in order to overcome this problem, it is often necessary to complement the system by adopting cathodic protection.

4.5 Cathodic protection

The principles and theory of cathodic protection have been extensively discussed in the corrosion literature Morgan, J.H. (1986) Ashworth, V. & Booker, C.J.L. (1987) and it is not intended to comment further here. Two techniques which are commonly used however involve the use of sacrificial anodes of magnesium, zinc or aluminium alloys which corrode preferentially and provide protection for the structure. The alternative is the use of impressed current via a rectifier using an inert anode assembly. The choice of method depends on the site conditions and the demands of the system. Cathodic protection can be applied not only to buried pipelines but also to the external sections of the vertical tubular materials for down-hole systems.

Experience has shown that in the absence of sulphate-reducing bacteria, protection of steel is achieved when the potential is depressed to -0.85 V (Cu - CuSO$_4$). Where microbial activity is high or the risk is known, the potential must be depressed to a value of at least -1.0 V (Cu - CuSO$_4$).

4.6 Alternative materials
Non-metallic materials such as composite glass fibre reinforced resin systems, PVC, polythene and polypropylene can be selected as an alternative to some of the more traditional alloys. In some instances, filter screens may be fabricated from fibrous non-metallic systems but their long-term performance has yet to be totally evaluated.

4.7 Biological control
Where control of the bacterial population is required, the use of biocide and bacteriostats is common. Biocides may act as either enzyme poisons or to disrupt the bacterial cell wall; bacteriostats inhibit bacterial growth and reproduction. Where these may be used for down-hole treatment or in reservoirs and storage tanks assessment of their chemical, biological and toxicological properties is important. The types of compound available range from phenolics, aldehydes quarternary ammonium compounds, chlorine and chlorine release agents, imidazoline, isothiazoline and bisthioyanate. Although many of these will inhibit or kill plantonic bacteria, few are able to penetrate the biofilm. This shortcoming has been recognised and the use of surfactants to disperse the film and improve biocide penetration, is now an accepted practice.

The use of UV irradiation in aqueous systems and borehole water and reservoir facilities has been successful. The system operates by passing the fluid to be treated through a narrow chamber in which a UV irradiation source is located. The disadvantage of the process is that it only kills the planktonic bacteria at the time of irradiation and hence reinfection may occur downstream of the UV chamber.

5 Case histories of biocorrosion

In discussion of practical situations the aim has been to assist the design, maintenance and operating engineer to become more aware of this type of problem and the phenomena associated with it thus allowing early and rapid counter-measures to be taken.

5.1 Example 1
Several vertical axial suction pumps had been installed in a river and were being used as part of a drainage scheme to control flooding during certain times of the year. During the first year, the operating conditions varied considerably from being intermittent to sequential and as a result the pumps were subjected to stagnant river water for long periods. Hence sludge build-up occurred on the internal surface of the components which encouraged the formation of corrosion tubercles.

Pitting attack having an etched and granular morphology developed on the phosphor-bronze impeller and in addition to this the wear ring and bell housing were similarly affected. In many cases, the pitted

regions had coalesced with the subsequent formation of grooves and
channels. In some areas, horse-shoe-shaped regions had developed
circumferential to the flow of water.

Examination showed sulphide present in both the pitted regions and
the corrosion products taken from several locations. Both the
corrosion products and slime films removed from the surface of various
components were found to contain a large population of active bacteria
with the predominant species being the sulphate reducers. Analysis of
the river water showed it had a variable BOD (biological oxygen
demand) value of 2-20 ppm with large quantities of nitrogenous
material and chloride and sulphate levels in the range 150-200 ppm
respectively.

The mechanism of the pitting corrosion in this instance is
considered to be due to the presence of microbial slimes with tubercle
formation. However, when the pumps were operated the shear forces
created by the water flowing in the space between the wear ring and
impeller removed the tubercles. These pitted regions then encouraged
additional turbulent flow conditions and eventually erosion corrosion
and some cavitation damage occurred leading to the formation of
grooves, channels and horse-shoe shaped areas.

The primary factor responsible for the onset of this corrosion was
the operating conditions adopted. Intermittent use allowing long-term
contact with stagnant water conditions exacerbated the situation. In
order to overcome this, an improved maintenance and monitoring
programme was implemented and the pumps operated more frequently.

5.2 Example 2
Approximately three years after 17 deep boreholes had been
commissioned and drilled in a region in the Middle East, a water
problem developed and souring of a number of wells occurred. During
the souring period, excessive sulphide production occurred which
caused serious internal corrosion of the down-hole tubular material,
suction pumps and ancillary materials. Subsequent chemical and
microbiological analyses confirmed the presence of high populations of
sulphate reducing bacteria. Operating conditions varied with some
boreholes operating for two or three months at a time whilst others
remained stagnant. In addition to this, subsequent water analyses
indicated changes in composition which were attributed to the external
corrosion of the downhole tubulars resulting in ingress of water from
other aquifiers. The primary cause of the external problem was due to
the combined effects of anaerobic corrosion and differential aeration
cells developing along the length of the well casing and varying with
depth. Substratum examination showed some were more anaerobic than
others. This is an example which illustrates that both the nature of
the operating procedures and an inadequate site survey, to assess the
corrosiveness of the environment, were responsible for the problem
experienced.

5.3 Example 3
A rotating stainless steel filter screen failed in the crevice regions
formed between the axial rods and the outer face of the stainless
steel mesh. The failures occurred within a three-year period

subsequent to commissioning.

Examination revealed that an extensive build-up of slime films had developed and pitting corrosion had occurred in many instances. Pit depths of up to 3 mm to 4 mm had developed and in many cases, these had penetrated longitudinally inside the axial rods creating hollow sections covered by a thin skin of metal. Subsequent metallographic investigations showed the axial rods had been fabricated from type 303 stainless steel, a sulphur bearing free machining alloy. The pitting corrosion had occurred at grain boundaries rich in maganese sulphide.

All the corrosion products and deposits including the water were heavily contaminated with both aerobic and anaerobic bacteria. Pseudomonads, fungi, bacilli, and sulphate reducing bacteria were all identified. This failure is a typical example of corrosion induced by bacteria developing a complex ecosystem within silt and sludge deposits present in the crevice regions formed between the axial rods and the stainless steel mesh. Corrosion occurred due to the formation of differential aeration and concentration cells exacerbated by the activity of the bacteria. End grain attack at the manganese sulphide stringers in the material continued causing the rods to become hollow.

6 Conclusions

Biocorrosion is a significant and persistent problem in many industries and is closely associated with aerobic and anaerobic bacteria and the presence of biofilms. The chemistry within a biofilm generated by the bacteria or the metabolic products generated by bacterial activity will encourage the corrosion of most of the engineering alloys now in use. In some cases the mechanism of corrosion is closely associated with the physiological and enzyme make-up of the organism and where this involves either the use of hydrogen or its liberation, corrosion problems associated with corrosion fatigue and hydrogen embrittlement can arise.

Although there has been a marked improvement in field testing techniques and the development of specific procedures for studying selective systems, problems still arise. Engineers still need better access and exposure to the data obtained from numerous studies which are now in progress and this requires improved technology transfer combined with availability of data-bases and the use of expert systems. In addition to this, it is important that the engineer acquaints himself adequately with other aspects which are likely to encourage this type of corrosion problem and, in particular, design features of the plant and operating conditions. Although each problem has to considered in context, there are a number of procedures and techniques available for control and prevention and if these are applied appropriately and monitored accordingly, a high degree of success will be achieved. Corrosion inspection and monitoring regimes are important features of good house-keeping and maintenance.

7 References

Ashworth, V. and Booker, C.J.L. (1986) **Cathodic Protection, Theory and Practice.** Ellis Horwood Ltd, Chichester, England.

Beugough, G.D. and May, R. (1924) 7th Report to the Corrosion Research Committee of the Institute of Metals, J. **Inst. Metals,** 32, 81-249.

Booth, G.H. et al (1968) Corrosion of mild steel by sulphate reducing bacteria: An Alternative Mechanism, **Brit. Corr. J.,** 3, 242.

Booth, G.H. (1971) Microbial corrosion. M & B Monograph CE/1 Mills and Boon, London.

Calderon, G.H. Stratfeld, E.E. and Coleman, C.B. (1968) Metal-organic acid corrosion and some mechanisms associated with these processes, in **Biodeterioration of Materials,** (eds. A.H. Walters and J.J. Elphick), 356 Elseveir, New York.

Dexter, S.C. (1983) (ed. Biologically Induced Corrosion), **NACE 8,** National Association of Corrosion Engineers, Houston, Texas.

Edyvean, R.G.J., Thomas, C.J. and Austen, I.M. (1986) Biologically Induced Corrosion, **NACE 8,** (ed. S.C. Dexter), National Association of Corrosion Engineers, Houston, Texas.

Ford, T.E., Maki, J.S. and Mitchell, R. (1987) Involvement of bacterial exopolymers in biodeterioration of metals, in **Biodeterioration 7,** (ed. D.R. Houghton, R.N. Smith and H.O.W Eggins), Elsevier Applied Science.

Gragnolino, G. and Tuovinen, O.H. (1984) The role of sulphate reducing bacteria and sulphur oxidising bacteria in the localised corrosion of iron-base alloys: A Review, **Int. Biodet.,** 20, 9-26.

Hamilton, W.A. (1968) Sulphate reducing bacteria and anaerobic corrosion. Ann. Rev., **Microbial,** 39, 195-217.

Iverson, W.P. (1987) Microbial Corrosion of Metals. Advances in **Applied Microbiology,** Vol 32.

King, R.A. and Miller, J.D.A. (1971) Corrosion by the sulphate reducing bacteria. **Nature,** 233, 491.

Kobrin, G. (1976) Corrosion by microbiological organisms in natural waters. **Materials Performance,** July .

Mara, D.D. and Williams D.S.A. (1972) The mechanism of sulphide corrosion by sulphate reducing bacteria, in **Biodeterioration of Materials,** Vol. 2, (ed. A.H. Walters and E.H. Hueck-van-der Plas), Applied Science, Barking, England.

Miller, J.D.A. (1971) (ed. Microbial Aspects of Metallurgy), MTP Aylesbury.

Mitchell, R. and Ford T.A. (1989) Surface Microbiology and Corrosion Processes, **Corrosion Research Sympsoium,** 17-19 April, New Orleans, Louisana, NACE.

Morgan, J.H. (1987) Cathodic Protection (2nd Edition NACE), National Association of Corrosion Engineers, Houston, Texas.

Newman, R.C., Wong, W.P. and Garner, A. (1986) A mechanism of microbial pittng in stainless steel, **Corrosion,** Vol 42, No 8, 489.

Pankhania, I.P. (1988) Hydrogen metabolism in sulphate-reducing bacteria oils role in anaerobic corrosion, **Biofouling,** Vol 1.

Purkiss, B.E. (1971) Corrosion in industrial situations by mixed microbial flora, Chapter 4. **Microbial Aspects of Metallurgy,** (ed. J.D.A. Miller), MTP Aylesbury.

Rickard, D.T. (1969) The microbiological formation of iron sulphides. **Stockholm Contrib. to Geology**, 20, 67.

Sequeira, C.A.C. and Tiller, A.K. (1988) (eds. Microbial Corrosion 1) Elsevier Applied Science, Barking, England.

Schaschle, E (1980) Elemental sulphur as a corrodent in deaerated neutral aqueous environments, **Mat. Ref.** 19(7)9.

Stoecker, J.G. (1981) Case 3 Penetration of stainless steel following hydrostatic tests in case histories: bacterial induced corrosion, **Mat. Perform.**, 20.

Tatnall, R.E. (1986) Biocorrosion of Metals in the Process Industries in **Biodeterioration 6**, CAB International UK.

Tiller, A.K. (1983) Chapter 3: Aspects of microbial corrosion in Corrosion Processes (ed. R.N. Parkins), Applied Science, Barking, England.

Tiller, A.K. (1988) The impact of microbially induced corrosion on Engineering alloys in **Microbial Corrosion 1** (Ed. C.A.C. Sequeira and A.K. Tiller) Applied Science, Barking, England.

Videla H. (1987) Electrochemical interpretation of the role of micro-organisms in corrosion, in **Biodeterioration 7**.

Walsh, M. and Mitchell, R. (1986) Hydrogen uptake by metals in the presence of bacterial films, in **Biodeterioration 6**.

White, D.C., Nivens, D.E., Nichols, P.D., Kerger, B.D., Hewson, J.M., Geesey, G.G. and Clarke, C.K. (1985) Corrosion of steels induced by aerobic bacteria and their extracellular polymers, **Proc. Fut. Workshop on Biodet.**, Held Univ. of La Plata, Argentina.

4 BIODETERIORATION OF MATERIALS USED IN CIVIL ENGINEERING

K.J. SEAL
Euro Laboratories Limited, Cranfield Biotechnology Limited,
Cranfield, UK

Abstract
A brief review of the principles of biodeterioration is
presented together with examples which are of relevance to
the civil engineering industry.
Keywords: Biodeterioration, Lubricants, Oils, Plastics,
Wood, Metals

1   Introduction

Biodeterioration may be described as a negative process in
the pragmatic sense of its definition. Implicit in the
study of biodeterioration processes, however, is the
positive aspect directed at controlling the problem
Biodeterioration has been defined as 'the study of the
deterioration of materials of economic importance by
organisms' (Hueck 1965). Eggins and Oxley (1980) widened
the definition to include structures and processes, whilst
an account of the meaning of biodeterioration and its
relationship with biodegradation have been described by
Eggins (1983). Eggins reminds us that the word
'deterioration' means 'to make worse' whilst 'degrade'
means 'to step down' or 'to break down'. Thus,
biodeterioration processes will include biodegradation
activity, but will also include a number of other perhaps
less direct effects of organisms on materials. We often
consider biodegradative processes as those which, by their
action, result in a product of improved quality. The
treatment of agricultural wastes to yield biomass or the
detoxification of waste pesticides are examples.
  Hueck (1965) has classified biodeterioration processes
into three types which reflect the wide meaning of the
definition.

1 **Mechanical**: insect and rodent attack on non-
   nutrient materials such as lead pipe and plastic
   cable.
2 **Chemical**: (a) assimilatory where the material is a

food source;   (b)   dissimilatory where waste products
or secondary metabolites are able to degrade the
material to no vital benefit of the organism.
3 **Fouling and soiling**: where the organism causes a
   worsening of the the material, structure, or process
   by its mere presence or the secretion of toxic
   metabolites. It may cause a stain, a blockage or
   foul the hull of a ship.

   This classification helps us to decide upon strategies
for the study and control of biodeterioration problems, but
does not necessarily reflect the metabolic activities or
ecological preferences of the organism. To a cellulose
utilising fungus there may be no distinction between a
natural cellulose fibre and a cotton fabric carefully
processed by man. The distinction is drawn with pathology
which is concerned with living organisms. Although there
may be some overlap biodeterioration does not concern
itself with disease, and this affects the control measures
which can be successfully used.
   The factors which encourage biodeterioration are legion,
and a combination of the inherent susceptibility of the
material by virtue of its chemical composition, surface
characteristics, or physical state.
The external environment and the extent to which
preventative measures are employed in an attempt to
control the biodeterioration processes must also be
considered. Environmental factors such as temperature,
humidity, pH, osmotic pressure and redox potential are very
important in initiating the problem and encouraging its
subsequent development. These factors have been reviewed
more extensively by Seal and Eggins (1981) and Onions et
al. (1981). The reader is also referred to Allsopp & Seal
(1986) for a readable introduction to biodeterioration.

2 The biodeterioration of materials of relevance to Civil Engineers

2.1 Wood
Living sapwood in any standing free, wherever it is growing
in the world, is more resistant to microbial attack than
heartwood. This is not the case, however, after the tree
is felled when the sapwood with its residual cell contents
and other food material in the parenchyma is subject to
microbial attack. There are several groups of micro-
organisms capable of colonising wood and in some cases
degrading it. They include Actinomycetes and other
bacteria, moulds, staining fungi, and soft, brown, and
white rot fungi (Levy 1969; 1971; Savory 1954; Liese 1970;
Carey 1975; Baecker and King 1981). Bacteria are not as
troublesome as fungi in the deterioration of wood which is

in service, i.e. wood which is not in contact with soil, although Kelly (1983) has recorded bacteria to be present in decaying in-service timber joinery in the UK. Actinomycetes are considered a significant group of organisms attacking wood when in soil contact (Baecker and King 1981).

The superficial decay of wood is known as soft rot in which Ascomycetes and certain Deuteromycetes are responsible for the destruction of cellulose in the wood cell wall (Savory 1954; Carey 1975). Brown and white rot fungi (Basidiomycetes) are responsible for causing a larger degree of degradation. Brown rot fungi attack the polysaccharide of the cell wall leaving the lignin unchanged. The wood becomes dark brown as decay proceeds with cracking along and across the grain (Carey 1975). White rot fungi are capable of attacking both lignin and cellulose, and may be generally distributed throughout the wood or localised into pockets of decay (Carey 1975).

In some areas of the world, however, decay agencies other than fungal ones are of importance. These include insects in the terrestrial environment, and crustaceans and molluscs in the marine environment.

2.2 Fuels and lubricants

Any country with a developed transport system and an engineering industry will use fuels and lubricants. Fuels will be stored prior to use and lubricants will be recirculated in closed or open systems. In both cases there will invariably be an ingress of water and the build-up of contaminant organic material leading to the growth of micro-organisms. The water may form a discrete layer in the bottom of a storage tank or it may be actively mixed with an oil to form an oil-in-water or water-in-oil emulsion. Such emulsions are used as lubricants and coolants in the metal-working industry to aid the machining or grinding or metal components, the rolling of aluminium, and the drawing of wires. Gasoline fuels used in piston driven internal combustion engines are generally made up of short-chain (less than C_9) volatile alkanes. These are not readily utilised by micro-organisms. Alkanes of chain length (C_{12}-C_{20}) are more susceptible to utilization, and this is reflected in the ability of a range of bacteria and fungi to grow at the interfaces of kerosene and water in the fuel tanks of jet aircraft, and diesel and water in gas turbine ship fuel tanks. The problems arising from microbial growth in fuels and lubricants fall into three categories.

1 The formation of troublesome bioslimes which may become detached and block pipework.
2 Losses in the useful properties of an additive.
3 The formation of metabolic products which directly or

indirectly contribute to corrosion of metal surfaces.

Fuels

Kerosene is widely used as an aviation fuel in jet engines.
For some time it has been known that in the presence of
water, which is an invariable consequence of temperature
changes during storage linked with solubility of water in
oil, fungal growth will occur at the fuel-water and water-
metal interfaces (Genner and Hill 1981). Fungal biomass
may become detached and drawn into fuel lines or block
filters resulting in fuel starvation to the engines.
Cladosporium resinae, the kerosene fungus, is most commonly
encountered in these situations although many other genera
have been isolated. The problem is normally associated
with grounded aircraft and is of particular importance in
hot climates where growth of the fungus is accelerated. In
flight temperatures may be as low as -40°C which, although
suppressing growth, may favour the survival of *C. resinae*
over other species in the fuel which are less resistant to
the low temperatures during flight.

At the water-metal interface, microbial colonies can
result in pitting corrosion as the consequence of a
differential aeration cell (see section on Metals). In
extreme cases, the pitting may cause complete penetration
of the aluminium fuel tank and fuel loss. The production
of acidic metabolic products and the utilization of
corrosion inhibitors may also help to control and delay
corrosion.

Diesel fuel used for driving marine gas turbine engines
can readily become contamination with fungi and bacteria.
Unfortunately, the use of sea water as a displacement as
the fuel is burnt ensures a constant microbial inoculum and
fresh nutrient supply for the micro-organisms. Fuel
blockages and corrosion problems have been observed to
occur resulting in reduced efficiency, which is of
particular concern to naval warships.

The prevention of microbial growths in fuels may be
achieved by adopting one or more of the following
practices.

1 The regular inspection of aircraft or ship fuel
  systems.
2 Monitoring the quality of fuel taken on board.
3 Adding biocides (organo-borons) to the fuel in order
  to suppress the growth of contaminants.
  Pasteurisation has also been suggested for ship fuels
  (Wychislik and Allsopp 1983)
4 Lining fuel tanks with materials to prevent corrosion.
  Butadiene nitrile rubber and polyurethanes are
  currently used in aircraft for this purpose.

Lubricants

Oils used as straight lubricants, where no water is present or at normal operating temperatures above 60°C, are not prone to biodeterioration problems. Two areas of usage, however, have received some attention: metal-working emulsions and lubricating oils used in slow-speed marine diesels. There are a range of emulsions used in the engineering industries for the drilling, cutting, grinding, and rolling of metals. Four classes may be distinguished.

1 **Neat oils**: straight mineral oils of varying viscosity.
2 **'Soluble' oil emulsions**: containing more than 50% by volume of oil in the undiluted form.
3 **Semi-synthetic 'soluble' oil emulsions**: containing less than 50% by volume of oil – perhaps as little as 5-10% – in the undiluted form.
4 **Chemical solutions**: containing no oil, but mixtures of synthetic compounds including esters and methyl silicones, which have lubricating and cooling properties.

It is the soluble and chemical solutions which are prone to contamination. A number of additives such as emulsifiers, corrosion inhibitors, extreme pressure additives, and coupling agents act as nutrients for the contaminants, resulting in a ion. The utilization of the emulsifier, accompanied by a reduction in pH can cause the emulsion to break, the oil separating out as a discrete layer. The development of anaerobic conditions in the sump (storage tank) can lead to the growth of sulphate-reducing bacteria which produce sulphide from sulphate in a dissimilatory fashion. The sulphide may combine with any iron present to generate a blackening of the emulsion in a highly reduced environment, or it may be released as hydrogen sulphide, usually first noticed when the emulsion recirculation system is turned on after shut-down at the week-end. Sulphides are, of course, highly corrosive and their effect may be detected on storage of the machined parts.

The conditions in which emulsions are used will often favour the growth and development of a microbial population. The concentrated soluble oil is mixed with water to give a final emulsion concentration of about 2% by volume. This is then continuously circulated around pipework, allowed to drip over machinery where it is reoxygenated, and often maintains a temperature of between 20 and 30°C depending upon the process and the climate. The system is continuously open to atmospheric contamination and to discarded wastes from the work-force. Bioslimes can develop on surfaces and microbes become

lodged in the maze of crevices and dead-ends which are
characteristic of many circulation systems.

Water can gain access to the oil in the crankcase of a
marine diesel engine from the water of the cooling system.
The water will then bring with it a microbial inoculum as
well as nutrients (Hill 1984). The oil is maintained at
35-45°C which is suitable for growth of many thermophilic
microbial species . The symptoms of an infection are
biomass (sludge) build-up, reduction in pH, malodours, and
oil emulsification. This results in corrosion problems and
the premature wearing of bearings. The use of various oil
cleansing systems such as filters, coalescers, and
centrifuges reduces contaminating water and particulate
matter, and this in turn reduces the likelihood of
microbial growth. The use of biocides is necessary only
periodically when problems arise.

2.3 Rubbers and plastics
Natural and synthetic formulations of rubbers and plastics
are extensively used in construction where they are often
exposed to microbial environments in water, and buried or
in contact with the soil. Many interacting forces may be
involved in a failure problem and it is often difficult to
assess the relative importance of micro-organisms in the
process. The recent review by Seal (1988) brings together
all the data available in this area.

The bulk polymers, polyethylene, polystyrene, and
polyvinyl chloride, in use today are regarded as
recalcitrant molecules and not, in their own right, subject
to biodeterioration. Workers have shown though that dimers
and short-chain oligomers of styrene are degraded
(Higashimura et al. 1983) and that unreacted low molecular
weight oligomers (less than 500 mol. wt) of ethylene
present in polyethylene can result in limited growth on the
polymer. However, it requires physical ageing to encourage
subsequent microbial utilization which then only occurs at
very low rates (figures in the literature vary between 0.3
and 3% weight loss per year; see Stranger-Johannessen
1979). Polymers based upon natural biopolymers, such as
proteins and cellulose, tend to be susceptible to
degradation and this limits their use to situations where
they never or only periodically come into contact with
conditions conducive to microbial growth. Thus, products
such as cellophane (packaging), cellulose ethers
(thickeners in emulsion paints and foodstuffs), casein-
formaldehyde (early type of plastic), crepe (latex rubber),
and rayon (regenerated cellulose), will all support the
growth of micro-organisms under suitable conditions (see
Allsopp and Seal 1986).

Natural rubber is a polymer of *cis* 1,4-isoprene and
contains between 2 and 3.5% protein. In its purest form
(pale crepe) it may still contain significant levels of

protein, lipid and carbohydrate, and is capable of absorbing up to 15% moisture. This can result in growth of micro-organisms on products made from crepe (Williams 1982), especially when they are exposed to high humidity conditions. The vulcanization process drastically increases the resistance of rubber to biodeterioration, but if subsequent oxidation takes place microbial growth may occur, resulting in cracking of the surface. Liquid latex is also prone to contamination and has to be protected using chemical preservatives.

There are a number of synthetic rubber formulations in use today including polyisoprenes, styrene-butadiene, neoprene, nitrile, and silicone rubbers. Of these only the polyisoprene and styrene-butadiene have been found to support microbial growth in any significant amounts. The other rubbers appear in both laboratory testing and in the field to be extremely recalcitrant. As we shall see below, it is often the impurities or additives which encourage biodeterioration.

The other main group of polymers worthy of mention are the polyurethanes. These are a diverse group of polymers based upon the reaction between an isocyanate and hydroxyl-containing compounds such as butan-1, 4-diol. In the presence of suitable catalysts, a long chain polymer is produced containing urethane groups which are similar to peptide bonds and subject to hydrolysis. There is much flexibility in the formulation of polyurethanes, and polyesters and polyethers can be inserted into the polymer to give the product elastomeric properties. The presence of ester bonds in a polyester polyurethane makes it more susceptible in the short term to biodeterioration than a polyurethane containing polyether chains (Seal and Pathirana 1982; Pathirana and Seal 1985). Within this generalization, it has been found that the different isocyanates and polyols used can also affect susceptibility (Darby and Kaplan 1968).

Cracking and embrittlement of cables, clothing, and automotive components containing polyester polyurethanes have been observed after relatively short in-service use (2-3 months in warm humid conditions). The storage of some car components in polythene packaging prior to use in the tropics has resulted in severe deterioration.

Many additives are incorporated into plastic formulations including plasticisers, ultra-violet stabilisers, antioxidants, pigments, fillers, processing aids, hydrolysis stabilisers, and fire-retardants. Individually, they are incorporated at low levels and most have been shown to have very little effect on the susceptibility of the formulation to biodeterioration. However, in total they may constitute 50 - 60% of the total weight and their combined utilization by micro-organisms may have a significant effect on the properties of the product. The additives most extensively studied have been plasticisers which confer flexibility on an otherwise brittle polymer, and may make up 40% of the total weight of the plastic. Plasticisers are extensively used in polyvinyl chloride (Seal & Pantke 1988), where a pliable product is required. They are esters of aliphatic (adipic, sebacic) and aromatic (phthalic) acids and as such may be subject to enzymatic hydrolysis.

The early literature recorded lists of susceptible and resistant plasticisers (see Berk et al 1957), but later research by Klausmeier (1966) showed that, using suitably isolated soil micro-organisms from enrichment cultures, degradation of previously resistant plasticisers could be observed. Work by Ribbons et al (1984), and Williams and Dale (1983) has further demonstrated that the phthalate plasticisers in particular are not as inert as was previously reported. It is, however, important to consider the physical aspect of the susceptibility of a plasticizer. It must be available at the surface of the plastic and not immobilized within the structure preventing its migration. Low migration rates reduce the rate at which biodeterioration occurs. Short-chain phthalates, such as dibutyl phthalate, are volatile and have been shown to have phytotoxic effects (Hardwick et al 1985). The effect may be extended to micro-organisms under confined conditions.

The presence of other additives such as organic fillers (cellulose based) or processing aids such as lubricants (based on vegetable oils, stearates) will all encourage surface growth which may then extend to other parts of a piece of equipment.

2.4   Surface coatings, sealants and adhesives
These products are formulations of a range of materials, some of which can encourage growth of micro-organisms. Paints contain polymers (acrylates, methacrylates, polyurethanes, and polyvinyl acetate), linseed oil, and cellulose ether thickeners. Sealants may be bitumen or synthetic polymer based, whilst adhesives may contain casein, starch, or cellulose. Water-based formulations of paint and adhesives may further encourage deterioration to occur during storage of the product prior to use (known as 'in-can spoilage'). This results in the loss in viscosity of an emulsion paint due to enzymatic degradation of the cellulose thickener (Tothill et al 1988) or the production of an odour in a casein-based adhesive, detected when the lid is first removed. While in service, the paint, sealant, or adhesive film may be subject to colonization by a range of micro-organisms which directly utilise the film causing cracking and loss of usefulness. The passive colonization of algae and fungi may, by the trapping of dirt and water, lead to aesthetic changes or accelerated environmental effects. Water may freeze and crack the

surrounding area, and the products of metabolism may accelerate hydrolysis and solubilization of the film (see Allsop and Seal 1986; Goll and Winters 1974; Upsher 1984).

2.5 Metals

The corrosion of metals is an electrochemical phenomenon. Whilst it is probably agreed that micro-organisms contribute to corrosion processes, the extent of the influence which micro-organisms exert on the corrosion process, and the mechanisms by which this occurs, are still under review (Miller 1981; Cragnolino and Tuovinen 1984). The literature cites a number of possible involvements which include the production of corrosive metabolic products such as acids, hydrogen sulphide, and ammonia, the formation of differential aeration electrolytic cells by fungal colonies on metal surfaces, and the assimilation of hydrogen produced at the cathode (termed cathodic depolarization). For a fuller description of these mechanisms the reader is referred to Miller (1981). The effects of these mechanisms are to enhance the rate of corrosion over that which would occur under the same conditions in the absence of micro-organisms. The corrosion results in either perforation (aluminium fuel tanks on aircraft are an example) or the formation of corrosion products (tubercles) which can reduce the diameter of cast iron water pipes and adversely affect flow rates. The acid product of the sulphur-oxidizing bacteria can be sulphuric acid capable of etching and perforating ferrous metals. Its greatest effect has been on concrete storage tanks and sewerage pipes in hot climates where, in the latter case, the lack of water reduced the volume of water flushed through the sewers each day so that they remained stagnant.

In recent years an economically important group of micro-organisms - the sulphate-reducing bacteria (SRB) (see Postgate's excellent monograph 1984) - has received much attention in connection with its involvement in metal corrosion. This has been due to two factors; the unique ability of the group to reduce sulphate to sulphide in a dissimilatory fashion, and the possession of a hydrogenase. Sulphide is corrosive in forming iron sulphide, which can function as a cathode, whilst hydrogenase is thought to be involved in removing hydrogen produced at the cathode, the presence of which can reduce corrosion (Parker et al 1988). SRB have been implicated in metal corrosion in anaerobic clays, and more recently in oil rigs involved in off-shore oil production where the use of sea-water injection systems introduces contamination into the well resulting in SRB activity. Hydrogen sulphide is produced and the oil becomes sour. Corrosion of pipework and storage areas may follow. Evidence is also accruing on the presence of SRB

in the fouling layers of the external oil platform
structures. Anaerobic conditions may be established and
the SRB then cause pitting corrosion beneath the fouling
(Costelow and Tipper 1984; Lewis and Mercer 1984;
Southwell et al 1974).

2.6 Other miscellaneous materials

Biodeterioration problems have been recognised in third
world countries in unusual cases such as where stone is
fashioned (Dukes 1972), or glass is being used for the
manufacture of microscope lenses (Nagamuttu 1967). Natural
animal products such as leather and wool will also
deteriorate under humid conditions. This is of particular
concern in the conservation of valuable objects of historic
interest. Algae are of particular importance as
biodeteriogens in connection with their soiling effects on
stonework and buildings in the terrestrial environment
(Grant 1982). Wee and Lee (1980) have reported that high
rise buildings in Singapore are badly affected by algal
growths. Yong et al (1972) have also noted that algae will
grow on exposed concrete and surfaces coated with cement
paints in Singapore. The use of fungicidal paints or
periodic cleaning with biocidal washes have been
recommended to contain the problem where this is practical
(Richardson 1973; Bravery 1981). Glass lenses may be
colonised by fungi under high humidity conditions, leading
to etching of the surfaces. Dry storage conditions and use
of volatile fungicides are recommended as control
strategies in such cases (Baker 1967).

3 Control methods

The specific control methods cited in the above sections
fall into two categories: the use of physical parameters
such as temperature pH, osmotic pressure, gaseous
atmospheres, and moisture content/relative humidity; and
the incorporation of chemical agents (biocides) to suppress
growth or kill the biodeteriogen outright. The choice of
method depends upon a number of factors which must be
assessed when a control strategy is sought. Biocides are
widely used in solvent- and aqueous-based systems where the
use of physical agents is not practical. Allsopp and
Allsopp (1983) have produced a list of biocides with
reference to their use and effective concentration ranges.
Proper control is only possible when the ecology and
mechanisms of the biodeterioration problem are fully
appreciated. It is all too easy to recommend a method of
control in the hope that it will be a panacea for all
problems. As this is never the case, it is necessary to
recognise the problem and evaluate it before instituting
control measures.

References

Allsopp, C and Allsop, D (1983). An updated survey of
commercial products used to protect materials against
biodeterioration. **Int. Biodeter. Bull.** 19, 99-146.

Allsopp, D and Seal, K J (1986) **An introduction to
biodeterioration.** Edward Arnold, London.

Baeeker, A A W and King, B (1981) Soft rot in wood caused
by Streptomyces. **J, Inst. Wood Sci,** 9, 65-71

Baker, P W (1967) An evaluation of some fungicides for
optical instruments. **Int. Biodeter. Bull.** 3, 59-64

Berk, S, Ebert, H, and Teitell, I (1957). Utilisation of
plasticisers and related organic compounds by fungi.
Ind. Eng. Chem. 49, 1115-24.

Bravery, A F (1981) Preservation in the Construction
Industry. In **Principles and practice of
disinfection** (eds A D Russel, W B Hugo and G A J
Ayliffe) pp; 379-402. Blackwell Scientific
Publications, Oxford.

Carey, J K (1975) Isolation and characterisation of wood-
destroying fungi. In **Microbial aspects of the
deterioration of materials** (eds D W Lovelock and
R J Gilbert) pp 23-37. Academic Press, London

Costlow, J D and Tipper, R C (1984) **Marine
biodeterioration: an interdisciplinary study.**
Naval Press, Annapolis, Maryland

Cragnolino, G and Tuovinen, O H (1984). The role of
sulphate-reducing and sulphur-oxidising bacteria in the
localised corrosion of iron-based alloys - a review.
Int. Biodeter. 20, 9-26

Darby, R T and Kaplan, A M (1968) Fungal susceptibility of
polyurethanes. **Appl. Microbiol.** 16, 900-5

Dukes, W h (1972) Conservation of stone: causes of decay.
Architects J. 156, 422-9

Eggins H O W (1983) Biodeterioration, past, present and
future. In **Biodeterioration** (ed T A Oxley and
S Barry) Vol 5 pp 1-9. J Wiley & Sons, Chichester

----- and Oxley, T A (1980) Biodeterioration and
biodegradation. **Int. Biodeter. Bull.** 16, 53-6

Genner, C and Hill E C (1981) Fuels and oils. In
Microbial biodeterioration (ed A H Rose) pp260-306.
Academic Press, London

Goll M and Winters H (1974) Pseudomonads and their
cellulases - mechanism of action and mode of detection.
J Paint Technol. 46, 49-52

Grant C (1982) Fouling of terrestrial substrates by algae
and implications for control - a review, **Int,
Biodeter. Bull.** 18, 57-65

Hardwick, R C, Cole, R A, and Fyfield T P (1985).
Plastics, phytotoxic vapours, and plant death.
Biologist 32, 22-4

Higashimura, T, Sawamoto, M, Hiza T, Karaiwa, M, Tsuchii,A and Suzuki, T (1983). Effect of methyl substitution on microbial degradation of linear styrene dimers by two soil bacteria. **Appl. Environ.Microbiol.**46, 386-91

Hill, E C (1984). Micro-organisms - numbers, types, significance, detection. In **Monitoring and maintenance of aqueous metal-working fluids** (eds K W A Chater and E C Hill) pp 97-112. J Wiley & Sons, Chichester

Hueck, H J (1965) The biodeterioration of materials as part of hylobiology. **Material und Organismen** 1, 5-34

Kelly, D M T (1983) Biotic and climatic factors affecting the colonisation of wood-invading fungi. PhD Thesis. Lancashire Polytechnic, Preston, UK

Klausmeier, R E (1966) The effect of extraneous nutrients on the biodeterioration of plastics. In **Microbiological deterioration in the tropics** pp 232-43. Society of Chemical Industry No 23, S.C.I, London

Levy, J F (1969) The spectrum of interaction between fungi and wood. **Rec. Ann. Conv. Br. Wood Preservers Ass.** 3, 81-97

----- (1971) Further basic studies on the interaction of fungi, wood preservatives and wood. **Rec. Ann. Conv. Br. Wood Preservers Ass.** 5, 63-75

Lewis, J R and Mercer, A D (1984) **Corrosion and marine growth on off-shore structures**. Ellis Horwood Ltd, Chichester

Liese, W (1970) The action of fungi and bacteria during wood deterioration. **Rec Ann Conv Br Wood Preservers Ass** 4, 1-14

Miller, J D A (1981) Metals. In **Microbial biodeterioration** (ed A H Rose) pp 149-202. Academic Press, London

Nagamuttu, S (1967) Moulds on optical glass and control measures. **Int Biodeter. Bull** 3, 25-7

Niloufari, P and Cockroft, R (1984) **Wood preservation in Iran**. Report no 412. National Swedish Board for Technical Development

Ocloo, J K (1978) The natural resistance of the wood of **Terminalia ivorensis**. A. Chev. (Idigbo, Emere) to both fungi and termites. **J. Inst. Wood Sci.** 8, 20-3

Onions, A H S, Allsopp D, and Eggins H O W (1981). **Smith's introduction to industrial mycology**, 7th edn. Edward Arnold, London

Parker, C H J, Seal, K J, and Robinson, M J (1988). Hydrogen absorption during the microbial corrosion of steel. **Biodeterioration** 7. (eds Houghton et al). Elsevier Applied Science, London

Pathirana, R A and Seal, K J (1985) Studies on polyurethane deteriorating fungi. Part 3. Physico-mechanical and weight changes during fungal deterioration. **Int Biodeter.** 21, 41-9

Postgate J R (1984) **The sulphate-reducing bacteria** 2nd edn. Cambridge University Press, Cambridge

Ribbons, D W Keyser, P Kunz, D A and Taylor, B F (1984) Microbial Degradation of Phthalates. In **Microbial degradation of organic compounds** (ed D T Gibson) pp 371-97. Marcel Dekker Inc., New York

Richardson, B A (1973) Control of biological growths. **Stone Industries** 8, 2-6

Savory, J G (1954) Breakdown of timber by ascomycetes and fungi imperfecti. **Ann. Appl. Biol.** 41, 336-47

Seal, K J (1988) The biodeterioration and biodegradation of naturally occurring and synthetic plastic polymers. **Biodeterioration Abstracts 2(4)** 295-317

Seal, K J and Eggins, H O W (1981) Biodeterioration of Materials. In **Essays in applied microbiology** (eds J Norris and M Richmond) pp 8/1-8/31. J Wiley & Sons, Chichester

Seal, K J and Pantke, M (1988) Microbiological testing of plastics. **International Biodeterioration** 24, 313-320

-----, and Pathirana, R A (1982). The microbiological susceptibility of polyurethanes - a review. **Int. Biodeter. Bull.** 18, 81-5

Southwell, C R, Bultman, J D, and Hummer, C W (1974) **Influence of marine organisms on the life of structural steels in seawater**. NRL Report 7672. National Technical Information Services: Springfield, Virgina

Stranger-Johannessen, M (1979) Susceptibility of photo-degraded polyethylene to microbiological attack. **J. Appl. Polymer Sci.** 35, 415-21

Tothill, I E, Best, D J and Seal, K J (1988) The isolation of **Graphium putredinis** from a spoilt emulsion paint and the characterisation of its cellulase complex. **International Biodeterioration 24,** 359-366

Upsher, F J (1984) Fungal colonisation of some materials in a hot-wet tropical environment. **Int. Biodeter.** 20, 73-8

Wee, Y C and Lee, K B (1980) Proliferation of algae on surfaces and buildings in Singapore. **Int. Biodeter. Bull.** 16, 113-7

Williams, G R (1982) The breakdown of rubber polymers by micro-organisms. **Int. Biodeter. Bull.** 18, 31-6

-----, and Dale, R (1983) The biodeterioration of the plasticizer diotyl phthalate. **Int. Biodeter.** 19, 37-8

Wychislik, E T and Allsopp, D (1983) Heat control of microbiol colonisation of shipboard fuel systems. In **Biodeterioration 5** (eds T A Oxley and S Barry) pp 453-61. J Wiley & Sons, Chichester

Yong, F N, Yeow, C T, Chua, N H and Wong, H A (1972) Method for screening and evaluating algicidal and algistatic surface coatings. **J. Singapore Inst. Architects** 53, 13-9

PART THREE
WATER SUPPLY ENGINEERING

5 BACTERIAL GROWTH POTENTIAL IN THE DISTRIBUTION SYSTEM

O. ADAM and Y. KOTT
Environmental and Water Resources Engineering Technion-
Israel Institute of Technology, Haifa, Israel

Abstract
Treated water was found to show bacterial aftergrowth in the distribution
system. Out of the various methods that are suggested to follow and measure
the bacterial regrowth capability in the distribution system is the method
that allows to measure the Maximum Growth Rate (MGR), that uses sensitive
turbidity meters measuring turbidity at 12° forward scattering light, attached
to a microcomputer. Water from the National Water Carrier was treated with
chlorine, chlorine dioxide, or ozone, was examined, for aftergrowth
capability. It was found that after ozonation, bacterial growth was enhanced.
Application of chlorine reduced regrowth capability of the water at 3-5 mg/l
applied chlorine. The same applies to chlorine dioxide at 3-10 mg/l applied
disinfectant. low chlorine or chlorine dioxide concentration of about 1mg/l
enhance aftergrowth of bacteria. Results from sampling of the distribution
system, indicated where potential aftergrowth of bacteria could occur and the
difference in quality between the various sampling points.
Keywords: chlorine, chlorine dioxide, regrowth, Maximum Growth Rate-
MGR

1 Introduction

There is a general belief that "water was created pure" (Chemviron Carbon
1987). This has been found to be erroneous already in Biblical times when
Elisha was called to heal the water source of Jericho, as a clear sign of enteric
disease in the water was mentioned [the Bible, Old Testament].
Currently in more developed societies, the basic bacterial criteria for water
quality is based, on one hand, on coliform fecal coliform or E. coli counts as
indicators, and total bacterial counts on the other. [APHA 1985, Cliver and
Newman 1987; C.E.C.E. 1980; W.H.O. 1982]. Often in these societies the water is
abstracted in one place and is transported hundreds of kilometres to other
places to open reservoirs which are subjected to dust, or micropollutants that
come out from industrial emission, which might dissolve in the water. In
addition, it is a very common practice nowadays to mix water which is
abstracted originally from different sources. The mixing might influence the
possibility of bacteria that exists in the water to grow in the distribution
systems.
Colony counts from water samples that are recommended in the U.K. are aimed
to indicate the efficiency of processes in the water treatment (Great Britain,
Dept. of Health 1969). However, the standard plate count in other countries
emphasize part of essential bacterial quality demanded by the authorities. This
might range from Poland 25/ml. Federal Republic of Germany 40/ml or 10/ml

if the water is chlorinated. (McNeill 1985), through the Canadian guideline which requires less than 500/ml bacteria based on geometric mean of the results from ten monthly samples (Cliver and Newman 1987) or 1000/ml in Israel (Israel Ministry of Health 1989).

The ability of bacteria to grow in water was noticed already as early as 1929 by Baylis (Baylis et al 1930). However, only in recent years due to long distribution systems mixing of various qualities, and additional knowledge concerning micropollutants in the water, more detailed research has been performed. The assimilable organic carbon (AOC) studies developed by Van der Kooij turned to be a key method in Holland and other countries (Van der Kooij et al 1982 a,b, Van der Kooij and Kijnan 1985). The use of ATP to measure bacterial regrowth in water is used as well as epifluorescence for direct count which is used by some others. The possibility to evaluate the maximum growth potential of bacteria in water was developed by Werner (Werner 1985).

The aim of the present study was to evaluate the regrowth potential of bacteria in water that was disinfected by various disinfectants.

2 Materials and Methods

Raw water from the Netufa Reservoir which supplies water to the national water carrier was used for the experiments. The laboratory experiments were performed using ozone generator Welsbach Model T-408. The ozone concentration was determined by the iodine titration method as mentioned in Standard Methods (APHA 1985). Regrowth experiments using sodium hypochlorite were performed in the laboratory as well as the chlorine dioxide. Few chlorine dioxide field samples were taken from pipelines which are branched from the national water carrier. Bacteria used for inoculation. Small activated carbon columns were fed daily with national water carrier water to enable indigenous bacteria to grow. The effluent water from the columns were centrifuged and the Pellet was resuspended in saline to a final concentration of 10^3 ml bacteria, which was used for the regrowth experiments.

3 Experimental Procedure

Sample of the water to be examined was inoculated to nutrient agar (Difco) incubated for 48h at 37°C, colonies were counted as mentioned in Standard Methods (APHA 1985). The main water portion was treated as follows: Filtration through 0.45μm pore size membrane filter (micropore). The water sample free from bacteria were seeded with 10^3 ml bacteria taken from the activated carbon columns, and divided to two parts. Part of it was incubated for 48h at 37°C. The major part was placed in quartz cells fitting Monitek model 251 turbidity meters, measuring turbidity at 12° forward scattering light. A magnetic stirrer was assembled which enabled continuous stirring. Each of the four turbidity meters was commanded by an HP 41 CX Hewlett-Packard minicomputer which measured the time interval and recorded the turbidity. A plotter attached to the system recorded the growth. Most of the experiments lasted 60 h at room temperature (18°C - 22°C). The recorded results enabled to calculate the Maximum Growth Rate (MGR) by using the programmed Hewlett-Packard calculator as was already published (Werner 1985; Adam and Kott 1989).

4   Results

Results from experiments performed with application of chlorine to the examined water as summarized in Table 1, from which it can be seen that at low concentrations of applied chlorine, the water enhanced bacterial growth. Application of 3mg/L and more of chlorine showed reduction in maximum growth rate (MGR) as compared with control water. When bacterial counts were compared, it can be seen that after 48 h all the examined water showed the same order of magnitude counts.

Table 1. Effect of Sodium hypochlorite on regrowth capability of bacteria in water

Chlorine added mg/L	Maximum Growth Rate (MGR)		Percent change in Growth		Bacterial Count/mL	
	Before	After	Reduction	Enhancement	0 h	48 h
1	0.1083	0.1234	-	13.9	8.0×10^2	3.9×10^6
2	0.1083	0.1146	-	5.8	8.0×10^2	3.8×10^6
3	0.2471	0.2347	5.0	-	4.1×10^3	4.0×10^6
5	0.2471	0.2397	2.0	-	4.1×10^3	3.0×10^6
10	0.2471	0.2253	8.8	-	4.1×10^3	2.3×10^6

Note: Results are an average of five experiments.

When chlorine dioxide was applied to the examined water the same trend was seen (Table 2). However, the percent of enhancement in the low concentrations of applied chlorine dioxide were higher, as well as reduction in MGR was higher. The bacterial counts after 48 h of incubation were the same order of magnitude. It was mentioned already that ozone might enhance bacterial growth. [Bancroft et al.,1984; Werner 1985 b; Adam and Kott 1989]. However, it seems that a comparison between MGR of other disinfectants and ozone was not done. Table 3 summarizes results that were received with the same water as the other experiments. It is worthwhile to note that at low applied concentration of ozone a reduction in MGR was noted. When higher ozone concentrations were applied, enhancement of MGR was noted. The plate counts showed growth of about two orders of magnitude after 48h.

Table 2. Effect of Chlorine dioxide on regrowth capability of bacteria in water.

Chlorine dioxide added (mg/L)	Maximum growth rate (MGR)		Percent change		Bacterial count/mL	
	Before	After	r.	e.	0 h	48 h
1	0.0542	0.0654	-	20.6	3.8×10^2	5.2×10^5
2	0.0542	0.0408	-	24.7	3.8×10^2	4.8×10^5
3	0.2232	0.1926	13.7	-	1.2×10^3	5.6×10^5
5	0.1727	0.1450	16.0[a]	-	2.0×10^2	3.8×10^6
10	0.2232	0.1855	16.9	-	1.2×10^3	4.8×10^5

a. A significant result that represents an average of five experiments.
r.- growth reduction ; e.- growth enhancement.

Table 3. Effect of Ozone on regrowth capability of bacteria in water

Ozone applied (mg/L)	Maximum growth rate (MGR)		Percent change in growth		Bacterial Count/mL	
	Before	After	r.	e.	0 h	48 h
1	0.1587	0.5590	1.76	-	$8.0x10^3$	$2.1x10^5$
2	0.0918	0.1026	-	11.7	$1.5x10^3$	$4.8x10^6$
3	0.1592	0.2271	-	42.6	$3.3x10^3$	$9.4x10^5$
5	0.1592	0.1823	-	14.5	$3.3x10^3$	$7.5x10^5$

Note: Results are based on an average of five experiments.
r.-growth reduction; e.- growth enhancement.

Table 4 summarizes results from water samples taken from the distribution system which orginially were national water carrier, water. The water is subjected at the exit of Point B.N. with chlorine dioxide, and again at Point KJ. It could be that this is the reason for the high MGR received at Point KJ. For an unknown reason sampling Point SH showed a very significant low MGR. Sampling Point NE showed an increase in MGR, which incidently fits complaints of the citizens in the area. The growth of bacteria in this research was followed up, up to 60 hours of incubation. However when the incubation was prolonged, a second MGR could sometimes be observed.

Table 4. Bacterial regrowth capability along the distribution system

Sampling point	Cholorine dioxide maximum residual	Regrowth (MGR)	Regrowth %	Bacterial count/ml 0 h	48h
B.N. (as control)	-	0.1409	-	$7.0x10^3$	$2.6x10^6$
KJ	0.4	0.2131	51.2	$7.0x10^3$	$2.5x10^6$[a]
KA	0.3	0.1489	5.8	$2.4x10^4$	$4.0x10^6$
SH	0.1-0.15	0.1079	-23.4[b]	$7.0x10^3$	$9.0x10^5$
NE	0.4-0.5	0.1743	23.7	$1.1x10^2$	$1.0x10^6$
KM	0.3[c]	0.1434	1.8	$7.0x10^3$	$2.4x10^6$

[a] after 24h.
[b] decrease in growth.
[c] chlorine gas applied.
Each figure represents an average result of two inoculations.

Figure 1 shows the two repitition experiments from sampling Point KM. It is worthwhile to mention that the MGR was after 20h and 26h of incubation, but when the water was incubated long enough a secondary MGR could be seen.

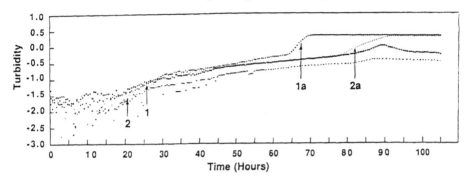

Fig. 1:   Two logarithmic growth phase in chlorinated water.

5   Discussion

Surface water might contain dissolved chemicals dust, bacteria and frequently, algae, protoza. In many countries, surface and other sources of raw water is treated, and by this, much of the organic material is removed. However, it was proven that despite of it, bacterial regrowth might occur, which effects the water quality [Olson 1982; Limoni and Teltsch 1985].In many countries where water temperature is low, bacterial aftergrowth would be very slow. Higher temperatures of the distribution system and long delivery pipes would enable better growth. The way to keep bacterial counts at low level would be disinfection. This study intended to evaluate regrowth capability of bacteria in water that was disinfected with chlorine, chlorine dioxide, and ozone. The results received showed that low concentration of chlorine enhanced the MGR. It is, though, that small particles of micropollutants were detached from each other, and became availabe to bacteria. While higher concentrations of applied chlorine caused a decrease in MGR, which indicated that chlorine had oxidized the available particles which were essential for bacterial growth. It might be interesting to mention that chlorine dioxide showed the same effect, although it is very well known that the chlorine dioxide has shown different oxidizing chemistry as compared with chlorine. Application of ozone to water is known to oxidize indifferently   any organic materials, thus it might produce inhibitors which might effect bacteria growth. The results in this study, where relatively small concentrations of ozone were applied, showed that at small concentration of applied ozone, the MGR decreased, as if inhibitors were created. However, at higher concentration, MGR was enhanced, just contrary to chlorine effect. It is interesting to mention that by using the growth of bacteria and being able to evaluate the maximum growth rate, it enables to follow, indirectly, their behavior of ecological effects in the system; whereas, exhaustion of growth materials essential to some bacteria, enables revival of others. This phenomenon seen, and not investigated yet, could lead to more detailed evaluation of bacterial growth in long pipes, where delivery from one point to another, might take several days.

6 Acknowledgment
The study was supported in part by the Mekorot Water Co. Ltd.

7 References

Adam, O. and Kott, Y. (1989) Evaluation of water quality as measured by bacterial maximum growth rate. **Water Res.** 23, 1407-1412.

Adam, O. ar.d Kott, Y. (1989). Bacterial growth in water. **Toxicity Assessment**, 4, 363-375.

American Public Health Association (1985). Standard Methods for the Examination of Water and Wastewater 16th ed. American Public Health Association, Washington, D.C.

Bancraft, K. Chrostowsk, Wright, R.L. and Suffet, H. (1984) Ozonation and oxidation competition values. Relationship to disinfection and micro-organisms regrowth. **Water Res.** 18, 473-478.

Baylis, John R., Chairman, E. Sherman Chase, C.R. Cox, J.W. Ellms, D.A. Eerson Jr., H.V. Knouse, H.W. Streeter (1930). Bacterial aftergrowth in water distribution systems. **Am. J. Pub. Health**, 20, 485-491.

Chemvivon Carbon Advertisment. **Water and Waste Treatment**, 30, (1987).

Cliver, D.O. and Newman, R.R (Editors) Committee on the challenges of modern Society (NATO/CCMS), 128, Special Issue, Drinking Water Microbiology. J. Environ. Pathol., Toxico. and Onco. 7, (5,6) pp. 282-287. Commission of the European Communities (1980). Council directives on drinking water.

Comission of the European Communities (1980). Council directive on drinking water.

Great Britain Department of Health and Social Security, Welsch office, Ministry of Housing and local Government. The Bacteriological Examination of water supplies. 4th ed. London. H.M. Stationery office (1969) Reports on Public Health and Medical Subjects (No.71).

Israel Ministry of Health (1989) Israel Water Quality Law.

Limoni, B. and Teltsch, B. (1984) Chlorine dioxide disinfection of drinking water. An evaluation of a treatment plant. **Water Res.** 9, 1489-1495.

McNeil, A.R. Microbiological water quality criteria: A review for Australia, Australian Government Publishing Service, Canberra. (1985) pp 274-277.

Olson, B.H. (1982) Assessment and implications of bacterial regrowth in water distribution systems. Project Summary/**USEPA** -600/S2-82-072.

The Bible - Old Testament - **Genesis** 26; 19-22.

Van Der Kooij, D. and W.A.M. Hijnen. (1985). Measuring the concentration of easily assimilable organic carbon (AOC) treatment as a tool for limiting regrowth of bacteria in distribution systems. Proc. Am.W.W.A. Technical Conf. Houston, Tex. American Water Works Association, Denver.

Van Der Kooij, D., A. Visser and J.P. Oranje. (1982a). Multiplication of fluorescent pseudomonas at low substrate concentrations in tap water. Antonie Van Leeuwenhoek. **J. Microbiol.** 48, 229-243.

Van Der Kooij, D., A. Visser and W.A.M. Hignen. (1982b). Determing the concentration of easily assimilable organic carbon in drinking water. **J. Am. W.W.A.** 74, 540-545.

Werner, P. (1985a) .Eine methode zur bestimnung der verkeimungsneigvng von Trinkwasser. **Von Vasser.** 65, 257-270.

Werner, P. (1985b). A new way for the decontamination of polluted aquifers by biodegradation. **Water Supply.** 3, Berlin. B. 41-43.

World Health Organization (1982). Guidelines for drinking water quality .Vol. 1. World Health Organization, Geneva.

6 FOULING AND CORROSION IN WATER FILTRATION AND TRANSPORTATION SYSTEMS

R.G.J. EDYVEAN
University of Leeds, UK

Abstract
Even the cleanest of natural waters will carry a biological loading of bacteria, microalgae and protozoa, as well as larger organisms. A filtration and transportation system provides a greatly increased surface area to the microorganisms and this has important implications in the formation of biofilms. Such biofilms may have significant effects on filtration, fluid flow, heat transfer, corrosion and the development of and release of, for example, pathogens and corrosive substances, further into the system. The environmental variables of such systems vary widely, according to the end use. These include the degree of filtration, use of biocides and additions of corrosion inhibitors. However, despite treatments, most systems will have active biofilms (although these need not necessarily be detrimental). The formation and effects of biofilms is described using systems in the off-shore oil industry as examples
Keywords: Filtration, Fouling, Corrosion, Water systems.

1 INTRODUCTION

The effects of fouling on filtration systems and water carrying pipework can be extensive and lead to increased costs due to lengthy shut-downs for repairs, filter changeouts *etc*. Filtration systems can become blocked while fouling will cause loss of efficiency in fluid flow and heat exchange, as well as corrosion problems, in pipework. There is a considerable body of information on this subject from the oil industry where, for use in reservoir pressure maintenance, large volumes of water are filtered and treated with the aim of removing all particles above 2-5 microns. Information is also available from cooling water systems in many industries. While every system has highly individual characteristics, the lessons learnt off-shore and in other industries have a general applicability.

2 TYPES OF WATER SYSTEM

There are two main types of water system, open or once-through, which tend to handle large volumes of water and to which the addition of persistent biocides and corrosion inhibitors would be prohibitive, and closed systems, which tend to have limited volumes of water and to which persistent biocides and corrosion inhibitors can be added. Examples of open systems include potable water supplies, sewers and industrial systems where water is used in large quantities, for example once-through cooling and the water handling systems on off-shore oil production platforms.

Examples of closed systems include recirculating cooling and many other industrial water systems. While biofouling can be found in both types of system and, indeed, biofouling problems can be greater in closed systems, open systems often provide the greatest challenge as they are constantly reinoculated and the fouling cannot be controlled by the continuous presence of biocides.

3 FILTRATION SYSTEMS

Filtration is used to control the size of particles in a fluid stream. This may seem an obvious statement but it is important to remember that filtration is not an antifouling or other treatment system, it only removes particles above a certain size and only in exceptional circumstances will that size include the smaller living organisms such as bacteria and the settlement stages of other fouling organisms. Thus a filtration system cannot be expected to prevent or cure a problem that is already established, or that may develop down-stream, in the system. Indeed filtration may actually enhance such problems.

Filtration systems are many and varied. However, they all tend to share the same principal features, that is they consist of a series of units removing progressively smaller particles until the desired "cut" is reached. These units may consist of coarse and fine screens, fibre bag or mixed media deep-bed filters, followed by finer sand, cartridge or more novel types. A typical multistage system used to filter seawater on off-shore oil production platforms prior to waterflooding the oil reservoir is shown diagrammatically in Fig 1. Seawater is lifted from the desired depth by pumps or gas lifts, screened to millimetres (not shown) and then to 80-100 microns, deaerated and fine filtered to 2-5 microns using disposable polypropylene cartridges. Guard filters may be used prior to the injection pumps. Typically, systems will be handling 70 million litres per day. However, one important point that the diagram does not show is the large amount of pipework, both up stream and downstream, associated with such a system. While the removal of suspended solids may result in a reduction in the inoculum for biofouling downstream, by no means all bacteria are removed and thus there is a considerable potential substratum for biofouling both in the filters and on the pipework.

4 THE DEVELOPMENT OF BIOFOULING

Providing that the substratum is not toxic and that high levels of biocide are not present, surfaces submerged in natural waters rapidly become covered by a biofilm (Characklis and Cooksey, 1983). Several stages in the formation of this biofilm have been identified:

1. Almost immediately on immersion, dissolved organic compounds are adsorbed onto the surface to form a "conditioning film" which may alter the surface charge, contact angle and other properties of the surface (Niehof and Loeb, 1973; Loeb and Niehof, 1977). Subsequent fouling thus "sees", and may be attracted to, a modified surface (although "conditioning" is not necessarily a prerequisite to biofouling).

2. The arrival, adhesion and reproduction of bacterial cells to the conditioned surface. Most of these colonising bacteria are probably carried to the surface on suspended particles (Wood, 1967). A typical developing biofilm is shown in Fig. 2.

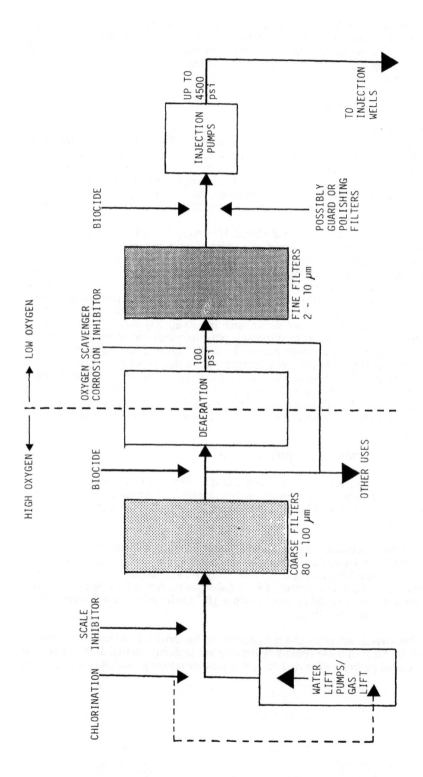

FIGURE 1. DIAGRAM OF A TYPICAL NORTH SEA WATERFLOOD PLANT.

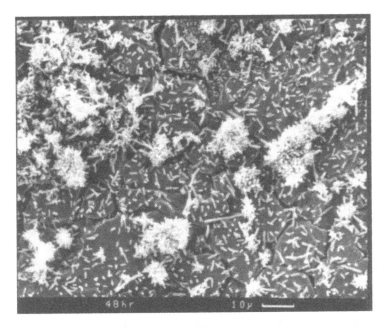

Fig. 2. A developing bacterial biofilm on stainless steel. Note the
clumps of bacteria, probably associated with particles, at the surface
and the beginning of a layered biofilm. (grown for 48h in nutrient
enriched water. Photograph courtesy of J.Little)

3. The development of interactions between individual colonies of bacteria and
the production of copious mucilage to produce a biofilm ("slime") on the surface.
The microbial cell content of a mature biofilm may only make up 10% or less of
its dry weight (Hamilton, 1985). Bacteria, especially when in a biofilm, are
highly resistant to biocide treatments. They can thus be found in most systems,
even with high biocide dosing. If treatment is not adequate then a typical reaction
to biocides is for the biofilm to produce more extracellular material for protection
and thus exacerbate the problems.

4. The arrival and development of organisms other than bacteria. This depends on
the environment, especially on nutrient and light availability, flow rates,
temperature, and the presence of biocides. Organisms such as protozoa,
microalgae and the settlement stages of larger fouling organisms can develop.
These may have arrived at the surface at the same time as the initial bacterial
inoculum but have longer development times or some may have the ability to
select settlement sites according to the underlying biofilm.

5. Growth of organisms and the development of a fouling community to the
limits imposed by the environment. A stable, but dynamic community will
develop with local areas of nutrient limitation leading to death and sloughing of
parts of the biofilm. These are then recolonised from the surrounding areas and
from the plankton.

65

5 EFFECTS OF BIOFOULING

Biofouling has three main effects in water filtration and transportation systems:

1. Altering the environment at the substratum/water interface.

2. Releasing detrimental metabolites into the water stream.

3. BLOCKING, either directly, by growth insitu, or indirectly, by the sloughing of slimes into the water stream, pipes, valves filters and control and monitoring equipment

6 EFFECTS OF FOULING ON FILTRATION SYSTEMS

The main effects of fouling on the filtration systems is to cause blockage, leading to a requirement for extensive shut-down times for cleaning or a decreased filter life. Filtration systems can become fouled in three ways:

1. Blockage by the bulk material to be removed.

2. The growth of fouling organisms on the inflow surface of the filter.

3. The growth of fouling organisms within the bulk of the filter

6.1 Blockage by bulk material
The idea of filtration is to remove suspended solids and thus filters are designed to be efficient at this without undue blockage problems. However, when the particulate material in the water is largely of planktonic organic origin, filter surfaces can rapidly become blocked. This is especially so during seasonal "blooms" (sudden increases in numbers) of planktonic organisms in natural waters.

Considerable problems have been experienced in the off-shore oil industry where plankton blooms often cause systems to be shut down (Mitchell and Finch, 1981; Edyvean and Sneddon, 1985; Edyvean and Lynch, 1990) and certainly reduce the working life of disposable filter elements (Figure 3).

It has been reported that more than 90% of suspended solids in seawater can be organic (Matthews *et al.*, 1984), with plankton, especially copepods,(minute shrimp-like animals), dinoflagellates and diatoms (armoured, single celled microscopic plants), making up a large proportion of the particles present. These plankton range in size from a few microns to a few millimetres and all have durable exoskeletons, together with considerable amounts of internally stored lipids and external mucilage (Edyvean and Sneddon, 1985). When the conditions of temperature, light and nutrients are adventitious, for example in the spring, then the phytoplankton (plants) grow at very high rates. These are fed on by the zooplankton (animals) which rapidly lay down lipids, both to aid buoyancy and to provide an energy store for the winter periods. In temperate waters the spring bloom is generally followed by a drop in numbers of plankton over the summer due to nutrient limitation and then another bloom in the autumn when storms mix the water layers and bring more nutrients to the surface. However, not only is there temporal variation in plankton abundance but there are also spatial variations, due to currents and diurnal migrations which can result in very high densities of plankton (and larger predators) drifting past water intakes

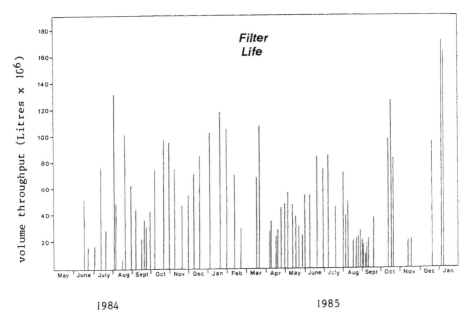

Fig. 3. Working life of 5 micron filter cartridges filtering seawater on an off-shore platform in the central North Sea. Reduction in life corresponds to spring and autumn blooms of planktonic organisms (Data courtesy of J.Lynch)

A typical sequence of blockage caused by planktonic organisms begins with the coating of the wires, threads or fibres of the filter material with sticky organic material, probably extracellular products of the plankton. Small particles and plankton such a coccoliths are trapped and this provides an anchorage for further blockage, especially by strands of organic material, which seem to be derived from copepods, and become wrapped around the wires or fibres of the filter material (a process often enhanced by backwashing procedures). The blockage rapidly builds up into a cake of amorphous material held onto a framework of planktonic skeletons by plankton derived lipids and mucilages (Fig. 4). The inherent movement of the wires or fibres that make up the filter elements enhances this process by trapping and crushing much of the plankton, providing more lipids, mucilage and skeletal remains to aid the blockage.

The tough external skeletons of many plankton can form a strong framework on which other blocking material can be trapped and it is this combination of hard, irregular shaped skeletal material and the sticky mass which, together with inorganic particulates produces such an effective blocking material.

Lipid and other debris from the crushed plankton will pass through the coarse filters and cause a similar sort (and often more severe) blockage on fine filter elements (Fig. 5). Again crushing within the filter elements results in more organic material being released, this time into the water system downstream of the filters. Such effects are carried on through the filters and into the pipework (see 7 below).

Fig. 4. Fouling of a woven wire coarse filter by plankton derived
material (North sea, spring bloom. Bar = 200 microns)

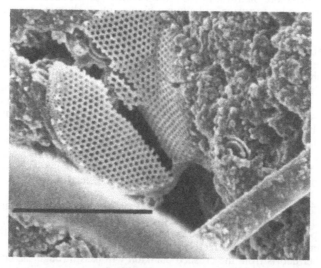

Fig. 5. Fouling of a polypropylene fine (5 micron rated) filter with
material that has passed through the coarse filter system. (North Sea,
spring bloom. Bar = 30 microns)

6.2 Fouling growth on and in the filters

Many types of filter provide an ideal substratum for the growth of fouling organisms, particularly bacteria. The flow of water over the surfaces of the filter provides considerable amounts of nutrients to the organisms. Unless a biocide is used, bacterial, and other, fouling of filters can result in blockage in a very short time. Fouling growth can be both on the surface and within the depth of the filter and a particular problem is the development of areas of anaerobiosis within sand filters which then become dominated by sulphate-reducing bacteria (SRB). These bacteria produce hydrogen sulphide and are characterised by the blackening they cause and a "bad egg" smell (see 7 below). The blackened area of the filter often has been consolidated by extracellular bacterial products and thus is ineffective as a filter.

7 THE EFFECTS OF FOULING ON WATER SYSTEMS

Fouling can be a considerable hazard in water systems. Upstream of the filters, in the water intakes, nutrients are high and light may be present, often allowing the growth of considerable amounts of macroscopic fouling. This may restrict flow, destabalise water movement patterns and will provide additional particulates to the filters.

The degree of fouling is considerably dependant on available nutrients and may be influenced by the conditioning film. Thus, organic compounds, fats, waxes and lipids, released from planktonic organisms by the crushing effects of filters, depositing onto downstream surfaces will encourage biofilm development.

Many papers have been published on the problems of bacteria in different water systems including in the off-shore oil industry (Lynch and Edyvean, 1988), cooling water (Sequeira et al., 1988) and heat exchangers (Gunn et al. 1987). The biofilms can produce corrosion, corrosive metabolites and extracellular products detrimental to the systems.

While much biofouling remains benign it can cause severe localised corrosion. Corrosion requires the presence of two types of reactive site on a metal surface, anodes, where metal ions are lost to the solution and cathodes, where a reduction reaction takes place to remove electrons left by the metal ions. In well aerated water this cathodic reaction is usually oxygen reduction and in deaerated water it is hydrogen reduction. On a uniformly corroding surface these anodic and cathodic sites are constantly changing. However, if conditions become such as to permanently separate the anode and cathode then the corrosion is localised. Oxygen concentration cells, for example, initiate localised corrosion by producing a difference in the dissolved oxygen level at two points on a metal surface; the area with the higher oxygen concentration becomes cathodic and the area with the lower oxygen concentration, anodic. Such cells can occur in oxygen-deficient zones beneath biofilms.

The activity of several bacterial groups can cause, enhance and localise corrosion. However, sulphate-reducing bacteria (SRB), a group of obligatory anaerobic organisms that reduce sulphate (from organic material) to sulphide are of particular concern. SRB generally grow in close association with other types of microorganisms and have been detected in biofilms within 3-20 days of exposure (Sanders, 1983; Hamilton and Sanders, 1984). The final product of their metabolic activities is hydrogen sulphide.

Hydrogen sulphide is not only highly toxic (deaths of personnel entering confined spaces in which bacterially generated hydrogen sulphide is present have been reported) but is also a corrosive agent in its own right and it can have an effect well away from the site of its generation. It is responsible for problems such as hydrogen blistering, hydrogen-induced cracking, sulphide stress cracking and corrosion fatigue of steels (Fontana and Greene, 1983. Thomas *et al.* 1988) and its presence can cause cracks in steel pipework to grow 25 times or more than they would do in unpolluted water (Thomas *et al.*, 1988).

SRB enhanced localised corrosion of metals is a well known phenomenon and there is an extensive literature on the subject (Miller and King, 1975; Tiller, 1982; Hamilton, 1983; 1985; Cragnolino and Tuovinen, 1984). It is characterised by extensive pitting under layers of black corrosion product. The classical theory of corrosion by these organisms is that of cathodic depolarisation which proposes that, under anaerobic conditions, SRB stimulate the cathodic corrosion reaction by utilising the hydrogen evolved there (Wolzogen Kuhr, 1961). Later evidence suggested that the main corrosion reaction is between metal anodic sites and a ferrous sulphide cathode formed from metal ions and the bacterially produced sulphide (Booth *et al.*, 1968). However, ferrous sulphide is not a permanent cathode and its regeneration is dependent on the SRB removal of hydrogen form its surface (King and Miller, 1971). This latter theory now forms the base line for most working hypotheses on the mechanisms of SRB mediated anaerobic corrosion (Hamilton, 1985). In addition there are proposals which stress the importance of bacterially produced sulphides and sulphur in the corrosion process (Salvarezza and Videla, 1980; Sanders, 1983) with the formation of sulphur concentration cells as a proposed mechanism (Schaschl, 1980).

Other bacterial groups are important, and the growth of iron bacteria, species of *Sphaerotilus*, *Gallionella* and *Crenothrix*, is commonly associated with corrosion of water distribution pipes (Tiller, 1982). It is postulated that, in areas of low oxygen availability these bacteria convert ferrous to ferric iron which precipitates as ferric hydroxide. This material shields the surface from oxygen and creates a localised differential cell (Sharpley, 1961a). Furthermore, these bacteria can produce large amounts of mucilage and can thus provide sites for other bacterial activity.

Sulphur-oxidising bacteria can be particularly important in some environments and microenvironments. Species of *Thiobacillus* and *Sulfolobus* are very diverse, being able to grow in a range of optimal temperatures (up to 90 C), oxygen concentrations from zero to saturated and pH values from neutral to very acidic (Bos and Kuenen, 1983). Corrosion is by acid production and also through the ability of some species to oxidise ferrous to ferric iron and destroy protective films.

There is an increasing realisation that different species of bacteria do not act alone but in "consortia" which can include both aerobic and anaerobic forms, providing mutual shelter and nutrients. An example of an interactive system on a macroscopic scale can be found in sluggishly moving or intermittently used piping systems which can be particularly susceptible to corrosion. Any water left in the bottom of a drained pipe, whether metal or concrete is a potential corrosion hazard, even without bacterial activity. However, a particular problem in concrete sewer pipes is the establishment of sulphur cycling, whereby bacterial degradation of organic compounds eventually results in the production of sulphides (by SRB) in stagnant water at the base of the pipe. Hydrogen sulphide then enters the damp air space at the roof of the pipe where bacterial oxidation (by sulphur-oxidising bacteria) converts it to sulphuric acid, causing considerable damage.

Another example of water left to stagnate in pipes is in hydrotesting. Pipelines are often integrity tested before use by filling them with water under pressure (hydrotesting) (Stoeker, 1984; 1986; Tatnall *et al.*, 1981; Kobrin, 1976). Often residual water form this process remains in the system for months or even years before the pipe is used for its intended purpose (King *et al.*, 1986). As this water is rarely treated (Stoeker, 1986), there have been many reported failures due to biologically induced corrosion even before the pipe is used (Stoeker, 1984; Tatnall *et al.*, 1981; Kobrin, 1976).

It has been recognised for some time that corrosion is often greater in systems containing large amounts of debris when compared to clean systems. The presence of rust, mill scale, welding consumables and general rubble (up to 4.4kg per linear metre has been reported, Broussard, 1982) provides both protection for microbial growth and enhances corrosion processes such as differential aeration cells.

Enhanced corrosion is not the only consequence of biofilm development. Other problems, such as the blockage of lines, valves *etc.* may occur. For example, the sulphide produced by SRB can react with iron to form a colloidal precipitate of iron sulphide which, in combination with bacterial cell mass can cause considerable blocking and malfunction of equipment (Crouch, 1982; Davis, 1967; Stranger-Johannessen, 1983). Other bacterial which cause plugging problems are the iron bacteria, which produce slimy precipitates of hydrated ferric oxide and can be found in both marine and fresh water systems (Sharpley, 1961a; 1961b) and a group of other organisms, known collectively as "slime-forming bacteria" which can produce considerable amounts of extracellular material under certain environmental conditions. These include members of the following genera: *Pseudomonas*, *Escherichia*, *Flavobacterium*, *Aerobacter* and *Bacillus* (Carlson *et al.*, 1961; Myers and Slabyi, 1962).

8 CONTROL METHODS

There are three important requirements to minimise biofouling problems in water filtration and transportation systems:

1. Awareness of potential problems.

2. Good integrated design

3. Provision for monitoring and action from the start (or even before) of operation.

To be aware, and have a knowledge of, the potential problems in water filtration and transportation systems is more difficult in reality that in theory. As all situations and environments are different, then the biofilm, its development and effects will also be different and, while generalities can be made, each system is unique even if physically identical.

Good integrated design must, for example, ensure compatibility between different parts of the system and ensure that the types of filtration equipment are suited to the particulates they are intended to remove. For example wedge wire screens are better than woven wire screens in the removal of planktonic organisms such as copepods as they present a smooth, rigid and easily backwashable surface to the inflow and thus do minimal damage to the organisms and reduce the release of lipids and other material.

Nearly all filtration systems require biocide addition (for example the chlorination shown in Fig. 1) to prevent insitu growth of fouling organisms. However, treatments such a chlorination have an effect on corrosion and this must be taken into account in the choice of pipework material. Deaeration will reduce normal corrosion but enhance the environment for anaerobic bacteria. Steps must be taken to ensure that particles are not recreated after filtration, for example by pH changes and corrosion (there have been examples in the off-shore industry where the filtered water at the final discharge point contained more particulates and of a larger size than the original supply). Tail end filtration, which comes at the end a "clean" water transportation is mainly used as a "polishing" system. If used without any additional biocide treatment great care has to be taken as they may easily become colonised and thus act as a bacterial reservoir and inoculum.

Provision for monitoring from an early stage is essential, as are planned treatment regimes. Biofilms are difficult to shift once established and only by monitoring at several points in the system can the effectiveness of control measures be assessed. It is not overstressing the case to say that if monitoring and control are introduced once a problem has been discovered then it is too late. There are now a number of descriptions of the use of monitoring schemes (Maxwell, *et al.*, 1987; Chen and Chen, 1984) and summaries of biocide types and usage (Bessems, 1983; Crouch, 1982; Stott, 1985; Lynch and Edyvean, 1988). It is often required to "slug dose" a system periodically and to change the biocides used so as not to build up a resistant population.

Awareness, design and monitoring are the keywords in controlling biofouling and biological corrosion in water filtration and transportation systems.

9 ACKNOWLEDGEMENTS

The author would like to thank J.Lynch and J.Little for the provision of some of the material used in this paper.

10 REFERENCES

Bessems, E. (1983) Biological aspects of the assessment of biocides, in **Microbial corrosion** (eds A.D.Mercer, A.K.Tiller and R.W.Wilson) The Metals Society, London, pp.84-89.

Booth, G.H., Elford, L. and Wakerly, D.S. (1968) Corrosion of mild steel by sulphate-reducing bacteria: an alternative mechanism. **Br. Corr. J.**, 3, 242-245.

Bos, P. and Kuenen, G. (1983) Microbiology of sulphur-oxidising bacteria, in **Microbial corrosion** (eds A.D.Mercer, A.K.Tiller and R.W.Wilson) The Metals Society, London, pp.18-27.

Broussard, D.E. (1982) Gel-plug technology used to clean FLAGS off-shore line. **Pipeline and Gas J.**, 209, 26-32.

Carlson, V., Bennett, E.O. and Rowe, J.A. (1961) Microbial flora in a number of oilfield water injection systems. **J. Soc. Pet. Engineers**, 1, 72-80.

Characklis, W.G. and Cooksey, K.E. (1983) Biofilms and microbial fouling, in **Advances in applied microbiology** (ed A.I.Laskin) Academic press, London, pp.93-108.

Chen, E.Y. and Chen, R.B. (1984) Monitoring microbial corrosion in large oilfield water systems. **J. Pet. Tech.**, 36, 1171-1176.

Cragnolino, G. and Tuovinen, C.H. (1984) The role of sulphate-reducing and sulphur oxidizing bacteria in the localised corrosion of iron-based alloys - a review. **Int. Biodeterioration**, 20. 9-27.

Crouch, B.A. (1982) Biocides in oilfield operations, in **Biocides in the oil industry**, Institute of Petroleum, pp.29-38.

Davis, J.B. (1967) **Petroleum microbiology**, Elsevier, New York.

Edyvean, R.G.J. and Lynch, J.L. (1990) Effect of organic fouling on the life of cartridge filters. **Filtration and separation**, 27, 114-117.

Edyvean, R.G.J. and Sneddon, A.D. (1985) The filtration of plankton from seawater. **Filtration and Separation**, 22, 184-189.

Fontana, G.M. and Greene, N.D. (1983) **Corrosion Engineering** McGraw Hill, New York.

Gunn, N., Woods, D.C., Blunn, G., Fletcher, R.L.. and Jones, E.G.B.. (1987) Problems associated with marine microbial fouling, in **Microbial problems in the Off-shore Oil Industry** (eds E.C.Hill, J.L.Shennan and R.J.Watkinson), Institute of Petroleum, John Wiley, Chichester, pp. 175-200.

Hamilton, W.A. (1983) The sulphate-reducing bacteria; their physiology and consequent ecology, in **Microbial corrosion** (eds A.D.Mercer, A.K.Tiller and R.W.Wilson) The Metals Society, London, pp.1-5.

Hamilton, W.A. (1985) Sulphate-reducing bacteria and anaerobic corrosion, **Ann. Rev. Microbiol.**, 39, 195-217.

Hamilton, W.A. and Sanders, P.F. (1984) Microbial corrosion; experience in the off-shore oil industry, in **Biodeterioration VI** (eds S.Barry, D.R.Houghton, G.C. Llewellyn and C.E.O'Rear) Jonh Wiley, Chichester. pp. 202-206.

King, R.A. and Miller, J.D.A. (1971) Corrosion by the sulphate-reducing bacteria. **Nature**, 233, 491-492.

King, R.A., Miller, J.D.A. and Stott, J.F.D. (1986) Subsea pipelines: internal and external biological corrosion, in **Biologically Induced Corrosion** (ed S.C.Dexter) NACE, Houston, pp.268-274.

Kobrin, G. (1976) Corrosion by microorganisms in natural waters. **Mat. Performance**, 15, 38-43.

Loeb, G. and Neihof, R.A. (1977) Absorption of an organic film at the platinum-seawater interface. **J. Mar.Res.**, 35, 283-291.

Lynch, J.L. and Edyvean, R.G.J. (1988) Biofouling in oilfield water systems - a review. **Biofouling**, 1, 147-162.

Matthews, R.R., Tunaal, T. and Mehdizadeh, P. (1984) Evolution of seawater filtration systems for North Sea applications. **Proc. 16 Ann. Off-shore Techn. Conf.** NACE OTC 4660. pp. 121-128.

Maxwell, S. McLean, K.M. and Kearns, J. (1987) Biocide application and monitoring in a waterflood system, in **Microbial problems in the Off-shore Oil Industry** (eds E.C.Hill, J.L.Shennan and R.J.Watkinson), Institute of Petroleum, John Wiley, Chichester, pp. 209-218.

Miller, J.D.A. and King,R.A. (1975) Biodeterioration of metals, in **Microbial Aspects of the Biodeterioration of Materials** (eds D.W.Lovelock and R.J.Gilbert), Academic Press, London, pp. 83-103.

Mitchell, R.W. and Finch, E.M. (1981) Water quality aspects of North Sea injection water. **J.Pet. Technol.**, 33, 1141-1152.

Myers, G.E. and Slabyi. B.M. (1962) The microbiological quality of injection waters used in Alberta oil-fields. **Producers Monthly**, 26, 12-14.

Niehof, R.A. and Loeb, G. (1973) Molecular fouling of surfaces in seawater, in **Proceedings of the III International Congress on Marine Corrosion and Fouling** (eds R.F.Acker, B.F.Brown, J.R.DePalma and W.P.Iverson), Northwestern University Press, Illinois, pp. 710-718.

Salvarezza, R.C. and Videla, H.A. (1980) Passivity breakdown of mild steel in seawater in the presence of sulphate reducing bacteria. **Corrosion**, 36, 550-555.

Sanders, P.F. (1983) Biological aspects of marine corrosion. **Metals World**, 2, 13-14.

Schaschal, E. (1980) Elemental sulphur as a corrodant in deaerated, neutral aqueous solutions. **Mat. Performance**, 19, 9-12.

Sequeira, C.A.C., Carrasquinho, P.M.N.A. and Cebola, C.M. (1988) Control of microbial corrosion in cooling water systems by the use of biocides, in **Microbial**

Corrosion 1 (eds C.A.C.Sequeira and A.K.Tiller) Elsevier, London, pp. 240-255.

Sharpley, J.M. (1961a) Microbiological corrosion in waterfloods. **Corrosion,** 17, 386-390.

Sharpley, J.M. (1961b) The occurrence of Gallionella in salt water. **App. Microbiol.,** 9, 380-382.

Stoecker, J.G. (1984) Guide for the investigation of microbiologically induced corrosion. **Mat. Performance,** 23, 48-55.

Stoecker, J.G. (1986) Overview of industrial biological corrosion: past, present and future, in **Biologically Induced Corrosion** (ed S.C.Dexter) NACE Houston, pp. 324-329.

Stott, J.D.F. (1985) Selection and use of biocides in seawater injection systems. Paper presented at the **2nd. Ann. workshop on secondary Recovery Technology,** London, June 1985 (Preprint).

Stranger-Johannessen, M. (1983) Off-shore structures at risk from sulphate-reducing bacteria. **Off-shore Engineer** (Feb.1983), 51.

Tatnall, R., Stoecker, J.G., Schultz, R., Kobrin, G., Pilusao, A. (1981) Case histories: bacteria induced corrosion. **Mat. Performance,** 20, 41-48.

Thomas, C.J., Edyvean, R.G.J. and Brook, R. (1987) Biologically enhanced corrosion fatigue. **Biofouling,** 1, 65-78.

Tiller, A.K. (1982) Aspects of microbial corrosion, in **Corrosion Processes** (ed R.N.Parkins) Applied Science, Barking, pp. 115-159.

Wolzogen Kuhr, C.A.H. von (1961) Unity of anaerobic and aerobic iron corrosion processes in the soil (reprint), **Corrosion,** 17, 293-299.

Wood, E.J.F. (1967) **Microbiology of Oceans and Estuaries,** Elsevier, London.

7 ENGINEERED WATER SYSTEMS AND WATERBORNE DISEASE

G.J. MISTRY and R. van WOERKOM
The Water Quality Centre, Thames Water Enterprises,
London, UK

Abstract
An overview of the processess controlling the growth of
microorganisms such as legionella within cooling, hot and cold
domestic water systems is examined. Broad outlines of the principles
of legionella colonisation are considered. Microorganisms will
colonise water systems and form biofilms which will protect the
colonising organisms in them from disinfection. Measures which reduce
the formation of biofilms and thus allow better disinfection are
discussed.
Keywords Legionella, biofilms, disinfection, temperature, control

1 Introduction

In recent years, the high profile of Legionnaires' disease has
caused panic in many quarters. Understandably so when one considers
the financial implications that an associated outbreak may have for a
business and the effect on employees and the public.

Often the maintenance of water systems, use of material that may
support growth, stagnant conditions and poor temperature regulation
create conditions where proliferation of microorganisms can occur,
especially opportunistic pathogens such as legionellae. These can
then be disseminated by aerosolisation via cooling towers, taps,
showers etc. and can lead to infection if sufficient numbers of
organisms from the water are inhaled. However if diagnosed in the
early stages the disease is easily treatable.

To aviod creating the type of environment that will encourage the
growth of opportunistic pathogens, a systems management checklist
should be drawn up where factors such as maintenance, materials,
stagnation and temperature control are regularly checked/reviewed.

Microbiologically, there are two principles to remember if growth
of bacteria such as legionella is to be controlled (Dennis, 1990):

(a) Prevention of surface fouling through good design, operation and
maintenance
(b) Control of water temperature such that conditions are hostile to
microbial growth.

2 Legionella

Legionellae are ubiquitous organisms known to proliferate in man made water systems. However they are fastidious and in the laboratory they need exacting conditions for growth.

The bacterium was recognised in 1977 (McDade *et al.*, 1977) following an outbreak of disease in Philadelphia at an American legion convention, which subsequently lent the organism its name. There are approximately 30 species of legionella, which are sub divided into serogroups and futher divided into subtypes. The majority (>90%) of cases of Legionnaires' disease are attributable to one species, serogroup and subtype, *Legionella pneumophila* serogroup 1 subtype Pontiac. Other legionellae have been implicated in disease, and therefore should not be overlooked (Fung *et al.*, 1989).

Infection can occur by inhaling aerosols (Baskerville *et al.*, 1981) containing the organism of 5um diameter or less. The dose needed to cause infection is not known, however the bacteria must be inhaled deep into the lungs of susceptible individuals before they can contract the disease. There is no evidence to show that the disease can be contracted by ingestion of water or that it can spread directly from person to person.

It can be argued that by preventing the formation of aerosols, infection will not occur. Under normal circumstances this is impractical since equipment such as water cooling systems are very efficient and cost effective. However, controls can be applied to these systems to reduce aerosol production and dispersion. It is important to remember that a simple act of running a tap can produce an aerosol.

Prevention of proliferation of legionella and other organisms in water systems must be the prime objective. We must look at the factors that influence growth.

3 Biofilms

Surface fouling especially that caused by biofilm formation creats problems such as poor heat exchange in cooling circuits and corrosion of pipes.

Biofilms are formed on surfaces by microorganisms in often response to the nature of environment surrounding them i.e it is a mechanism of survival in hostile environments such as those which may have low nutrient status or unfavourable temperatures.

The formation of biofilms can be described in three stages (Costerton *et al.*, 1988);

a. Absorption - an organism adheres to a surface, this can be either permanent or temporary. If permanent then this involves chemical attachment to the surface by polymers etc. which act as bridges.
b. Microcolony fomation - the attached colony subdivides with the formation of a small colony of one species.
c. Biofilm formation - this occurs by internal replication and by recruitment of other microorganisms from the bathing (surrounding water) phase.

Many microenvironments may exist within these films, which bear little resemblence to the bathing waters. Microorganisms are able to change themselves phenotypically to such an extent that it is impossible to define them chemically, structurally or functionally without reference to the growth environment (Herbert, 1961). Hence with organisms such as legionella there may be more than one species, serogroup and subtype present.

4 Cases

Outbreaks at Wadsworth Memorial Hospital between 1977 and 1980 and Kingston District Hospital in the early 80's, established two features of the ecology of legionella;

a. They establish themselves in piped water systems and survive even when water conditions are hostile
b. the organism is associated with surfaces or particulate matter.

Studies of a passenger ship and two cooling towers, both associated with cases of Legionnaires' Disease, confirmed that *Legionella pneumophila* colonised these systems forming part of the biofilm (Colbourne and Dennis, 1988) . Slime in the fresh water tank and foam/slime in the cooling towers indicated the presence of legionella with only low levels present in the water occasionally.
It was also noted that the effectiveness for example of chlorine would be reduced as a result of formation of diffusion barriers across the biofilm which would restrict the ingress of the disinfectant.

5 Other factors

Water temperatures between $20°C$ and $45°C$ encourage growth, but organisms may survive for sometime, at temperatures above and below this range. Nutrients can be provided from the supply water and ingress of foreign materials such as dust and debris through for example, open access to cisterns. Nutrients may also be provided by the use of inappropriate construction materials such as, some washers, coatings and jointing/sealing compounds encouraging growth.
Scales, sludges, and scum deposits may contain nutrients. Also they may protect legionellae from adverse conditions. For example, during chlorination, if all scale is not removed from a cistern, then penetration of scales will be low therefore preventing the disinfection being effective.
Growth can also be accelerated in the presence of slow moving, stationary or stagnant waters held at ambient temperatures, for example long lengths of pipework supplying a infrequently used tap.

6 Control

To control some of the factors that influence growth of legionella within water systems, measures should be taken such as:

(a) Ensuring that operating temperatures of water systems prevent multiplication. For drinking and domestic cold water water systems temperatures must be maintained below 20°C. For hot water systems temperatures should be kept at or above 60°C with all outlet temperatures being above 50°C, but below 55°C.

(b) In cooling water systems routine maintenance must be carried out and an appropriate chemical treatment regime should be in place. Regular monitoring for chemical and microbiological parameters should identify an potential problems that may arise.

(c) Periodic cleaning and disinfection of cooling (bianually) and domestic water systems (annually). For cooling systems, the tower and associated distribution pipework, including any water supply cistern should be disinfected, cleaned and disinfected again. For water systems calorifiers and cisterns should be inspected, drained, pasteurised and/or disinfected, annually. Water softeners, showers etc. should be regularly inspected and properly maintained.

(d) Records of all actions must be kept, so that planned maintenance can be reviewed, modified if necessary in a single water system logbook.

(e) Inspection and assessment of existing water systems should be undertaken to assess problems that may exist. This should be done visually initially, and backed up with targetted sampling at strategic points to indicate overall water condition.

7 References

Badenoch, Sir John. (1988) Second Report of the Committee of Inquiry into the outbreak of Legionnaires Disease in Stafford in April 1985. HMSO, London. 1987.

Baskerville , A., Fitzgeorge , R.B., Broster, M., Hambleton, P., & Dennis, P.J., (1981). Lancet, ii, 1389.

Chartered Institute of Building Services Engineers. (1988) Code of Practice, Minimising the risk from Legionnaires Disease, CIBSE.

Colbourne, J.S. and Dennis, P.J. (1988) Legionella: a biofilm organism in engineered water systems. In Proceedings of 7th International Biodeterioration Symposium Sept, 1987, J. Int. Biodeterioration

Colbourne, J.S. and Dennis, P.J. (1989) The ecology and survival of Legionella pneumophila J IWEM Aug 1989.

Costerton, J.W., Cheng, K.-J., Geesey, G.G., Ladd, T.I., Nickel, J.C., Dasgupta, M. and Marrie, T.J. (1987). Bacterial biofilms in nature and disease. Ann. Rev. Microbiol., 41, 435-464.

Dennis, P.J. Bartlett, C.L.R. & Wright, A.E. (1984). In Legionella Proceedings of the 2nd International Symposium, ed. C. Thornsberry, A. Balows, J.C. Felley & W. Jakuboski. American Society for Microbiology, Washington, D.C.

Department of Health and Social Security (1989) Code of Practice for the Operation and Maintenance of Cooling Towers and Hot and Cold Water Services in Health Service Buildings, HMSO, London,

Fang Guo-Dong., Victor, L., Vickers, R.M., (1989) Disease due to the Legionellaceae (Other than *Legionella pneumophila*) Medicine Vol 68, no2.

Health & Safety Executive, Guidance Note EH48. Legionnaires Disease. HMSO, London. 1987.

Herbert, D. (1961). The chemical composition of microorganisms as a function of their enviroment. Symp. Soc. Gen. Microbiol., 11, 391-416.

Keevil, C.W., West, A.A., Walker, J.T., Lee, J.V., Dennis, P.J.L., Colbourne, J.S. (1989) Biofilms: Detection, implications and solutions. Watershed 1989 Edts., Wheeler, D., Richardson, M., Bridges, J. Pergamon Press

McDade, J.E., Sheppard, C.C., Fraser, D.W., Tsai, T.R., Redus, M.A., Dowdle, W.R. and the laboratory investigation team (1977). New England J.Med., 297, 1197-203.

8 STUDY ON BIOFOULING FORMING IN INDUSTRIAL COOLING WATER SYSTEMS

Gy. LAKATOS
Ecological Institute of Kossuth University, Debrecen, Hungary

Abstract
In industrial cooling water systems the formation of biofouling on the heat exchange surfaces is a frequent
phenomenon and the elimination of this deposit causes many
difficulties. This paper presents results of algological
investigations and possible methods for the elimination of
biofouling in cooling water systems of seven Hungarian
industrial plants. Results of laboratory, pilot-scale and
full-scale eliminating measures are dealt with.
Keywords: Biofouling, Algal-fouling, Industrial Cooling
Water Systems, Elimination.

1 Introduction

The formation of biofouling on heat exchange surfaces of
industrial cooling water systems and its elimination
arise frequently as serious water management problems
(Goysich and McCoy, 1989; Holmes, 1986; Lakatos and Tokár,
1977; Maguire, 1956; Sladeckova, 1961; Smith et al.,1988
etc.).
    The formation of biofouling involves the danger of
increased corrosion besides decreasing cooling efficiency
and excess energy demand (Birchall, 1979; Kirkpatrick et
al., 1980; Mansfield, 1978). The proliferation of micro-
organisms can be traced back to the fabourable thermal,
light conditions and nutrient supply (Characklis, 1981).
    Biofouling forming on the heat exchange surfaces in the
open cooling water systems can be divided into inorganic
and organic fractions. While the former one is caused by
the sedimentation of suspended matter from the cooling
water and by the precipitation of mainly ferric compounds
produced in the course of scaling and corrosion processes,
the latter one consist of living and dead algae, fungi and
slime-forming bacteria (Characklis et al., 1982; Sladeckova,
1969). The mud part of fouling is on the one hand a
nutrient pool for the microorganisms, and on the other hand

it provides a substratum for futher settling. The quickly
and massively proliferating microorganisms clog up the
water distribution and aeration troughs and form a several
centimeter thick biofouling deposit on the heat exchange
surfaces.

In order to develop and apply the effective protection
methods against this phenomenon in industrial cooling water
systems, it is, first of all, necessary to study the
species composition, the qualitative and quantitative
features of biofouling forming on the heat exchange
surfaces.

To prevent the formation of biofouling on the heat
exchange surfaces or to eliminate it the following measures
are available:
- By adding selective biostatic or biocide chemicals
  into the feed water to prevent the proliferation of
  microorganisms in recycling water systems.
- By treatment of substrates preventing the settlement
  and proliferation of microorganisms, for example
  antifouling paints are applied to reduce surface
  biofouling.
- By making circumstances unfavourable for micro-
  organisms (e.g. selective discolouring, decreasing of
  the nutrient supply etc.).
- By applying biological measures (e.g. cyanophages
  Barnet   et al.,1981; Safferman and Morris, 1964;
  saprophytic fungi, etc.).

The full-scale biocide treatment according to the
efficient working program of industrial cooling water
systems must always be preceeded by a series of laboratory
and pilot-scale measurements (Bussy, 1949; Fitzgerald,
1962; Jones et al. 1987; Lakatos and Tokár, 1977; Lutz and
Merle, 1983; Nakayama et al., 1981). In these preliminary
measurements a special emphasis is laid on the
determination of the required quantity of biocide, the
duration of the exposition and the factors influencing the
treatment.

This paper summarizes the results of algological
investigations of biofouling forming on the heat exchange
surfaces in different Industrial Works, of NE Hungary. We
compare the composition of algal-fouling forming on
vertical cooling plates consisting of different material in
the David type cooling towers of TVM. In the laboratory
comparative studies on the efficiency of some biocides were
carried out. This paper reports data on the application of
Bonion trademark biocide in full-scale water systems of
TIFO.

2 Material and methods

The composition of biofouling on heat exchange surfaces was

studied in the cooling water system of seven Hungarian
industrial plants between 1975 and 1989. The qualitative
biofouling samples were taken in the "Tiszai" Oil Refinary
Works (TIFO), "Tiszai" Petrochemical Works (TVK). "Tiszai"
Chemical Industries (TVM), "Jászberényi" Refrigerator Works
(JH), the "Lenin" Metallurgical Works (LKM); Hungarian
Bearing Works (MGM) and "Ózdi" Metallurgical Works (ÓKÜ)
from their vertical plates of David-type cooling towers and
cooling basins.

The chemical (organic and inorganic), algological and
hydrozoological investigations of biofouling were carried
out to reveal the circumstances of the biofouling formation
and its structural composition. This paper presents results
of the algological analysis.

The algal species composition of the biofouling forming
on different vertical cooling plates, such as glass (g),
asbestos slate (a), aluminium (Al) and polyesther (p)
plates has been studied in the "Tiszai" Chemical Industries
(TVM).

In laboratory experiments we have not worked with
individual algal species, but with algal-communities that
have already formed biofouling deposits and whose dominant
genus were as follows: Lyngbya, Oscillatoria, Phormidium.
From the biofoulig deposits we have cut little discs of
known area and weight, similar to the method Goldman et al.,
(1963), and the discs have been kept in Gorham alkali
nutrient solution (Hughes et al., 1958).

The laboratory experiments have been made with the
solutions of different concentrations of the following
biocides or biostatics: Busan 77, Copper-ion, Sterogenol,
$KMnO_4$ and Hypo. the photosynthesis-respiration activity of
biofouling discs were made by dark and light bottle method
(Gaarder and Gran, 1927); the disolved oxygen was
determined by the Winkler method (Felföldy, 1974). We have
refered the quantity of oxygen produced in an hour to the
unit quantity to 10 cm^2 of the discs area.

For the chemical treatment of cooling water in
recirculation systems the Bonion biocide was studied in
full-scale in TIFO. To measure the efficiency of biocide
the chemical composition and a-chlorophyll content of
biofouling were used as a basis. We have also performed
taxonomical analysis of algal-fouling.

3 Results and discussions

3.1 The algal species composition of biofouling
In the samples collected during 15 years from the seven
Works, 139 taxa were identified (Table 1), which belong to
three phylla. The most species (71) were present in the
samples had been taken from the exchange surface in the TVM
and the least species (20) were found in MGM. The cooling
water system provide special circumstances for the algae,

Table 1. Algal taxa in the biofouling on heat exchange surfaces of different industrial cooling water systems

	TIFO	TVK	TVM	JH	LKM	MGM	ÖKÜ
Cyanophyta							
Aphanocapsa biformis A. Br.					+		
Chrooccoccus minutus (Kg.) Nag.			+	+			+
Lyngbya epiphytica Hieron.						+	
L. aerugino-coerulea (Kg.) Gom.			+				
L. cryptovaginata Schkorb.					+		
L. kützingiana Kirchn.			+				
L. kützingii Schmidle			+				
L. lagerheimii (Möb.) Gom.			+				
L. limnetica Lemm.					+		+
L. martensiana Menegh.			+				+
L. putealis (Mont)			+	+			
Microcoleus lacustris (Rabh.) Fsrl.			+				
M. subtorulosus (Bréb.) Gom.			+				
Oscillatoria agardhii Gom.		+	+		+	+	+
O. angusta Koppe.							+
O. chalybea (Mert.) Gom.	+	+	+	+	+	+	+
O. chlorina Kg.							+
O. cortiana Menegh.			+				
O. formosa Bory	+		+				
O. geminata Menegh.			+				
O. irrigua (Kg.) Gom.			+	+			
O. limosa Agh.	+		+				+
O. princeps Vauch.					+		
O. pseudogeminata G. Schmid.			+				+
O. subtilissima Kg.					+		
O. tenuis Agh.					+		
Phormidium favosum (Bory) Gom.			+				
Ph. foveolarum (Mont.) Gom.		+				+	+
Ph. molle (Kg.) Gom.					+		
Ph. retzii (Agh.) Gom.		+	+				
Ph. subfuscum Kg.			+				
Ph. tenue (Menegh.) Gom.		+			+	+	
Romeria sp.							+
Synenchococcus sp.							+
acillariophyceae							
Achnanthes hungarica Grun.			+				
A. grimmeri Krasske			+				
A. lanceolata Bréb.	+		+				+
A. linearis W. Sm.			+				
A. microcephala Kütz.	+		+	+	+		
A. minutissima Kütz.	+						
A. plönensis Hust			+				
A. sp. I.	+	+			+		
A. sp. II.		+	+	+			
A. sp. III.	+	+				+	+
Amphora veneta Kütz.							+
Bacillaria paradoxa Gmelin	+	+	+		+		

Table 1. (cont.)

	TIFO	TVK	TVM	JH	LKM	MGM	ÖKÜ
Caloneis amphisbaena (Bory) Cleve			+				
Clampylodiscus clypens Ehr.			+				
C. hybrenicus Ehr.	+						
Ceratoneis arcus Kütz		+	+				
Cocconeis hustedtii Krasske	+	+	+		+	+	+
C. placentula (Ehr.)	+						+
C. sp.			+		+	+	+
Cyclotella comta (Ehr.) Kütz.							
Cymbella caespitosa (Kütz.)			+				
C. cistula (Hemprich) Grun.	+						
C. gracilis (Rabh.) Cleve	+						
C. ventricosa Kütz.	+	+	+	+	+	+	+
C. sp.			+				
Denticula thermalis Kütz				+			
Diatoma anceps (Ehr.) Grun.							+
D. hiemale (Lyngb.) Heiberg					+		+
D. vulgare Bory	+			+	+		
Diploneis oculata (Bréb) Cleve							+
Epithemia argus Kütz.							+
Fragilaria capucina Desmaz.	+						
F. construens (Ehr.) Grun.	+					+	+
F. nitzschioides Grun.							+
Frustulia rhomboides Ehr.	+				+		
F. vulgaris Thwaites					+		
Gomphonema acuminatum Ehr.			+	+			
G. constrictum Ehr.							+
G. longiceps Ehr.		+					
G. olivaceaum (Lyngb.) Kütz.	+	+	+		+	+	+
G. parvulum (Kütz.) Grun.			+			+	+
Gyrosigma spencerii (W. Smith) Cleve				+			
G. sp.		+					
Hantzschia ampioxys (Ehr.) Grun.			+				
Mastogloia smithii Thwaites	+	+			+	+	+
Melosira granulata (Ehr.) Ralphs.			+				
M. varians Agh.			+	+			+
N.cryptocephala Kütz.							
N. cuspidata Kütz.					+		
N. exiqua (Gregory) O. Müller	+						
Navicula gracilis Ehr.	+						
N. halophila (Grun.) Cleve							+
N. heufleriana (Grun.) Cleve	+						
N. hungarica Grun.		+			+	+	+
N. minima Grun.							+
N. muralis Grun.							+
N. mutica Kütz.	+	+	+	+	+		
N. placentula (Ehr.) Grun.	+						
N. protracta Grun.							+
N. pygmaea Kütz.			+	+			
N. radiosa Kütz.			+			+	

Table 1. (cont.)

	TIFO	TVK	TVM	JH	LKM	MGM	ÖKÜ
N. simplex Krasske	+			+			
N. tuscula (Ehr.) Grun.			+				
N. viridula Kütz.			+				
N. sp. I.	+	+		+	+	+	+
N. sp. II.	+				+		
Nitzschia acicularis W. Sm.			+	+			+
N. amphibia Grun.			+				+
N. clausii Hantzsch.	+		+				
N. commutata Grun.					+	+	+
N. dissipata (Kütz.) Grun.			+				
N. kützingiana Hilse			+	+			
N. linearis W. Sm.		+	+		+		
N. palea (Kütz.) W. Sm.	+		+				+
N. parvula Lewis	+	+	+	+		+	+
N. pseudoamphioxys Hust.			+				
N. recta Hantzsch.			+				+
N. sigma (Kütz.) W. Sm.			+				
N. sygmoidea W. Sm.	+	+	+	+			
Pinnularia interrupta W. Sm.	+						
P. microstauron (Ehr.) Cleve			+				
P. viridis (Nitzsch.) Ehr.		+	+	+			+
Rhopalodia gibba (Ehr.) O. Müller	+						
Stauroneis anceps Ehr.			+				
S. phoenicenteron Ehr.			+				
Stephanodiscus astraea (Ehr.) Grun.		+			+		
Surirella biseriata Bréb.			+				
S. ovalis Bréb.			+				
S. ovata Kütz.		+	+	+			+
S. robusta Ehr.			+				
Synedra affinis Kütz.			+		+	+	
S. ulna (Nitzsch.) Ehr.	+	+	+	+			+
S. vaucheriae Kütz.	+						

Chlorophyta							
Chlamydomonas sphaeroides Gerloff	+						
Chlorella pyrenoidosa Chick.	+						
Crucigenia tetrapedia (Kirchn.)	+	+	+			+	+
C. qadrata Morren				+			
Hyaloraphidium contortum Pascher et Kors.	+	+			+		+
Scenedesmus acuminatus (Lagh.) Chod.	+	+		+			
S. acutus Meyen	+	+			+	+	
S. ecornis (Rolfs) Chod.					+		+
S. quadricauda (Turp.) Bréb.				+	+		+
Tetraedron sp.						+	
Ulothrix sp.							+
Vaucheria sessilis Dekand			+				

the composition of algal-fouling are often influenced by
seasons, the water supply and the treatment.
It is characteristic in general that the algal flora
consists of particular species occuring in a mosaik-like
fashion which is visible to the naked-eye on the basis of
different colours of biofouling. Although the species
number of diatoms was the highest (Table 2), their
estimated biomass was only very small in comporasion with
the biomass of filamentous Cyanophyta algae.

Table 2.   The distribution of taxa by phyllum

	Number of species	%
Cyanophyta	34	24,5
Chrysophyta	93	66,9
Chlorophyta	12	8,6
	139	100,0

Two species were present in all the studied industrial
works:
    Lyngbya chalybea
    Cymbella ventricosa.
Diatoms occurring in the biofouling belong mainly to the
Pennales, but some Centrales species were also identified.
The constant species which were present and abundant in the
biofouling of five or six Works can be systematized to
Centrales and only two species belong to the Cyanophyta or
Chlorococcales:
    Oscillatoria agardhii
    Cocconeis hustedtii
    Gomphonema olivaceum
    Navicula mutica
    Nitzschia parvula
    Synedra ulna
    Crucigenia tetrapeadia.
The above-mentioned dominancy of diatoms is similar to the
results of some authors (Characklis and Cooksey, 1983;
Cooksey and Cooksey, 1986; Escher and Characklis, 1983).
Beside the typical eu-fouling species (e.g. Achnanthes sp.,
Gomphonema sp., etc.) some eu-planktonic species also
occurred many times (Crucigenia sp., Scenedesmus sp., etc.)
which might suspend into the biofouling from the supplemental water.
To assess the similarity of the algal-fouling in
different Works Sörensen's index was used and cluster
analysis based on WPGM method was applied (Fig. 1.).
The algal-fouling collected in TVK and LKM are found to
be the most similar, although these two Works use different
water sources as supplemental water. The MGM joins to this

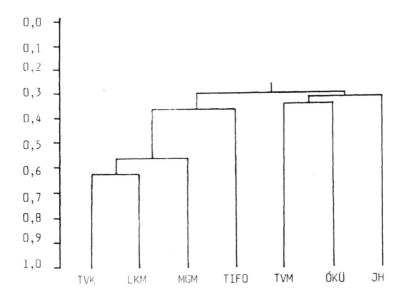

Fig.1. Dendrogram on the basis of species composition
in algal - fouling found in different Works.

block of dendrogram, but this Works is mainly supplied by
drinking water. The Works (TVK and TIFO) using the same
supplemental water source differ considerably in their
algal-fouling from each other. This provides evidence that
the algal colonization on the cooling exchange surfaces
depends only partly on the type of supplemental water, it
may also be influenced significantly by other factors e.g.
the spreading of spores by air.

3.2 Biofouling formation on different substrates
Table 3   contains the number of taxa and their percentage
distribution by phyllum of biofouling on different
substrates in TVM.
      The cluster dendrogram was also constructed on the basis
of taxonomical data (Fig.2.). The glass and the asbestos
slate plates show the largest similarity, the aluminium
plate occupies a middle position, while the poliesther
substrate differs greatly from them. On the glass and
asbestos slate plates the species number of diatoms was
twice as much as that of blue algae. The large part of
biomass is produced by filamentous algae. On the aluminium
plates the biofouling is - unlike all the other cases - not
contiguous it can only be found in "spots". The number of
blue-green algae is minimal and the basis population

Table 3.   Number of taxa and their percentage distribution
            by phyllum of biofouling on different substrates

	Glass		Asbestos		Aluminium		Polyester	
	No	%	No	%	No	%	No	%
Schizomycophyta	0	0,0	0	0,0	1	3,3	0	0,0
Cyanophyta	17	36,2	14	35,0	2	6,7	3	50,0
Bacillariophyceae	28	59,6	26	65,0	27	90,0	3	50,0
Chlorophyta	2	4,2	0	0,0	0	0,0	0	0,0
	47	100,0	40	100,0	30	100,0	6	100,0

consists of diatoms, but even the latter are found in
remarkably smaller quantities than on the plates made of
glass and asbestos slate for example. Very small numbers
of algal taxa were found on the polyester cooling plates
and it is very promising from the point of view of
preventing the formation of biofouling. Of the different
substrates, such as glass, asbestos slate, aluminium and
polyester, applied under identical circumstances,
aluminium and polyester showed the most favourable
operation, that is virtually no blue-green algae and only
few diatoms adhesed to these.

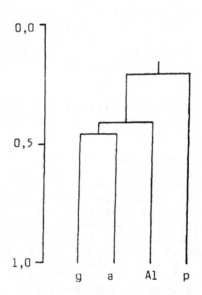

Fig.2. Dendrogram for different substrates on the
       basis of algal-fouling.

Table 4.   The effect of Busan 77, copper-ion, sterogenol KMnO$_4$ and chlorine (Hypo) on the gross photosynthesis of biofouling

Concentration ppm	Gross photosynthesis mg O$_2$/10 cm^2/h				
	Busan 77	Copper	Sterogenol	KMnO$_4$	Hypo
0	0,550	0,550	0,550	0,550	0,550
2	0,545	0,550	0,550	0,496	0,412
4	0,540	0,453	0,496	0,438	0,357
10	0,528	0,398	0,287	0,307	0,250
20	0,495	0,230	0,075	0,095	0,020

Duration of treatment: 12 hours
Light intensity: 2000 Lux
Temperature: 20,0 C

3.3 Laboratorical study of the effect of some biocides on biofouling

Results of laboratorical DLB experiments are summarized in Table 4   and Figure 3. The applied chemicals can be devided into two groups on the basisof their effect mechanisms. Chemicals with algicide effect kill the algaedefinitely, and the algistatic compounds only hinder the proliferation of algae, while they are contact with them.

Busan 77 had an algicide effect only in a great concentration (30 ppm) and after a long period of treatment, but it is possible that the lower concentration is satisfactory for the prevention of the formation of biofouling in the recirculating water system.

The copper ion is generally of biostatic effect (Fitzgerald, 1964), but cheap, is not corrosive and can be easily added from its water solution. To kill algal-fouling formed, based on DLB experiments copper is needed in 20 ppm concentration in fhe case of 12 hours exposition. The lakes were treated with copper in the 1980'ies (Moore and Kellermann, 1904) and we can read about the toxic effect mechanisms in the paper of Cairns et al. (1972). The effect of copper on hindering the growth and photosynthesis of diatoms is reported by Steemann Nielsen and Andersen (1970) and Stauber and Florence (1987).

In longer pilot-scale experiments Cu concentration proved to be lower, it was first 10 ppm required for inhibiting the proliferation of algae, but then 1-5 ppm was sufficient to preserve the protection anti-fouling layer on the surface. Beside the advantageous features of copper as an algicide, the resistance of algae to this element is well known from the references (Fitzgerald and Faust, 1963; Kuwabara and Leland, 1986).

Good results were obtained with sterogenol, which proved

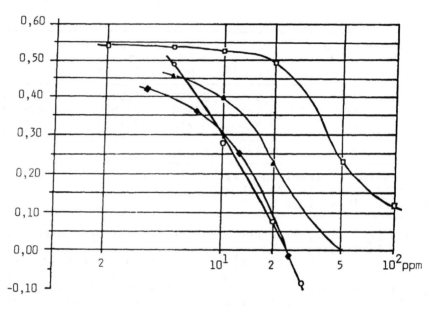

O$_2$ mg/g dry weight/h

Fig. 3.
The effect of Busan 77 (—□—), coper-ion (—▲—), Sterogenol (—●—), chlorine (Hypo,—■—) on the gross photosynthesis of biofouling.

an effective algicide in laboratorical experiments. It eliminates the algal-fouling in 10 ppm concentration under longer exposition. Because or its intensive foaming it cannot be used without an appropriate anti-foaming agent.

Potassium-permanganate has also been examined, the effective concentration of which was found between 10-20 ppm, but its using is uneconomical because of its unfavourable corrosive character and, even in laboratorical experiment the MnO_2 derived from oxidation, formed a thick blackish-brownish layer on the bottom of the glass basin.

Free chlorine has been used for a long time (Stratton and Lee, 1975; Holzwarth et al., 1984; Ibrahim et al., 1982; Soracco et al., 1985), and is often added to water in Hypo form. According to our findings it can act as a biocide or biostatic material depending on the concentration applied. The necessary concentration to prevent the formation of biofouling is less than 5 ppm, to eliminate the filamentous blue-green algae is about 15 ppm. Its disadvantage is that, the maintenance of an exact concentration is difficult and in case of overdose it corrodes the structural materials.

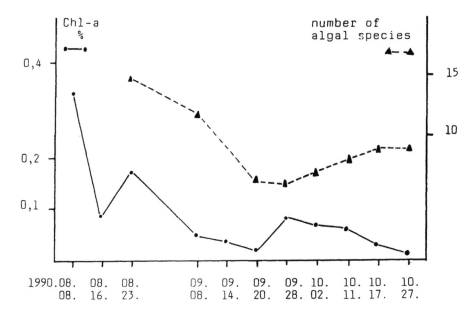

Fig.4. Chlorophyll a % and number of algal species in biofouling of TIFO.

3.4 Full-scale study of biofouling treatment in TIFO's recirculating water systems by using Bonion biocide

For the chemical treatment of industrial recirculating cooling water systems in NE Hungary to the middle of 1980'ies copper sulphate and the products trademarked Busan were applied successfully.

In the years past the products bought for foreign currency (Busan, Drew, Nalco, etc.) were replaced by biocides produced in Hungary, e.g. by Bonion products. Among them Bonion 605 biocide was added into the recirculating cooling water system of TIFO in concentration of 20 ppm twice a week beside the application of corrosion inhibiting and dispersing agents. Under the influence of treatment with Bonion biocide the organic matter content of biofouling on heat exchange plates changed to a slight degree, but it remained above 50 %. The chlorophyll a content of biofouling decreased significantly at the beginning of the treatment, but later it showed an increasing and then again a decreasing tendency (Fig. 4.). The number of algal species changed partly similarly to the chlorophyll a content, it decreased to half following the treatment but two months later it increased slightly due to the appearance of some diatom species again on the cooling surfaces. Our algological, bacteriological and zoological

studies suggest that the applied Bonion biocide has a
biostatic effect, since in the biofouling not only algal
species but Nematoda taxon were also obsorved. In spite of
the aplication of biocide treatment two diatom species
could always be identified:
Fragillaria construens
Navicula placentula.
The adaption and resistance of algae to biocides are
widely reported (Fitzgerald and Faust, 1963; Genter et al.
1988; Kuwabara, 1986; Rai et al., 1981; etc.). The
adaption mechanisms involve the intra- and extracellular
exculsions (Silverberg et al., 1973). It has been shown
that certain algal species release organic that may
significantly decrease biologically available copper
(McKnight and Morel, 1979).
From the results of our full-scale experiments with
Bonion biocide it can be concluded that this organic
product possesses a definite biostatic character and the
species occurring in the algal-fouling become adapted and
resistant to it. Our results also confirm that studies on
the taxonomical composition of biofouling and the resistance of algal species are needed to control the operation
of cooling systems and provide essential background
information to the solution of problems resulting from the
formation of biofouling.

4 References

Barnet, Y.M. Daft, M.J. and Stewart, D.P. (1981) Cyano-
    bacteria-cyanophage interactions in continuous culture.
    Journal of Applied Bacteriology, 51, 541-552.
Birchall, G.A. (197⁹) Control of fouling within cooling
    water systems. **Effluent and Water Treatment Journal,** 19,
    571-578.
Bussy, I.J. (1949) Growth and control of algae in openair
    swimming pools. **Research Institute for Public Health
    Engineering,** The Hauge, Netherlands, 1, 1-78.
Cairns, J.Jr. Lanza, G.R. Parker, B.C. (1972) Pollution
    related structural and functional changes in aquatic
    communites with emphasis on freshwater algae and
    Protozoa. **Procedings of the Academy of Natural Sciences
    of Philadelphia,** 124, 79-127.
Characklis, W.G. (1981) Fouling biofilm development: A
    process analysis. **Biotechnology and Bioengineering,** 23,
    1923-1960.
Characklis, W.G. Trullear, M.G. Bryers J.D. and Zelver, N.
    (1982) Dynamics of biofilm processes: methods. **Water
    Res.,** 17, 1-10.
Characklis, W.G. and Cooksey, K.E. (1983) Biofilms and
    microbial fouling. **Adv. Appl. Microbiol.,** 29, 93-138.
Cooksey, K.E. Cooksey, Barbara (1984) Adhesion of fouling
    diatoms to surfaces: some biochemistry, in **Algal Bio-**

fouling (eds L.V. Evans and K.D. Hoagland), Elsevier
   Science Publishers B.V. Amsterdam, pp. 41-53.
Escher, A. and Characklis, W.G. (1982) Algal-bacterial
   interactions within aggregates. **Biotech. Bioeng.**, 24,
   2283-2290.
Felföldy, L. (1974) A biológiai vízminősítés. (Method of
   biological water quality). in **Vízügyi Hidrobiológia,**
   VIZDOK, Budapest, 3, 1-242.
Fitzgerald, G.P. (1962) Bioassay for algicidal chemicals in
   swimming pools. **Water and Sewage Works,** 109, 361-363.
Fitzgerald, G.P. (1964) Factors in the testing and applica-
   tion of algicides. **Applied Microbiol.,** 12, 247-253.
Fitzgerald, G.P. and Faust, S.L. (1963) Factors affecting
   the algicidal and algistatic properties of cooper.
   Applied Microbiol., 11, 345-351.
Gaarder, T. and Gran, H.H. (1927) Investigations of the
   production of plankton in the Oslo Fjord. **Rapp. Proc.
   Verb., Cons. Perm. Int. Explor. Mer.,** 24. 1-48.
Genter, R.B. Cherry, D.S. Smith, E.P. and Cairns, J.Jr.
(1988) Attached-algal abundance altered by individual and
   combined treatments of zinc and pH. **Environ. Toxic.
   Chem.,** 7, 723-733.
Goldman, C.R. Mason, D.T. and Wood, B.J.B. (1963) Light
   injury and inhibition in Antarctic freshwater
   phytoplankton. **Limnol. Oceanogr.,** 8, 313-322.
Goysich, M.J. and McCoy, W.F. (1989) A quantitative method
   for determining the efficacy of algicides in industrial
   cooling towers. **Journ. Indust. Microbiol.,** 4, 429-435.
Holmes, P.E. (1986) Bacterial enhancement of vinyl fouling
   by algae. **Applied and Enviromental Microbiology,** 52,
   1391-1393.
Holzwart, G. Balmer, R.G. and Soni, L. (1984) The fate of
   chlorine in recirculating cooling towers. Field results.
   Water Res., 18, 1429-1435.
Hughes, E.O. Gorham, P.R. and Zehnder, A. (1958) Toxicity
   of unialgal culture of Microcystis aeruginosa. **Can. J.
   Microbiol,** 4, 226-236.
Ibrahim, J. Squires, L. Mitwalli, H. and Taha, M. (1982)
   Chlorine as an algicide in a conventional water treat-
   ment plant. **Intern. J. Environmental Studies,** 20, 44-46.
Jones, C.A. Leidlein, J.H. and Grierson J.G. (1987) **Methods
   for evuluating the efficacy of biocides against sessile
   bacteria.** Cooling Tower Institute 1987 Annual Meeting,
   Technical Paper Number TP87-6. New Orleans, LA.
Kirkpatrick J.P. McIntire, V. and Characklis, W.G. (1980)
   Mass and heat transfer in a circular tube with
   biofouling. **Water Res.,** 14, 117-127.
Kuwabara, J.S. (1986) Physico-chemical processes affecting
   copper, tin and zinc toxicity to algae: A review, in
   Algal Biofouling (ed L.V. Evans and K.D. Hoagland),
   Elsevier Science Publishers B.V. Amsterdam, 129-144.
Kuwabara, J.S. and Leland, H.V. (1986) Adaptation of
   Selenastrum capricornutum (Chlorophyceae) to copper.

Environ. Toxicol. Chem., 5, 197-203.

Lakatos, Gy. and Tokár, M. (1977) Effect of some biocids on attached algae (biotecton). **Acta Biol. Debrecina, 14,** 105-114.

Lutz, P. and Merle, G. (1983) Discontinuous mass chlorination of natural draft cooling towers. **Wat. Sci. Tech.,** 15, 197-213.

Maguire, J.J. (1956) Biological fouling in recirculating cooling water systems. **Ind Eng. Chem.,** 48, 2162-2167.

Mansfield, G.H. (1978) Some aspects of cooling water treatment. **Effluent and Water Treatment Journal,** 17, 552-555.

McKnight, D.M. and Morel, F.M.M. (1979) Release of weak and strong copper complexing agents by algae. **Limnol. Oceanogr.,** 24, 823-837.

Moore, G.T. and Kellerman, K.F. (1904) A method of destroying or preventing the growth of algae and certain pathogenic bacteria in water supplies. **U.S. Dep. Agr. Bur. of Plant Ind. Buil.,** 64, 15-44.

Nakayama, S. Tanaka, M. Yamauchi, S. and Tahata, N. (1981) An antibiofouling ozone system for cooling water circuits. I - Application to fresh water circuits. **Ozone: Science and Engineering,** 2, 327-336.

Rai, L.C. Gaur, J.P. and Kumar, H.D. (1981) Phycology and heavy metal pollution. **Biol. Rev.,** 56, 99-151.

Safferman, R.S. and Morris, M.E. (1964) Control of algae with viruses. **Journal of the American Water Works Association,** 56, 1217-1224.

Silverberg, B.A. Stokes, P.M. and Fernstenberg, L.B. (1976) Intracellular complexes in a copper tolerant green alga. **J. Cell Biol.,** 69, 210-214.

Sladeckova, A. (1961) Fouling of the cooling equipment of steam power plants. **Energetika,** 7, 327-329.

Sladeckova, A. (1969) Control of slimes and algae in cooling systems. **Werh. Internat. Verein. Limnol.,** 17, 532-588.

Smith, A.L. Muia, R.A. and Clancy, M.O. (1988) **New application technology for controlling algal fouling in recirculating cooling water systems.** Cooling Tower Institute 1988 Annual Meeting, Technical Paper Number TP88-15. Houston, TX.

Soracco, R.J. Wilde, E.W. Mayack, L.A. and Pope, D.H. (1985) Comparative effectiveness of antifouling treatment regimes using chlorine or a slow-releasing bromine biocide. **Water Res.,** 19, 763-766.

Stauber, J.L. and Florence, T.M. (1987 Mechanism of toxicity of ionic copper and copper complexes to algae. **Marine Biology,** 94, 511-519.

Steemann-Nielsen, E. and Wium-Andersen, S. (1971) The influence of Cu on photosynthesis and growth in diatoms. **Physiol. Plant.,** 24, 480-484.

Stratton, C.L. and Lee, G.F. (1975) Cooling towers and water quality. **J. Wat. Pol. Cont. Fed.,** 47, 1901-1912.

PART FOUR
ENGINEERING MATERIALS

9 MAJOR BIODETERIORATION ASPECTS OF BUILDINGS IN ROMANIA

A. POPESCU
The Building Research Institute (INCERC), Bucharest, Romania
T. BESCHEA
The Research, Engineering and Design Institute for Oil Refineries (IOITPR), Ploieşti, Romania

Asbtract
The ensuring of a good building durability is not possible without a profound knowing of the natural or industrial aggressive factors that can contribute to its determination, During the last time the biotic factor, represented by the action of the various beings on buildings, became very important as mechanical and physico-chemical factors. This paper presents a planning attempt of buildings microbiological determination in Romania, detailing their negative aspects. The knowing of these aspects means a first step in taking some adequate measures for prevention and controlling of the various biodeterioration phenomena in buildings.
Keywords: Biodeterioration, bacterial attack, microbial corrosion, biologic fouling, fungic attack, moulding, buildings, sewerage systems, soils, wood structures.

1 Introduction

The importance of the aspects related to buildings biodeterioration has been minimized, if not ignored, for a long time. However, during the last 5o years, the literature in the field has succeeded in gradually changing these opinions. Today, one can admit, with good reason, the acknowledgement of a real pathology of buildings besides other mechanical,physico-chemical and climatic environment factors are considered to have a significant weight in the civil and industrial buildings deterioration. From bacteria to man, through their activity, a huge number of living beings are involved, to a greater or smaller extent, in the extremely varied forms in which the environmental conditions act upon buildings. And, while for the large organisms the problems have been solved, for the microorganisms of the unseen, microscopic world, the hypotheses needed a long time to become certitudes, as has also happened in the case of human or animal pathology.

The difficulties encountered in the correct assessment of these problems are numerous. The development of the ecological studies, especially during the last 2o-3o years, has shown the extend and complexity of the biotic and abiotic factors interference, particularly at a microscopic level. The high variety of interrelations, in the mutual conditioning or intensification of living and non-living factors, and especially their negative influence on the elements of a building, has given rise to highly difficult problems that affected the technical conception of the civil engineer, dominated for centuries by what could be directly seen, "de visu".

Undoubtly, the complex mechanisms by which the microorganisms act directly or indirectly in buildings deterioration, are still to be studied in interdisciplinary researches for their thorough understanding. In this respect, the exchange of scientific information and documentation between research laboratories in the field from various countries, can have significant, favourable results. Gaining knowledge and determining the actual extent of the microbiological aggressiveness and explaining the modalities by which microorganisms negatively act on buildings elements, contribute decisively to developing and applying the methods for the prevention and control of this type of biodeterioration, as well as to increasing buildings durability.

In Romania, the study of the problems related to microbiological biodeterioration in buildings, represents a significant object of the research themes approached by the buildings biodeterioration team within the Building Research Institute - INCERC - and by the Applied Microbiology Laborators in the Research, Engineering and Design Institute for Oil Refineries - ICITPR - in Ploiești. A number of other laboratories have studied or are studying aspects related to the biodeterioration of historical monuments, especially the stone ones, wood biodeterioration by macromycetae, implications of moulds growth in dwellings on the tenants health.

In the following sections, the types of microbiological deterioration occurring in the Romanian buildings will be calssified according to a general taxonomic criterion.The paper also presents some methods or procedures dealt with in the above mentioned institutions for the prevention and control of the aggressive microorganisms affecting various types of civil and industrial constructions.

2 Bacterial biodeterioration of buildings

Microbial aggression upon buildings or building elements is usually determined by the growth possibilities offered by the surface or structure of these buildings to the microorganism populations.

The microorganisms attached to these surface will act negatively on the structures acting as a growth support, determining zhem. The aggressive microorganisms are usually both bacteria and microscopic fungi, but one of these two population types is always prevailing. In this chapter, we grouped those types of biodeterioration with prevailing bacteria populations.

2.1 Microbial corrosion of sewerage systeme

The sewerage pipes and especially the main concrete ones undergo a particular type of bacterial biocorrosion having sometimes severe results. There are cases when the concrete wall was destroyed up to a 4 cm depth only after 4-5 years of service, or some other cases when, due to this biocorrosion entire sections had to be replaced after a lo year service. The biocorrosion process has different characteristic features on the two surface of the pipes. On the internal surface the biocorrosion mechanism is provided by several bacteria populations acting sequently. Inside the pipe, due to the occurrence of anaerobic zones, bacterial populations cf the DESULFOVIBRIC DESULFURICANS type develop, that produce hydrogen sulphide by reducing the sulphates. The presence of hydrogen sulphide favours the development of aerobic bacteria. some of them autotrophic, especially of the THIOBACILLUS species. Some populations, such as TRIOBACILLUS THIOPARUS and TRIOBACILLUS DENITRIFICANS develop under the alkaline pH offered by concrete. These populations oxidize the hydrogen sulphide slower, but the concrete pH gradually decreases, thus favouring the growth of another bacterial group whose main representative is THIOBACILLUS THICOXIDANS or CONCRETIVOURS that produce sulphuric acid. The growth of this last group causing the concrete pH to reach values of 1.5-2.5 determine, naturally, a strong corrosion of concrete. Once penetrating into the pipe wall material, the biocorrosion process can rapidly develop, extending in area as well as velocity. The physico-chemical abiotic factors, represented by the water and acidity of the circulating liquids prolong and intensify the bacterial corrosion. Once the biocorrosion processes penetrate deeper, the possibility of anaerobic areas occurrence increases, therefore increasing the production of hydrogen sulphide due to DESULFOVIBRIO. The metabolic acid products give rise to a number of other bacterial aerobic and anaerobic populations that can combine with the destructive action produced by the THIOBACILLUS

populations. Nevertheless, these interrelations and interferences still need complex studies.

On the external surfaces, deterioration is represented
by a particular aspect of biocorrosion in soil.

For cast iron sewer pipes, the corrosion processes on
the internal surface are also electrochemical in nature,
especially due to hydrogen sulphide. However, the decisive factors are still the bacterial ones, represented by
DESULFOVIBRIO growth and iron bacteria. The external surface shows a biocorrosion process similar to the one
affecting concrete conduits.

In Romania, the researches focused to a greater extent
on the intimate mechanisms of these bioaggression complex
processes, with special emphasis on the protection means,
such as:

Execution of precast concrete units for sewer pipes
using special sulphates resistant cements,
Replacement, under certain conditions, of the concrete
or cast iron conduits with fused or recrystallized
basalt conduits or plastic ones;
Conduits protection by coatings or by impregnation with
bituminos materials or with by-products in the coke-
chemical industry modified against bacterial aggression;
The possible future use of coatings for cast iron or
plastic conduits or of chemical compounds with antibacterial properties introduced in the concrete mass.

2.2 Biocorrosion in soils

The concrete or steel structural members, as well as the
adequate protection materials can undergo biological dete-
riorations when placed in aggressive soils. Soil aggressivity depends on an intricate group of factors such as:
moisture, aeration and chemical composition degree number
and variety of aerobic and anaerobic bacterial populations
of microscopic fungi populations and actinomycetas. This
biodeterioration type is mainly provided by bacteria;
however, the microscopic fungi populations also have a
great contribution. The complexity of deteriorating factors in the aggressive soils accounts for the possible e-
xistance of some interfering mechanisms difficult to
assess, that still need thorough studies for their understanding. This is the type of biocorrosion with most diverse possibilities, with the formation of microbiocenoses
and trophic chains. Among bacteris, the most significant
ones are the PSEUDOMONAS, FLAVOBACTER, BACILLUS, MICROCO-
CUS species, as well as the sulphate reducing and sulphur-
oxidizing populations, especially in the soils with a high
pyrite content. Among the microscopic fungi, the most significant ones are ASPERGILLUS, CHAETOMIUM, FUSARIUM,
PENICILLIUM and TRICHODERMA species, Obviously it is also
worth mentioning the action of plant roots.

Concerning biocorrosion in soil, the researches carried

out in the Romanian laboratories focused on the following aspects:

   Setting up of some criteria and methods for estimating soil aggressivity, researches carried out in collaboration with institutions in Hungary;
   Studies concerning bacterial biodegradability of bitumens, especially for those used in hydroinsulations and pipes;
   Studies concerning bacterial and fungal biodegradability for a number of film-forming protection systems based on acetalic, perchlor-vinylic or petroleum polymers, as well as on by-products in the coke-chemical and petroleum industry;
   Obtaining of bitumens or other bioresistant protection systems;
   Methodology for film application using the above mentioned systems and for providing concrete surfaces impregnation using these protection systems by coatings;
   Development of methodologies for laboratory testing or for "in situ" exposure of some models protected with such bioresistant protection systems.

2.3. Biodeterioration by biologic fouling

The biologic fouling phenomenon is caused by biological growth developing on water immersed buildings, such as water cooling industrial systems, hydraulic engineering structures and waste water treatment systems. The fouling film acts destructively upon the supporting structure, first chemically by metabolic acid products or by products resulting from dead decomposing organisms. Due to such mechanisms, the microorganisms can penetrate the supporting structure mass, their destructive action being favoured by water penetration and by the freezing-thawing phenomenon. These depositions burden the supporting structure, finally destroying it. The main interfering organisms, particularly during the first stage, are the bacteria, that, by synthesizing the exocellular polymeric compounds form gelatinous adjacent films. Afterwards, this film has a different evolution. Thus, in the case of industrial or soft waters, some other organisms attach to the bacterial layer such as algae, protozoa, worms, hydrocarbons.

   In Romania, the studies related to this type of biodeterioration focused on the recycling water cooling systems, where the biological fouling, consisting especially of bacteria, has extremely severe consequences, such as the decrease in heat transfer. This has negative effects on the technological processes, diminishes the pipes diameter up to their obstruction, thus causing an increase in power consumption, and strongly corrodes the internal surface of metallic or concrete conduits, as well as of the tanks and systems for water dispersion.

The studies resulred in:
- development and use of complete technologies for re-cycled water treatment with antibacterial, antifungi-cidal and anticorrosive effects;
- manufacturing of chemical products used for these technologies, dispersers, crust and corrosion inhi-bitors;
- development of technologies for anticorrosive pro-tection by film-forming and impregnation of the concrete suriaces of tanks and water dispersion ele-ments.

Also, besides these research works, laboratory studies are performed, these being initiated by INCERC - Bucharest and ICITPR - Ploiești in collaboration with the Romanian Institute for Marine Researches (IRCM) - Constanza, for developing antifouling biocide aditives. By means of such substances, anticorrosive protection systems will be pro-vided for structural members located in seawater.

3 Fungic biodeterioration of buildings

These processes can be grouped according to a systematic criterion into biodeterioration processes caused by micro-mycetae - microscopic fungus - and especially by moulds, and biodeterioration processes caused by a macromycetae acting especially on wooden structures.

3.1 Buildings Moulding
The extremely low food requirements and the impressive adap ting capacity account for the ubiquitous nature of the mi-croscopic fungi populations that cause the moulding pro-cess. Moulds can develop on any solid support that provides them with a minimum of organic substances and mineral salts on condition that this support also provides a minimum hu-midity level, initially 70% for spores germination and then even 40-50% for mould growth (usual 75-95%). Thus, it be-comes abvious why moulds represent one of the major fac-tors in the constructions biodeterioration processes.

The types of fungi developing on building walls do not represent typical populations but ordinary ones with huge spreading possibilities.

The geographical sprend of moulds in constructions has almost no limits. In the tropical areas, due to the air humidity that reaches almost 100%, the moulding process is a common one, in temperate cold areas, the condense occur-rence on cold walls in the decisive factor of thin process.

In Romania, the inadequate heating conditions of rooms, due to the poorly understood power saving policy, particu-larly during the last 10-15 years, caused the development of condensation in residential buildings and in social and industrial spaces, as well as the corresponding increase in moulding processes.

Besides humidity, which is the main factor the moulding of walls is also facilitated by the use of some finishing materials and by the presence of some aerosols with biodegradable organic substances in the rooms atmosphere. These growths cause the deterioration of finishings, of plasterings and, sometimes, even of the concrete structures of the buildings masonry.

Most moulds act upon buildings mechanically, producing micro-cracks or enlarging the existing ones. At the same time, moulds also act chemically, their metabolic products, carbon dioxide and organic acids having a highly aggressive behaviour. The direct action of mould mycelium combines with the destructive actions of other microclimatic physical and chemical factors.

Beisdes microscopic fungi (moulds and yeasts) representing the majority mould films also contain bacteria, and sometimes algae, lichens and moss. All these populations are grouped due to trophic relations into complex associations - microbiocenoses - whose composition differs, function of the finishing chemical composition and environmental conditions.

Moulds in building habe negative effects on the tenants, especially on children, create an unpleasant phychic feeling of discomfort, and cause a rise in the expenses of the tenants who are forced to have treatment of biocided walls performed at a shorter time interval. These negative economic effects can amplify under certain conditions, as the mould films represent a contamination source for the objects inside rooms, for example in storage for food or industrial products, in museums, libraries, archives.

Moulding usually occurs on the internal walls of the building and especially in those areas affected by condensation. More recently, moulding can be seen on the external walls, on surfaces oriented to north or on industrial units with external emission of nutritive aerosols, such as the sieving sections within the grain silos or some bread manufacturing units.

As the main factor that favours moulding is humidity with its various manifestation types, the simplest way to prevent the occurrence of the moulds is to eliminate all causes that lead to relative humidity increase in rooms and to reduce condenation. However, there are situation when these requirements cannot be met due to numerous causes. Under these circumstances, the use of special chemical products, toxic for the moulds occurring in buildings, should be considered. These should not affect the health of the people dwelling or performing activities in the rooms whose walls have been subjected to such treatments.

In Romania, within the collaboration between INCERC -

Bucharest and ISITPR-Ploieşti, a number of products amd technologies were developed and applied for the prevention and control of moulds growth on walls and ceilings. These are registered as FUNGOSTOP trade mark.

These products and technologies can be used for water soluble finishings or for aqueous dispersions. At present, researches are being carried out for perfecting some efficient biocide additives for paints soluble in organic solvents (paints based on flax oil and perchlorvynilic paints). Also, a number of complex laboratory methodologies have been developed for testing and checking the biocides activity, the resistance of plasterings, of polishing plaster coats and biodidated finishings against fungi action, for checking the possible influence of biocides and biocidated finishings on the people health state.

The fungicizing technologies differ according to the purpose of the treatment that can be curative for controlling the already existing moulding phenomenon, or prophylactic for preventing the occurrence of the moulding phenomenon.

For specific situations when not all requirements can be met for performing curative or preventive treatments, such as rooms with low temperatures and excessive humidity of the supporting layer, some temporary treatments can be prevented.

The FUNGOSTOP technologies for controlling moulding phenomeny have in view a complex treatment that should include not only finishings, but the plastering and polishing plaster coat as well, a method that can provide a higher, long-term efficiency of the antifungical protection. An exaggerated increase of the biocide concentration only in the finishing cannot provide this efficiency, sometimes causing the negative modification of the mechanic characteristics of the finishing complex system and especially the attachment to the supporting layer. Function of the composition and thickness of the layer in which the biocide additive is incorporated, a 2–4 year antifungical protection can be provided if only the finishing is treated, while a 5–1o year antifungal protection can be provided if both the finishings and their supporting structures are treated.

Laboratory tests were performed for the following main categories of finishing systems: based on lime, on clay, on acrylic, and vynilic aqueous dispersions, emulsified paints, coloured cements. Most of these systems have already been long applied in industrial units and dwellings, thus confirming the laboratory test results.

The microorganisms tested by laboratory methods for moulds occurrence were selected populations of Aspergillus flavus, Penicillium citrium, Alternaria tenuis,

Paecilomyces varioti, Trichoderma viride, Fusarium roseum and Aureobasidium polluans. The antifungi resistance was estimated visually while the extent to which test samples previously infected with mist sprayed or immersed inoculatipns were covered with moulds, was estimared microscopically.

The specific microbiological testings such as microbiostatic or microbiocide activity were proceeded by testings on laboratory large scale models, Due to these testings, the compatibility between various biocide additives and building materials was established, alternatives for application technologies were obtained and the physicalmechanical parameters of finishings were determined such as: attachment to the support and friction resistance, these models being subjected to a strong fungal attack.

It is worth mentioning that, for checking the FUNGOSTOP technologies, an experimental, uninhabited apartment was used various alternatives were applied, while the relative humidity of the air inside the rooms was permanently controlled. In these rooms, under certain experimental conditions, laboratory animals were introduced. After a predetermined period of exposure, anatomical-pathological testings on their various organs were performed-

3.2 Biodeterioration of wood structures
Although used in constructions since ancient times, wood can be easily deteriorated by a large number of living organisms, from bacteria and fungi to insects and rodents. The numerous studies carried out in this field showd that the most aggressive populations are the fungi the Macromycetae group, both Ascomycetae and Basidiomycetae, and the xylophagous instects.

The microscopic fungi - moulds and yeasts - as well as bacteria have a low significance.

In Romania, the fundamental researches in this field are carried out within the Research and Lesign Institute for Wood Industry (ICPIL), Bucharest. There is a wide range of fungicidal and insecticidal products, products used against rotting, that are applied by brushing or spraying technologies or by vacuum or normal pressure impregnation.

4 Conclusions
- The microbiological aggressivity, both bacterial and fungal, represent one of the major factors causing civil and industrial constructions biodeterioration.
- An attempt for classifying the biodeterioration types in buildings is presented, based on the systematic criterion of majority populations.
- The civil and industrial buildings in Romania are subjected to various types of microbial aggressions, the moulding process being the most significant one due to

its extent and consequences.
- The researches carried out in the Romanian laboratories
in the field especially those in INCERC - Bucharest and
ICITPR - Ploieşti,   led to the development of complex
technhologies and of products for controll buildings biodeterioration.
- The complexity of the biological, physico-chemical and
ecological mechanisms in buildings biodeterioration require an international collaboration of the laboratories
working in this field for solving the problems related
to buildings biodeterioration.

5 References

Beşchea, T. and Ionescu-Homoriceanu, S. (1979) Bacterial
    deterioration of the bitumens chromatographic frac-
    tions, in the papers of the II-nd Symposium on Indus-
    trial Microbiology, Jassy, Romania
Beşchea, T. and Ionescu-Homoriceanu, S.(1981) Chemical
    composition modifications of the petroleum bitumens
    bacterially deteriorated, in the papers of the II-nd
    Symposium on Industrial Microbiology, Bucharest,
    Romania
Beşchea, T., Ionescu-Homoriceanu, S. and Popescu, A.(1981)
    Control of petroleum bitumens biodeterioration, on the
    papers of the III-rd Symposium on Industrial Microbio-
    logy, Bucharest, Romania
Berinde, F.(1986) Prevention and control of fungi attack-
    ing wood in structures, Ceres Printing House, Bucharest,
    Romania
Bravery, A.F.(1985) Moulds and mould control, in Informa-
    tion paper of Building Research Establishment, U.K.
    no.11
Bravery, A.F.(1987) Methods for estimating the fungicides
    efficiency for wood and wood products, in SAB Techni-
    cal Services, U.K. no23
Cornish, P.(1987) Condense and moulds growth, in Building
    Services, U.K. no.9
Fesus, I.(1986) Analysis of soil biocorrosive action, on
    the papers of the IV-th Symposium on Soil Corrosion,
    Siozok, Hungary
Fesus, I. (1982) Assessment of buildings plastics resis-
    tance against moulds, an FTV paper, Budapest, Hungary
Fesus, I.(1986) Moulding of dwellings internal walls
    an FTV paper, Budapest, Hungary
Fesus, I. and Frish, M.(1987) Control of moulds affecting
    structures members and buildings, an FTV paper, Budapest
    Hungary
Foldes, A and Fesus, L,(1984) Anticorrosive protection of
    underground urban steel pipes, in the joint publication
    INCERC-Bucharest - FTV Budapest

Gilchrist, F.(1959) Microbiological aspects concerning
    sewerage corrosion, paper edited by the South-African
    Council for Scientific and Industrial Research, Cape
    Town, the South-African Republic
Gorog, J., Vanes, F and Labory, I.(1972) Deterioration of
    bituminous in active microbiological soils, in
    Korröziös figyelö no.6
Grant, O. Hunter, C.A. and Flannigan, B(1989) The need for
    humidity of izolated moulds in inhabited rooms, in
    International Biodeterioration, U.K.
Hunter, A.C. Grant, O. Flarrigan, B and Bravery A.F.
    (1988) Mould in buildings: asores in the atmosphere of
    inhabited rooms, in International Biodeterioration,U.K.
Ionescu-Homoriceanu, S. and Popescu, A. (1987) Biodeterio-
    ration in buildings, in Science and Technique Almanach
Ionescu-Homoriceanu, S., Popescu, A. and Tomescu, I.(1986)
    Aspects concerning the biodeterioration of emulsified
    paints based on co-polymers used in buildings, in the
    papers of the XVIII-the FATIPEC Congress, Venice, Italy
Medgyesi, I. and Fesus, I. (1985) biocorrosion in buil-
    dings. Methods for the analysis of the building mate-
    rials biostability, an FTV paper, Budapest,Hungary
Medgyesi, I., Foldes, A. and Bleuer, M.(1986) Aspects re-
    lated to the selection of materials for pipes placed
    in the soils, i.e. anticorrosive protectiob of some
    types of pipes, in the papers of the IV-th Symposium
    on soil corrosion, Siofok, Hungary
Popescu, A., Teodorescu, D. and Ionescu-Homoriceanu, S.
    (1984) Anticorrosive protection by impregnation of some
    reinforced concrete members, in the papers of the III-
    rd Conference "Concrete structures statics and behaviour
    in time", Bratislava, Czechoslovakia
Popescu, A., Teodorescu-Homoriceanu, S. and Beschea, T.
    (1983) Behaviour of some hydroinsulating materials
    towards natural aggressivity, in the paper of the III-rd
    RILEM Symposium "Testing of bituminous coatings and
    materials",Beograd, Yugoslavia
Popescu, A., Ionescu-Homoriceanu, S and Beschea, T. (1986)
    Behaviour of some biocidated bitumens exposed in
    aggressive soils, in the papers of the IV-th Symposium
    on soil corrosion, Siofok, Hungary
Popescu, A., Ionescu-Homoriceanu, S. and Beschea, T.(1980)
    Influence of technological in-service conditions on the
    structure of some water cooling towers, in the papers
    of the III-rd Symposium on "in-situ behaviour of
    buildings", Tg.Mures, Romania
Popescu, A., Ionescu-Homoriceanu, S. and Gomoiu, M.(1981)
    Study of the methods related to the antifouling pro-
    ducts testing for marine water immersed concretes, in
    the papers of the III-rd INCERC Symposium "Anticorro-
    sive protection of buildings structures",Mamaia,Romania

Popescu, A. and Ionescu-Homoriceanu, S. (1988) Prevention
and control of moulds in buildings, in "Construcţii"
journal, no-3-4, Bucharest-Romania

Popescu, A. and Ionescu_Homoriceanu, S.(1989) A new
Romanian biocide for water borne coatings, in"Polymers
Paint Colour Journal", no.4237, U.K.

Samson, R,A. (1985) Mould occurrence in modern dwellings
and working spaces, in "European Journal of Epidemio-
logy"

Tonk, E. and Seidl, A.(1984) Moulding of masonry st uc-
tures, in "Szakiparitechnika", no.5, Budapest,Hungary

Verhoeff, A.Attwood, A. Brunekreef, B and Samson, R.A.
(1988) Specification and identification of mould spores
in dwellings, in "Study of the Wagenigen University",
the Netherlands

Wolkober, Z. and Gyorgy, M.(1986) Rate of soft PVC
corrosion in soil, in the papers of the IV-th Symposium
on soil corrosion, Siofok, Hungary

10 THE BIODETERIORATION OF POLYESTER POLYURETHANE IN SOIL/MARINE CONTACT

M.J. KAY, L.H.G. MORTON and E.L. PRINCE
School of Applied Biology, Lancashire Polytechnic,
Preston, UK

Abstract
Plastics are used widely in many aspects of civil engineering and
often come into contact with hostile environments. Such materials
need to be rigorously tested in order to evaluate their potential
usefulness in their end-use environment, and this should include an
assessment of their susceptibility to microbial biodeterioration. The
methodology for plastics testing is currently being assessed by the
Plastics Project Group of the International Biodeterioration Research
Group. In the investigations presented here, polyester polyurethane
foam test pieces, both with and without the incorporation of the
formulation biocide Vinyzene B.P. (10,10'-oxybisphenoxarsine) were
found to be susceptible to degradation in both soil and marine
environments.

1 Introduction

Plastics possess a broad range of chemical and physical properties
which may be tailored to meet the particular requirements of
industry. Specifically, they have been formulated for durability, and
are able to resist weathering and/or microbial biodeterioration. In
addition they are both cheap and relatively easy to produce.
Consequently they have replaced many traditional constructional
materials such as wood, metal and rubber.

In 1989 the construction industry used 20% of the plastics
produced in the U.S.A., representing some 5,180,000 tonnes of
material (Modern Plastics International, 1990). The range of plastics
applications is shown in Table 1. Plastics may be exposed to hostile
soil or marine environments in applications such as subterranean and
submarine pipelines. Of the various types of plastics used in the
manufacture of pipelines, the most common is unplasticised polyvinyl
chloride (P.V.C.) which accounted for 80% of this market in Western
Europe in 1984. The popularity of this material for pipeline
applications has increased in recent years, and between 1982 and
1983, 44% of the water mains which were renewed were replaced with
unplasticised P.V.C. pipeline. Four other plastics are also generally
used for underground transport of water; polyethylene, polybutylene,

Table 1. Plastics commonly used for construction in the United States of America (1989)

Application/material	1000 tonnes (1989)
Flooring[a]:	
Epoxy	11
Polyvinyl chloride (PVC)	173
Polyurethane foam	146
(rug underlay)	
Insulation:	
Phenolic (binder)	236
Polystyrene foam	115
Polyurethane foam (rigid)	205
Panels and siding:	
PVC	349
Pipe, fittings and conduit:	
Acrylonitrile-butadiene-styrene (ABS)	73
High density polyethylene (HDPE)	223
Low density polyethylene (LDPE)	61
Polypropylene	15
Polystyrene	10
PVC	1452
Reinforced polyester	67
Profile extrusions[b]:	
PVC (incl. foam)	150
Resin-bonded woods:	
Phenolic	863
Urea and melamine	441
Vapour barriers:	
LDPE	82
PVC	25
Wall coverings[c]:	
Polystyrene	12
PVC	34

a: Excluding bonding or adhesive materials
b: Including windows, rainwater systems
c: Including swimming pool liners

After **Modern Plastics International** (1990).

acrylonitrile-butadiene-styrene and, for use in hot water systems, chlorinated polyvinyl chloride (Memmott and Beardsell, 1985). Plastics are also used as anti-corrosion coatings for steel pipes used for transporting commodities such as water, oil and gas. During 1975, Western Europe used 15,000 tonnes of low density polyethylene for this purpose (Aukes, 1979).

The exposure of plastics to hostile environments has led to reports of in-service failures due to microbial attack, and several workers have attempted to devise methods to evaluate the susceptibility of these materials to such degradation (Baskin and Kaplan, 1956; Darby and Kaplan, 1968; Seal and Pathirana, 1982). Plasticised P.V.C. is widely used for electrical insulation, floor coverings and wall coatings and this material has been reported to be susceptible to biodeterioration by both fungi and bacteria (Hitz et al., 1967; Booth and Robb, 1968). These workers reported that the plasticisers incorporated into the formulations were the components susceptible to attack in these cases. Other synthetic polymers used widely in civil engineering include polystyrene for insulation, and polyamides such as nylon-6,6 which are used in fibre form as geotechnical fabrics for soil stabilisation in structures such as roads and dams. Conflicting evidence has been presented regarding the susceptibility of these materials to microbial attack, since they are used in a wide variety of forms and both the test methods used to determine susceptibility and the parameters employed to assess degradation have varied widely. In studies using a polyethylene and a polyamide (designated P2010-V and PK-4 respectively), Allakhverdiev et al. (1967) found that only the polyamide was susceptible to biodeterioration. Degradation was assessed both by recording the presence of fungal growth on the material and by comparing the physical properties of degraded test pieces with those of uninoculated controls. However, Potts et al. (1973), found that neither polyethylene nor the polyamides nylon-6, nylon-6,6 or nylon-12 were susceptible to fungal deterioration when exposed to a mixed fungal inoculum as specified in a standard test method (ASTM Method D-1924-63). These workers used a visual rating of the extent of fungal contamination of the test material in order to indicate its susceptibility to microbial deterioration. In contrast to the previous workers, Colin et al. (1981) recorded that nylon-6,6 and both low density and high density polyethylene were significantly degraded within 32 months in soil contact. Degradation was assessed by measuring both physical and chemical changes. The degradation of nylon-6,6 was considered to be an oxidative process, while the mechanism involved in the degradation of polyethylene was unclear, since little evidence of either oxidation or microbial attack was found. Polystyrene has been reported to be resistant to microbial degradation (Potts et al., 1973), although Sielicki et al. (1978), using ^{14}C labelled polystyrene, detected a level of $^{14}CO_2$ release on soil burial amounting to 1.5-3% of total carbon over a period of 4 months, which was taken as an indication of some degree of degradability. These figures were acknowledged to be 15 to 30 times higher than values previously reported.

The polyurethanes are versatile plastics and are used for a

variety of applications in the construction industry. The
susceptibility of polyester polyurethanes to microbial
biodeterioration has frequently been reported, but the polyether
types have been found to be completely or almost completely resistant
(Ossefort and Testroet, 1966; Darby and Kaplan, 1968; Martens and
Domsch, 1981).

A number of standard methods are available to evaluate the
susceptibility of polymeric materials to microbial biodeterioration.
These methods involve the exposure of test pieces either to a defined
group of micro-organisms (usually fungi) in a petri dish test
(American Society for Testing and Materials, 1980; International
Organisation for Standardization, 1978) or to an undefined mixture of
fungi and bacteria by burying them in a standard soil (Deutsches
Institut fur Normung, 1984). Seal and Pantke (1986) recorded the
results of an interlaboratory collaborative investigation carried out
by the Plastics Project Group of the International Biodeterioration
Research Group (I.B.R.G.) in which a petri dish test based upon DIN
53 739 (Deutsches Institut fur Normung, 1984) was evaluated. It was
concluded that visual assessment alone was an unreliable parameter to
assess biodeterioration of polyester polyurethane and that weight
losses were not statistically acceptable over the 4 week period of
this test. Shuttleworth and Seal (1986) considered that because of
the physical dimensions of the test pieces specified in many standard
tests, several months may be required to obtain visible signs of
biodeterioration and that even longer periods may be necessary to
record changes in their physical properties. They therefore proposed
a rapid test method to determine the biodegradability of polyester
polyurethane which involved microscopic observation of the
penetration of thin films of the test material by the fungus
Gliocladium roseum. Penetration was recorded within 4 to 7 days. In
an alternative series of tests using soil burial, Pommer and Lorenz
(1985) observed that degradation proceeded more rapidly if the
polyester polyurethane test pieces were buried in soil in a stressed
attitude. Bentham **et al.** (1987) confirmed this result quantitatively,
recording a significant reduction in the tensile properties of
degraded stressed test pieces compared to degraded unstressed
material. These results led them to propose a test method which
involved stressing the test pieces in order to rapidly determine
their susceptibility to microbial degradation.

In-service, polyester polyurethanes may be exposed to a range of
temperatures whilst in contact with their end-use environment.
Varying temperatures due to differences in insolation or from the
thermogenic activities of thermophilous micro-organisms may affect
the deteriogenic activity of the microbial population (Morton and
Eggins, 1976). Conflicting evidence has been recorded regarding the
susceptibility of polyester polyurethane foams to microbial
degradation at elevated temperatures. Martens and Domsch (1981) and
Filip (1985) assessed the degradation of polyurethane foams incubated
in leachate from garbage landfill for 3 months at temperatures of 22
and 50°C. The former authors labelled polyurethane foams with ^{14}C and
evaluated degradation both by recording visible signs of structural
damage to the material and by measuring the quantities of ^{14}C-

labelled products released on degradation. The latter author used the criterion of weight loss to evaluate degradation. Both groups of workers recorded markedly more degradation when the test material was incubated at 50°C rather than at 22°C. Martens and Domsch (1981) attributed the degradation to chemical hydrolysis but Filip (1985) found no significant weight loss of test materials when incubated in sterile leachate. It was considered that the observed degradation in non-sterile leachate was due to the action of bacterial hydrolases and was therefore a biological effect. Other biologically active environments such as compost have been reported to rise in temperature due to the thermogenic activities of thermophilous micro-organisms (Fergus, 1964). These micro-organisms have been found to be widespread in soils, particularly in the upper layer and in plant debris near the soil surface (Apinis, 1963) but have been reported to be active only where soils were insolated (Eggins et al., 1972). The deteriogenic activity of this group of micro-organisms in terms of their ability to degrade polyester polyurethane has not been previously investigated.

During exposure to such environments, these materials may be stressed by compression, torsional or shearing forces. It was therefore considered that an investigation into the effects of soil temperature on the biodeterioration of stressed polyester polyurethane test pieces should be undertaken.

Polyester polyurethanes may also be exposed to hostile marine environments. Jones and Le Campion-Alsumard (1970) qualitatively assessed the extent of microbial attack on polyester polyurethane sheets subjected to submergence in the sea for 44 months. Marked degradation of the material was recorded, and a number of marine micro-organisms were implicated. Cable sheathing, used for seismic measurements in the Atlantic Ocean, was damaged after a number of years in-service (Stranger-Johannessen, 1985). Star-like fissures accompanied by a dark brown discolouration were observed in the material. Fungal degradation of the material was confirmed but the extent of degradation was not quantified. Materials exposed to marine environments may also be subject to physical stresses. The effects of stress on the extent of degradation of polyurethane in the marine environment have not however been investigated, and attempts were therefore made to quantify this relationship.

The reported failure in-service of polyester polyurethanes as a result of microbial degradation has led to the use of biocides such as Vinyzene B.P. which contains 10,10'-oxybisphenoxarsine (OBPA) as its active ingredient. OBPA has been reported to be active against a wide spectrum of fungal and bacterial micro-organisms (Versteegh, 1983) and is thus incorporated into materials during formulation with the intention of preventing biodeterioration. Bentham et al. (1987), found that the incorporation of Vinyzene B.P. protected polyester polyurethane test pieces buried in soil for 28 days at 25°C. There are, however, no reports of the effectiveness of Vinyzene B.P. in protecting polyester polyurethane from attack by either marine micro-organisms or thermophilous soil micro-organisms. The present investigation therefore includes an assessment of the performance of Vinyzene B.P. as a formulation biocide under these conditions.

2 Materials and methods

2.1 Test materials
Polyurethane formulations prepared both with and without the addition of the formulation biocide Vinyzene B.P. were available for this work. These were supplied as dumbell shaped test pieces with gauge dimensions of 60x10x10 mm.

2.2 Soil burial
The test pieces were buried in soil and maintained at temperatures of 25, 30, 37, 42 or $50^{\circ}C$. At each soil temperature six test pieces containing the formulation biocide Vinyzene B.P. and six test pieces without the biocide were provided. Unplasticised P.V.C. clips were used to fasten both ends of each set of six test pieces together, in order to place them under stress. Sets of six test pieces were buried for four weeks in plastic boxes each containing 1 kg of John Innes No. 2 compost with a moisture content of 35% (w/w). The activity of the soil micro-flora was confirmed by burying unprotected cotton fabric (Shirley Test Cloth) in each soil for up to 2 weeks prior to use. The destruction of this material by cellulolytic organisms was taken as an indication that the activity of the soil micro-flora was suitable for the purposes of the test. Changes in the physical properties of the test samples due to the incubation temperatures themselves were determined by incubating six test pieces of each material at the specified temperatures as unexposed controls.

2.3 Submergence in sea water
Stressed test pieces were placed in weighted sacks and submerged in sea water ponds at the Port Erin Marine Laboratory, Isle of Man. Six test pieces containing the formulation biocide together with six test pieces without the biocide were exposed to these conditions. In order to determine the effect of sea water itself on the physical properties of the test pieces, a separate laboratory experiment was also performed using sterile sea water. A set of six stressed biocide-free test pieces and a set of six stressed pieces containing Vinyzene B.P. were each submerged in 800 ml aliquots of sea water sterilised by the addition of 0.1% $HgCl_2$, and incubated at ambient temperature. In addition to the above, six stressed test pieces of each material were also incubated at ambient temperature as unexposed controls. All test pieces were incubated for 12 weeks.

2.4 Evaluation of physical properties
Tensile strength at break ($N mm^{-2}$) and percentage elongation at break of the test pieces were measured using a vertical tensometer (J.J. Lloyd Instruments Ltd.) with a load cell of 2000 N and a cross head speed of 100 mm min^{-1}.

3 Results

3.1 Physical properties of test pieces after soil burial
Figures 1 and 2 show the effect of soil burial on the physical properties of stressed polyester polyurethane test pieces maintained

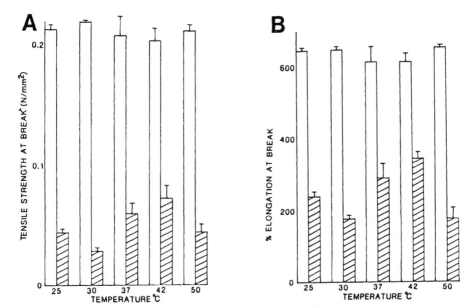

Figure 1 Physical properties of stressed biocide-free test pieces maintained at various temperatures, either buried in soil for 4 weeks ▨, or left unexposed as controls □.

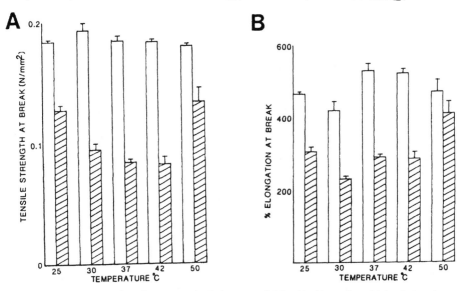

Figure 2 Physical properties of stressed test pieces, containing biocide and maintained at various temperatures, either buried in soil for 4 weeks ▨, or left unexposed as controls □.

     A) tensile strength at break

     B) % elongation at break

     Error bars indicate the standard error of the mean.

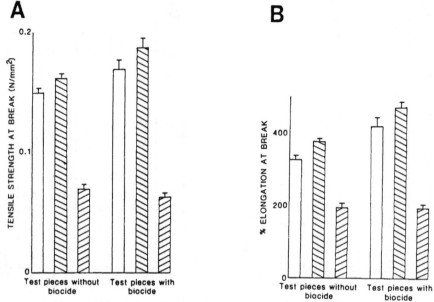

Figure 3 The effect of three months submergence on the physical properties of test pieces, as assessed by,

A) tensile strength at break

B) % elongation at break

Incubation conditions : sea water ▨, sterile sea water ▨, unexposed controls □,

Error bars indicate the standard error of the mean.

at various temperatures. Also shown are the physical properties of appropriate unexposed control test pieces. All of the test pieces exposed to soil burial showed a significant decrease in the values of the physical properties measured (P<0.001), compared to appropriate control replicates, with the exception of those containing biocide and buried in soil at 50°C, where no significant reduction in % elongation at break was recorded.

The incubation temperatures used were found to have no significant effect on the physical properties of unexposed control test pieces, with the exception of those containing biocide and assessed using the parameter of % elongation at break, where there was some evidence (P<0.05) of variability with temperature.

3.2 Physical properties of test pieces after submergence in sea water

Figure 3 shows the physical properties of polyester polyurethane test pieces submerged either in seawater or in sterile seawater, together with those of unexposed control replicates. All of the test pieces submerged in seawater showed a significant decrease in the values of the physical properties measured (P<0.001), both compared to appropriate unexposed controls, and also compared to replicates exposed to sterile sea water.

116

The only evidence for a difference between the physical properties
of test pieces submerged in sterile sea water and those of
appropriate unexposed controls was found in the case of biocide-free
test pieces, where a slight increase (P<0.05) in the % elongation at
break was recorded.

4 Discussion

Soil burial is an accepted method of assessing the susceptibility of
polymeric materials to microbial biodeterioration (Deutches Institut
fur Normung, 1984), and to a considerable extent simulates conditions
which may be encountered in-service by subterranean structures. It is
somewhat more difficult to simulate in the laboratory the conditions
which may be encountered by structures in marine contact, although in
the present investigation facilities were available to expose test
pieces to submergence in sea water ponds, with suitable laboratory
experiments being provided as supplementary controls.

As a result of this investigation data are available on the
effects of soil burial and exposure to sea water on the physical
properties of flexible polyester polyurethane foams. Significant
degradation of biocide-free test pieces buried in soil was recorded
at all the incubation temperatures used, with the most marked
reductions in physical properties being recorded at incubation
temperatures of 30 and 50°C. These latter findings support those of
Martens and Domsch (1981) and Filip (1985) in that there seems to be
evidence of an increased degradation at higher temperatures.

Bentham et al. (1987) recorded that Vinyzene B.P. provided
protection of stressed polyester polyurethane test pieces over 28
days of soil burial, and considered that such test material
constituted an abiotic control since the presence of the biocide
prevented microbial attack. In their investigations no degradation
was recorded in the case of these abiotic controls, suggesting that
the physicochemical effects of the soil were insignificant. In
contrast to these findings, in the present investigation significant
degradation of test pieces containing the biocide Vinyzene B.P. was
recorded at all of the incubation temperatures used, although in each
case the extent of such degradation was markedly lower than was
recorded for biocide-free test pieces incubated under similar
conditions. These results suggest either that the degradation
recorded was a result of both biological and physicochemical effects,
or that the biocide was ineffective under the experimental conditions
employed.

The findings presented on the effects of the exposure of
polyurethanes to sea water are particularly interesting. The data
provide relevant supplementary information to augment the
observations of Jones and Le Campion-Alsumard (1970) and of Stranger-
Johannessen (1985), in that the susceptibility of polyester
polyurethane test pieces to degradation in the marine environment was
confirmed. Significant reductions were recorded in the physical
properties of test pieces submerged in sea water, but not in those of
test pieces submerged in sterile sea water as compared to unexposed

controls. The chemical action of the sea water alone over the test period can therefore be considered insignificant and thus the degradation of the material may be attributed to the biological or physical action of the marine environment. No significant differences in the extent of degradation were recorded between test pieces containing the formulation biocide Vinyzene B.P. and biocide-free test pieces, and it can therefore be concluded that the biocide was completely ineffective in protecting the test pieces from degradation in the marine environment.

These findings are of interest to those environmentally aware scientists who are concerned with the fate and effect biocides the environment. The prospect of plastics being disposed of in high moisture landfill sites or in the marine environment justifies further investigations in this field.

5 Acknowledgements

The authors thank Caligen Foam Limited for providing the polyurethane foam test pieces used for this work; Mr. C. Bridge, Port Erin Marine Laboratory, Isle of Man, for providing access to sea water ponds and Ms. E. Giblin for secretarial assistance.

6 References

Allakhverdiev, G.A. Martirosova, T.A. and Tariverdiev, R.D. (1967) Change in the mechanical properties of polymeric films under the action of micro-organisms in soil. **Soviet Plastics**, 2, 22-23.

American Society for Testing and Materials (1963) Recommended practice for determining resistance of plastics to fungi (ASTM-D1924-63).

American Society for Testing and Materials (1980) Standard practice for determining the resistance of synthetic polymeric materials to fungi (ASTM G21-70).

Apinis, A.E. (1963) Occurrence of thermophilous microfungi in certain alluvial soils near Nottingham. **Nova Hedwigia**, 5, 57-78.

Aukes, M. (1979) Protecting the pipelines. **Plastics Today**, 6, 1-5.

Baskin, A.D. and Kaplan, A.M. (1956) A study of mixed spore culture and soil burial procedures in determining mildew resistance of vinyl-coated fabrics. **Applied Microbiology**, 4(6), 288-293.

Bentham, R.H. Morton, L.H.G. and Allen, N.G. (1987) Rapid assessment of the microbial deterioration of polyurethanes. **International Biodeterioration**, 23, 377-386.

Booth, G.H. and Robb, J.A. (1968) Bacterial degradation of plasticised P.V.C. - effect on some physical properties. **Journal of Applied Chemistry**, 18, 194-197.

Colin, G. Cooney, J.D. Carlsson, D.J. and Wiles, D.M. (1981) Deterioration of plastic films under soil burial conditions. **Journal of Applied Polymer Science**, 26, 509-519.

Darby, R.T. and Kaplan, A.M. (1968) Fungal susceptibility of polyurethanes. **Applied Microbiology**, 16(6), 900-905.

Deutches Institut fur Normung (1984) Prufung von Kunststoffen.
Einflub von Pilzen und Bakterien. Visuelle Beurteilung. Anderung
der Massen order der physikalischen Eigenschaften. DIN 53 739.

Eggins, H.O.W. von Szilvinyi, A. and Allsopp, D. (1972) The isolation
of actively growing thermophilic fungi from insolated soils.
International Biodeterioration Bulletin, 8(2), 53–58.

Fergus, C.L. (1964) Thermophilic and thermotolerant molds and
actinomycetes of mushroom compost during peak heating. **Mycologia**,
56, 267–284.

Filip, Z. (1985) Microbial degradation of polyurethanes, in
**Biodeterioration and Biodegradation of Plastics and Polymers.
Proceedings of the Autumn Meeting of the Biodeterioration Society.
Occasional Publication No. 1** (ed. K.J. Seal), Publications
Service, Lancashire Polytechnic, Preston, pp. 51–55.

Hitz, H.R. Merz, A. and Zinkernagel, R. (1967) Determination of the
resistance of plasticised P.V.C. to attack by fungi and bacteria
by the weight loss method and evaluation of mechanical properties.
A report on the co-operative test of the Task Group A (Biological
Attack) of the ISO Technical Committee 61, Working Group 6.
Material und Organismen, 2(4), 271–296.

International Organisation for Standardization (1978) Plastics –
determination of behaviour under the action of fungi and
bacteria – evaluation by visual examination or measurement of
change in mass or physical properties, Geneva. ISO 846-1978 (E).

Jones, E.B.G. and Le Campion-Alsumard, T. (1970) The
biodeterioration of polyurethane by marine fungi. **International
Biodeterioration Bulletin**, 6(3), 119–124.

Martens, R. and Domsch, K.H. (1981) Microbial degradation of
polyurethane foams and isocyanate based polyureas in different
media. **Water, Air, and Soil Pollution**, 15, 503–509.

Memmott, C. and Beardsell, K. (1985) The hidden giant. Major national
assets increasingly depend on P.V.C. pipe systems. **Plastics
Today**, 23, 16–19.

Modern Plastics International Special Report (1990) Materials ´89
United States of America. **Modern Plastics Inernational**, 20(1), 35–
43.

Morton, L.H.G. and Eggins, H.O.W. (1976) The influence of insolation
on the pattern of fungal succession onto wood. **International
Biodeterioration Bulletin**, 12(4), 100–105.

Ossefort, Z.T. and Testroet, F.B. (1966) Hydrolytic stability of
urethan elastomers. **Rubber Chemistry and Technology**, 39, 1308–
1327.

Pommer, E.H. and Lorenz, G. (1985) The behaviour of polyester and
polyether polyurethanes towards micro-organisms, in
**Biodeterioration and Biodegradation of Plastics and Polymers.
Proceedings of the Autumn Meeting of the Biodeterioration Society.
Occasional Publication No. 1** (ed. K.J. Seal), Publication Service,
Lancashire Polytechnic, Preston, pp. 77–86.

Potts, J.E. Clendinning, R.A. and Ackart, W.B. (1973) The effect of
chemical structure on the biodegradability of plastics, in
**Degradability of Polymers and Plastics. Institute of Electrical
Engineers,** The Plastics Institute, London, pp 12/1–12/9.

Seal, K.J. and Pantke, M. (1986) An interlaboratory investigation into the biodeterioration testing of plastics, with special reference to polyurethanes. Part 1: Petri Dish Test. **Material und Organismen**, 21, 151–164.

Seal, K.J. and Pathirana, R.A. (1982) The microbiological susceptibility of polyurethanes. A review. **International Biodeterioration Bulletin**, 18(3), 81–85.

Shuttleworth, W.A. and Seal, K.J. (1986) A rapid technique for evaluating the biodeterioration potential of polyurethane elastomers. **Applied Microbiology and Biotechnology**, 23, 407–409.

Sielicki, M. Focht, D.D. and Martin, J.P. (1978) Microbial degradation of (^{14}C) polystyrene and 1,3-diphenylbutane. **Canadian Journal of Microbiology**, 24, 798–803.

Stranger-Johannessen, M. (1985) Microbial degradation of polyurethane products in-service, in **Biodeterioration and Biodegradation of Plastics and Polymers. Proceedings of the Autumn Meeting of the Biodeterioration Society. Occasional Publication No. 1** (ed. K.J. Seal), Publications Service, Lancashire Polytechnic, Preston, pp. 93–102.

Versteegh, N. (1983) Protecting plastics from microbial attack. **Plastics and Rubber International**, 8(3), 96–97.

11 INFLUENCE OF MATERIALS ON THE MICROBIOLOGICAL COLONIZATION OF DRINKING WATER

D. SCHOENEN
Hygiene-Institut, Bonn, West Germany

Abstract
Drinking water is influenced intensively by materials with which it comes into contact. The products of corrosion of metals and cement mortar for example, are released into the water. Materials of organic basis (coatings, films, sealants, plastic pipes and hoses) or with organic additives (cement mortar with organic compounds) can lead to an intensive growth of microorganisms. Field observations with bituminous and epoxy resin coatings, PVC films, polyamide pipes and a plastic-containing cement mortar show, for example, that there is a considerable increase of microorganisms in the water and a visible microbial growth upon the surface of the materials. In experimental investigations on some materials growth could always be shown. This growth only developed on materials which liberate organic microbiologically usable components such as solvents or plasticizers. Purely mineral products such as glass, enamel and cement mortar or metals, and also some plastic materials without the release of organic compounds, do not lead to an increase in microorganisms. A microbial surface colonization can be established with the help of culture methods on these materials as well.
    To avoid deterioration of the drinking water it is necessary to examine such materials - a test procedure is given. The growth of specific hygienically relevant germs, e.g. Legionella pneumophila, can be supported by materials, as demonstrated by investigations on pipes and hoses.
Keywords: Drinking Water, Materials, Microbial Colonization.

1 Introduction

Drinking water undergoes many changes during transportation. Some of these are influenced by constituents of the water and interactions between the water and the materials with which it comes into contact. Corrosion is a well known problem involving destruction of

the materials and contamination of the water by products
of corrosion. Materials, especially plastics and coatings,
can contaminate the water without visible corrosion,
thereby impairing its taste and flavour Campbell (1940);
de Jong (1973); McFarren et al. (1977); Kreft et al.
(1981); Karrenbrock and Haberer (1982); Miller et al.
(1982); Larson et al. (1983); Montiel and Rauzy (1983);
Schoenen and Karrenbrock (1984); Yoo et al. ((1984); Bell
and Sorg (1985); Burlingame and Brock (1985); Krasner and
Means (1985); Sorg and Bell (1986); Alben et al. (1987);
Frensch et al. (1987); Bernhardt and Liesen (1988); Mäckle
et al. (1988). This paper confines itself, however, to a
description of the growth of microorganisms these
materials can support, which will be demonstrated by field
observations and experimental investigations.

The first cases of microbiological impairment of
drinking water due to materials were described in 1912 and
1916. Gärtner (1912) described colony count increases in
the water from a drinking water reservoir which had
previously been coated with a bituminous paint. He made
some laboratory tests and confirmed that the bituminous
paint caused the growth of the microorganisms. The next
recorded case of drinking water quality impairment due to
the material of a storage tank was that of Jones and
Greenberg (1964). The tank was built of redwood.

Houston (1916) reported an event involving a colony
count increase and the colonization of the water with
coliform organisms. The multiplication, including that of
coliforms, was caused by the leather seals in a water
pump. The growth-promoting effect of the leather was also
confirmed by controlled tests.

The microbial contamination of the water is not limited
to the public water supply. The colonization of the water
is possibly caused, above all, by hoses joined to the
installation system either permanently or temporarily.
Further field observations described in the literature are
listed in table 1. The type and place of the alteration as
well as the materials leading to the growth of
microorganisms are summarized.

2 Field observations

In the water of a rural community a colony count
increase occurred Schoenen and Schöler (1983). The maximum
was 20,000 cfu/ml (colony forming units). The colony count
increases only appeared in that part of the water system
which was supplied from a particular service reservoir. In
the other areas of the distribution system the colony
count remained consistently low. The chlorine content was
0,1 mg Cl_2/l in the initial section of the distribution
system but there was no free chlorine in the water of the
reservoir. The two compartments of the reservoir had been

Table 1. Field observations from literature.

Observed impairments	Place of observation	Material	Author(s)
increase of microorganisms	water reservior	bituminous painting	W.Gärtner 1912
microbial colonization	water reservoir	painting	Aug.Gärtner 1915
increase of cfu and Bacillus coli	sealing of water pump	leather	A.C.Houston 1916
increase of Coli aerogenes bacteria	sealing of pipe joints	jute	G.O.Adams, F.H.Kingsbury 1937
increase of cfu and coliform bact.	sealing of pipe joints	hemp	C.K.Calvert 1939
increase of cfu	water pipe	PE-, PVC pipe	W.Zimmermann 1956
increase of cfu	water pipe	plastic pipe	W.Ahrens, C.Siegert 1957
microbial corrosion, surface growth	rapid sand filter	PVC-painting	A.Schwartz, W.Schwartz 1962
increase of coliform bact., surface growth	reservoir	redwood	F.E.Jones, A.Greenberg 1964
increase of psychrophile microorganisms	reservoir	plastic painting	L.Wolff, A.Heintz 1966
surface growth	reservoir	epoxy resin	H.Barth 1972
increase of cfu, pellicle on the surface of the water	reservoir	epoxy resin	B.deJong 1973
increase of cfu, surface growth	reservoir	PVC film	P.Groth 1975

increase of cfu, growth of P.aeruginosa, coliform bact., mould	plumbing work	plastic pipe, fluxes	A.P.Burman, J.S.Colbourne 1977
increase of cfu, surface growth	reservoir	epoxy resin	H.Klopfer 1976
growth of coliform bact., surface growth	reservoir	redwood	R.J.Seidler, et al. 1977
increase of cfu	reservoir	coaltar epoxy resin	W.E.Ellgas, R.Lee 1980
increase of cfu, surface growth	reservoirs, concrete filter, pneumatic steel storage tank	bitumen, chlorinated rubber, epoxy resin	K.-H. Roggenkamp 1982
growth of Legionella pneumophila	plumbing work	sealing	J.S.Colbourne et al. 1984
growth of Legionella pneumophila	warm water supply	rubber	C.J.Niedeveld et al. 1986
increase of cfu	pneumatic steel storage tank	bitumen	H.Bernhardt, H.-U.Liesen 1988
increase of cfu	reservoir, water pipe line	bitumen, chlorinated rubber	H.Mäckle et al. 1988

cfu = colony forming units

given a coating of bituminous paint. The compartment of
the reservoir coated first was emptied and examined after
a 5-month working period. On the surface of the coating,
microbial growth was evident. The growth was removed with
a window scraper and collected (120 ml/m² growth).
Following this the reservoir was cleaned. Colony count
increases exceeding the guideline of 100 cfu/ml no longer
appeared. The microbial growth, however, was observed up
to 1 1/2 years after the bituminous coating had been
applied (Fig. 1).

Pronounced colony count increases were repeatedly
observed in the water from a newly constructed service
reservoir with a lining of epoxy resin. There was no
disinfection of the water at all. The reservoir consisted
of two chambers each of 5000 m³ capacity. The reservoir
was repeatedly drained and thereby a surface growth was
observed. Even after a period of ten years the epoxy
resin coating produced a microbial growth on its surface
(Fig. 2). Remarkable colony count increases no longer
appeared after the first inspection, after which the
reservoir was intensively cleaned.

In the water of a service reservoir coated with
PVC film, colony count increases in the water and a growth
on the surface of the PVC film were evident. Figure 3
shows the results of the bacteriological examination of
the non-disinfected water from the well and from the
reservoir. On average, the water remained in the reservoir
for two days. Figure 4 shows the amount of growth on the
PVC film in the period of 2 years.

In the following example, no colony count increases
occured after disinfection of the water. The chlorine

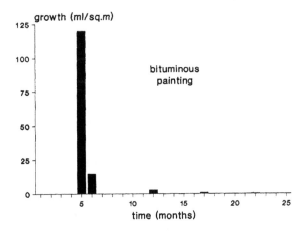

Fig. 1.   Amount of surface growth of a bituminous painting
in a drinking water reservoir at 5 inspections in
different temporal intervals.

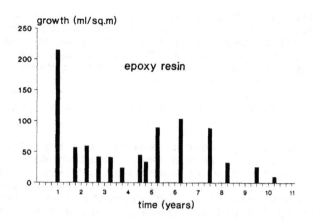

Fig. 2. Amount of surface growth of an epoxy resin
painting in a drinking water reservoir at 14 inspections
in the period of 11 years.

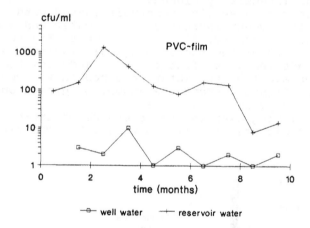

Fig. 3. Number (monthly average) of colony forming units
(cfu) in the water of a drinking water reservoir with a
lining of PVC film and in the water of the well running to
the reservoir.

content of the water leaving the reservoir was 0,5 mg
Cl_2/l. The reservoir was lined with tiles embedded in a
plastic-containing mortar. Microbial growth in the form of
macrocolonies was invariably situated over a joint with an
easily recognizable pore. The largest colonies had a
diameter of approximately 8 cm and had formed principally
in those regions in which the tiles did not adhere firmly
to the underlying surface. The colonies also formed

repeatedly at the same points.

Undesirable multiplication of microorganisms can also occur in pipes. A colony count increase was observed in the water of a pipe of 800 mm nominal diameter. The pipe conveyed non-contaminated groundwater that was introduced into the system in a bacteriologically acceptable condition. Increases of the colony count were found in the water at the end of the 7 km-long water pipe. Macrocolonies of about 1 cm diameter were apparent on the surface of the line. The macrocolonies had developed solely on those parts of the pipe in which the cement mortar lining contained plastic additives. Thereafter the water entering the pipe was continuously disinfected with chlorine. Colony count increases no longer appeared, although at the annual inspections the surface growth could be observed further on in the pipe.

High colony counts appeared in the water from a well in an uncharged drainage area (Fig. 5). In order to improve the bacteriological conditions, concentrated disinfectant solutions were added to the well repeatedly . This only led to a temporary improvement of the bacteriological quality of the raised water (Fig. 5). The water was raised by a pump over a 45 m-long polyamid rising-pipe line. The polyamide rising-pipe was replaced by a high-grade steel pipe, as a polyamide rising-pipe had already produced a microbial growth in a previous incident. There was also a visible growth on the inner and outer surface of the pipe. After the replacement of the pipe, colony count increases were no longer observed in the water from this well.

The surface growth was repeatedly collected at its point of origin and microbiologically and microscopically

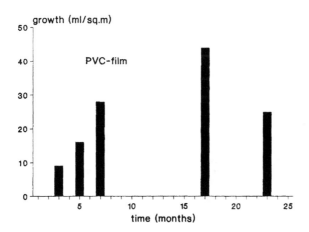

Fig. 4. Amount of surface growth of the PVC film in a drinking water reservoir at 5 inspections in the period of 2 years.

examined. Bacteria and mould were always found and nematodes, ciliates and amoebae were often present. The microscopically detected number of cells ranged from about 10^{10} to 10^{12} /ml. Bacteriological cultivation showed colony counts between 10^6 and 10^8/ml. The identified germs may be seen in table 2.

3 Microbial testings

From field observations it could never be conclusively established whether the growth of microorganisms in the water and adhering to the surface of the materials was solely due to the materials or whether or not the nature of the water might have made a significant contribution. Time and again it was surmised, principally by operators of waterworks, that the adverse effects were caused either by the water and its constituents, or else by undesirable interactions between the materials and the water. In particular, these conjectures were reinforced by the fact that the materials which led to a striking degree of bacterial multiplication in one place did not give rise to similar impairment in others. In order to elucidate some of the questions arising from field observations, most authors who observed these untoward incidents carried out controlled tests Gärtner (1912); Houston (1916); Gerstein (1928); Adams and Kingsbury (1937); Calvert (1939); Ahrens and Siegert (1957); Jones and Greenberg (1964); Böing (1957); Barth (1972); de Jong (1973); Groth (1975); Burman and Colbourne (1977); Seidler et al. (1977); Ellgas and Lee (1980); Schoenen and Schöler (1983); Colbourne et al.

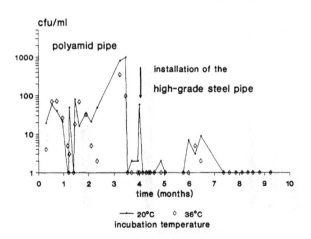

Fig. 5. Number of colony forming units (cfu) in the water of a well with a polyamid rising pipe and after exchange by a high-grade steel pipe.

(1984); Colbourne (1985); Niedewald et al. (1986);

Three methods can be distinguished:

1   increase of the number of microorganisms in the water;

2   visible growth of miroorganisms on the surface of the materials;

3   cultivation in petri dishes.

3.1   Increase in the number of microorganisms in the water

To assess the materials the method already described by Gärtner in 1912 was generally used. The materials were exposed to the water and the colony count increases in the water were observed. Meanwhile different conjugations of this method had been developed Gärtner (1912); Calvert (1939); Ahrens and Siegert (1957); Böing (1957); Schmidt (1960); Burman and Colbourne (1977); Althaus et al. (1979); Ellgas and Lee (1980); Kooij et al. (1982). Essential differences are produced by the times of observation (between 5 days and 20 months), the water with which the materials came into contact (aqua dest., buffered aqua dest., mineral salt solution, ground water, unchlorinated water, dechlorinated water, sterile water, diluted waste water), and the primary contamination of the test water (Escherichia coli, Enterobacter aerogenes, Clostridium perfringens, Klebsiella oxytoca, K. pneumoniae, Pseudomonas aeruginosa, P. fluorescens, P. species, Salmonella enteritidis, S. infantis, S. paratyphi B, S. soneii, S. typhi, S. typhimurium, Staphylococcus albus, S. aureus, mixed flora from drinking water and surface water). The experiments without materials or with inert materials also led to an increase of microorganisms, hence it was decidedly difficult to determine whether or not the materials caused the increase of microorganisms.

A distinctive change in the investigation method was described by Calvert (1939). The water was changed at intervals of 3 days ("three day cycle method"). This method of investigation was adopted by Burman and Colbourne (1977); Ashworth and Colbourne (1981); Colbourne et al. (1984); Colbourne (1985) and further developed into a routine test protocol for the U.K. anonym (1983). The materials were exposed in 2-litre flasks, filled at the start with a mixture consisting of 100 ml untreated river water (River Thames) and dechlorinated tap water. The water was changed twice a week. The experiments lasted for a period of at least 8 weeks. Data obtained included the colony count after incubation at 22 and 37°C, the populations of mould, P. aeruginosa and coliform bacteria.

In addition, the oxygen concentration and therewith the biochemical oxygen demand (BOD) was determined.

This method was used by the author in a varied form to prove the influence of pipe- and hose-materials on the microbial colonization of the water (see below).

3.2 Visible growth of microorganisms on the surface of materials

Groth (1975) first adopted formation of a surface growth as assessment criterion. He performed these tests following the discovery of colony count increases in the water of a reservoir lined with PVC film, associated with the appearence of growth on the surface of the film. 12 PVC films used for lining service reservoirs were tested. Only two of the materials showed no evidence of growth, but the film used in the reservoir showed surface growth. This method, using surface growth as criterion for assessing materials, was developed as a standard method in West Germany anonym (1984).

The comparable investigations carried out by the author with different materials in clear water reservoirs and later on in a laboratory test plant with basins having continuous through flow are described below.

3.3 Cultivation in petri dishes

The exposition of samples in petri dishes on oligotrophic nutrient agar was first described Böing by (1957) and taken further by Barth (1972). The surfaces of the materials and the agar were then contaminated with specific microorganisms. Materials, which released substances capable of utilizing bacteria, should then cause the formation of surface colonies.

4 Comparing investigations

Growth on the surface was used to assess the materials in the following investigations Schoenen and Schöler (1983). The test plates (30x90cm) of different materials were exposed to the water in drinking water reservoirs for 5 months. Thus the test materials were continuously wet. In order to detect surface growth, the test plates were gently scraped with a window scraper. The surface growth or the liquid film adhering to the specimen was then collected in a measuring cylinder. These studies were carried out in 15 service reservoirs with water of different origin, treatment, quality and chlorination. In table 3 the results of four chosen materials are shown. Bituminous painting and chlorinated rubber always gave

Table 2   Microorganisms found in surface growth of
different materials Dott and Schoenen (1985).

Genus/species/other groups	frequency
Gram-negative bacteria	+++++
P. fluorescens	++++
P. putida	++++
P. aeruginosa	+
P. cepacia	+++
P. diminuta	+++
P. acidovorans	++
P. putrefaciens	+
P. alcaligenes	++
P. pseudoalcaligenes	++
P. spec.	+++++
Plesiomonas spec.	+
Comamonas terrigena	+
F. devorans	++
F. breve	+++
F. aquatile	+++
F. rigense	++
F. spec.	+++++
CDC-Gr. VeI/VeII/IIk	+++
Cytophaga spec.	++
Caulobacter spec.	++
Hyphomicrobium spec.	++
Sheathed bacteria	+++
Chromobacterium spec.	+
Acinetobacter calcoaceticus	++++
Alcaligenes spec.	+++
Achromobacter spec.	++
Moraxella like bacteria	+++
Pasteurella spec.	+
Citrobacter spec.	+
Klebsiella spec.	+
E. agglomerans	++
Vibrio spec.	++
Spirillum spec.	++
Zoogloea spec.	++
Acetobacter spec.	+
Aeromonas spec.	++
unindentified gram-negative rods	+++++
Gram-positve bacteria	++
Bacillus spec.	+
Micrococcus speç.	+
Planococcus spec.	+
Arthrobacter spec.	++
Streptomyces spec.	+
Actinoplanes spec.	+
Nocardia spec.	+
Pasteurella spec.	+

occurrence

+++++	always	P.	= Pseudomonas
++++	most frequent	F.	= Flavobacterium
+++	frequent	CDC	= Center of Disease
++	occasional		Control (Atlanta)
+	sporadic		

Table 3. Results of 4 selected materials from 15 reservoirs with water of different characteristics. The dose of chlorine is mentioned because it is the only component of the water with an influence to microbial behaviour.

Reservoir	Chlorine mg/l	Bit. paint.	Chlor. rubber	Asbestos cement	Polym.-metha.
1	0	+	+	-	-
2	0	+	+	-	-
3	0	n.e.	+	n.e.	-
4	0	+	+	-	-
5	0	+	+	-	-
6	0,1	+	+	n.e.	-
7	0,1	+	+	-	-
8	0,15	+	+	-	-
9	0,15	+	+	-	-
10	0,15	+	-	-	-
11	0,2	+	-	-	-
12	0,2	-	-	-	-
13	0,2	-	-	-	-
14	0,4	-	-	-	-
15	0,4	-	n.e.	-	-

+ = surface growth;   - = no surface growth
n.e. = not examined
Bit. paint = bituminous painting
Chlor. rubber = chlorinated rubber
Polym.-metha. = polymethylmethacrylat

rise to a surface growth, when the dose of chlorine in the water was < 0,15 mg/l. Asbestos-cement and polymethylmethacrylate never showed any surface growth. The development of a surface-attached growth on materials in contact with water was thus seen to be a material-specific phenomenon not affected by the nature of the tap water. Surface growth can be suppressed by disinfection. All materials, including those without visible growth, had a microbial colonization on the surface that was detectable by contact cultures or by scanning electron microscopy Schoenen and Tuschewitzki (1982).

Having proved that the components of the water had no influence upon the bacterial growth, it was possible to test the materials in one type of water. In the following experiments the size of the test plates was reduced and they were exposed in 100-litre tanks. A treated, non-disinfected groundwater was used for the through flow in the tanks. The exchange of the water was 4 to 6 times the capacity per day. The time of observation was shortened from 5 to 3 months. For the assessment of materials in accordance with the DVGW-Arbeitsblatt W 270 anonym (1984) two examination periods of three months are required.

Microbial testing of materials cannot yet be replaced by chemical or physicochemical investigations Anonym (1977) and testing with TOC (total organic carbon) is insufficient.

In order to discover the influence of pipe- and hose-materials on the microbial colonization of the water, pipes and hoses with narrow diameters were well regulated by a through flow of water and samples were taken and examined Schoenen and Wehse (1988). The pipes and hoses had a length of 10 m and an inner diameter of 3 mm and they were flushed through 5 days a week with 5 l of water each time.

In corresponding investigation plants, the influence of the materials on the colonization of the water with Legionella pneumophila kept at temperatures of 35 - 40°C was tested and with E. coli, K. pneumoniae and P. aeruginosa at room temperature.

5 Results

Results are shown of some materials often used in drinking water supply systems and of comparable materials. Finally, the results of the contamination of the water by pipes and hoses are shown.

5.1 Pure mineral materials and metals

Materials which would not be expected to exhibit any tendency to support microbial growth, by reason of their composition (and for which this has been confirmed by testing) are: glass, enamel, cement mortar, asbestos-cement and metals (high-grade steel, copper). On high-grade steel plates, an intensive growth was observed during the initial stages of experiments. This growth, however, was invariably eliminated if the test pieces were given a sufficiently thorough cleaning beforehand. It can thus be inferred that the protective coating, which is sometimes applied to high-grade steel sheets during manufacture, had caused the growth formation.

5.2 Coatings

Solvent-containing coatings used as corrosion inhibitors in drinking water have frequently given rise to microbial growth of an undesirable nature. Specific examples of such behaviour in the field are known for bituminous painting, chlorinated rubber, PVC, and epoxy resins. It is possible that the growth-promoting effect was caused by the solvents. Growth formation, however, declines fairly rapidly. After 1 1/2 to 2 years these materials no longer give rise to a detectable growth of bacteria. In addition, the processing and especially the curing of the coating affects the microbiological behaviour. If solvent-containing coatings are cured for a couple of days after application at elevated temperatures of about 50 - 80°C, they lose so much solvent that no detectable growth of bacteria occurs.

Solvent-based coating compositions are thus unsuitable for drinking water supplies, if they are applied and cured under the temperature conditions normally prevailing for structures and technical equipment of drinking water supply and distribution facilities. They may, however, be acceptable provided they are held for a certain time (a few days) at an elevated temperature (>50°C).

In addition to the solvent-containing epoxy resins mentioned above, solvent-free epoxy resins are also employed in drinking water applications. The solvent-free epoxy resins contain a so-called accelerator. In the products tested, benzylalcohol was added as an accelerator. In the tested compositions, up to 5% of accelerator proved acceptable, while materials with a higher proportion caused bacterial growth. In contrast to the solvent-based epoxy resins, curing at higher temperatures had no effect on microbial behaviour.

Epoxy resins, both of the solvent-based and solvent-free types, must be compounded from two components directly prior to use. Both components must be mixed together in prescribed stoichiometric proportions, so that a chemical reaction between them occurs. If the mixing ratio is not correct, there may also be evidence of bacterial growth. As has been shown by experiments, both under-use and over-use of hardener leads to bacterial growth. Coatings not exceeding a 20% over-use or under-use of hardener do not exhibit any technical deficiencies but result in microbial growth.

5.3 PVC materials

Soft PVC films for the lining of service reservoirs have repeatedly led to microbiological impairment of water quality. In addition, a surface-attached growth has been observed time and again. The microbial growth is caused by additives, especially phthalates that are used in the preparation of soft flexible films. Some PVC film types which get their flexibility through

copolymerization do not cause bacterial growth.
The rigid PVC pipes tested showed no evidence of
growth formation.

5.4 Polyamid pipes

A single sample of a polyamid pipe showed an intensive
growth, both during laboratory tests and also earlier in
the field. No information is available concerning the
nature of the constituents which the bacteria can
utilize.

5.5 Cement mortar

Cement mortar is employed for the lining of water
mains and reservoirs. Additives are frequently introduced
to improve the ease of application. Cement mortar without
plastic additives does not give rise to microbial
impairment of water quality. In contrast to this,
cement mortars containing plastics as bonding agents or
for similar purposes have repeatedly resulted in colony
count increases.
Which constituents are released by the additives and
function as substrates for microbial growth is not known
at present as there is no information available on the
composition of the various products on the market. Thus
it is not possible to indicate which plastic or
combination of plastics may be acceptable in practice.
Besides materials in direct contact with water, those
with merely indirect contact can also lead to bacterial
growth. For example, plastic-containing cement mortar is
also used in tile laying, and may then cause bacterial
growth, as shown by a field observation. The organic
components reach the surface by pores and cause bacterial
growth. This influence of plastic-containing
cement mortar could be proved by the test plates.

5.6 Rubber

Rubber is widely used in drinking water supply
facilities for the sealing of pipe joints as well as for
the surface coating of metallic components as an
anti-corrosive agent. All the materials examined except
one exhibited an undesirable growth in the course of our
own experiments. Colbourne (1985) however, found some
rubber products which did not produce an unwanted
bacterial growth.

5.7 Non-hardening sealants

Non-hardening sealants are frequently used for the
sealing of expansion joints in drinking water reservoirs.
Time and again they display some growth formation. The
intensity of surface growth differs greatly. It is not
known which components of the silicons and polysulphides
cause the bacterial growth.

5.8 Reference materials

To prove the microbial behaviour of materials it is
necessary to include in different test batches materials
of a well-known microbiological behaviour. Glass and
high-grade steel proved to be suitable negative controls,
which do not lead to bacterial multiplication. When using
high-grade steel it is important to clean it intensively
before exposure.

A material of well-defined composition which is easily
utilizable by bacteria (positive control) is paraffin.
Solvent-based coatings, such as bituminous paint and
chlorinated rubber, may also serve as positive controls.

For microbiological testing there are two acknowledged
methods; those of the UK anonym (1983) and West Germany
anonym (1984). In order to test the conformity of the two
methods PVC films were exchanged and examined in London
and Bonn. The investigations gave corresponding results
Schoenen and Colbourne (1987).

6 Contamination of the water by pipes and hoses

The influence of the pipes and hoses on
microbiological contamination of the water is
demonstrated in figures. Figure 6 shows the results of
the first water sample from PVC-, PE-, PTFE-, silicone-
and rubber hoses during an observation time of half a
year. The number of colony forming units (cfu) decreased
with the amount of water passing through. There was no,
or hardly any, colony increase in the water of copper and
glass pipes. The water of the insufficiently cleaned
high-grade steel pipe had high initial colony counts
(Fig. 7). There was only a slight colony increase in a
second high-grade steel pipe which had been intensively
cleaned with benzine and   acetone beforehand (Fig. 7).

The contamination of the water with Legionella
pneumophila from pipes and hoses is demonstrated in Fig.
8. Materials such as rubber and PVC which promote the
growth of the unspecific flora do so to L. pneumophila
Schoenen at al. (1988). On the other hand, in the water
of the PA pipe there was no growth of L. pneumophila but
an intensive growth of unspecific microorganisms and in

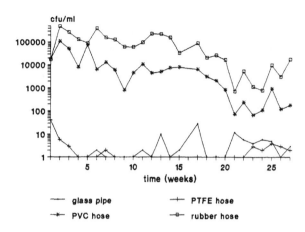

Fig. 6. Colony forming units (cfu) in the water of pipes
and hoses when the water was exchanged regularly over a
period of 6 months Schoenen and Wehse (1988).

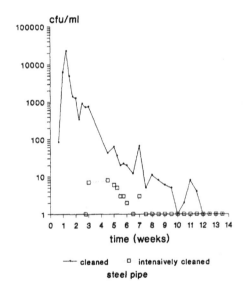

Fig. 7. Colony forming units (cfu) in the water of two
high grade steel pipes, one insufficiently and one
intensively cleaned.

the water of the PTFE hose there was a growth of L.
pneumophila but no increase of unspecific organisms.
However, there are still no satisfactory explanations for
this reaction of PA- and PTFE hoses.

The results of pipes and hoses contaminated with E. coli, Klebsiella pneumoniae, Citrobacter freundii and Pseudomonas aeruginosa (detectable in 100ml samples) are shown in figures. Only the water of the rubber hose had a long lasting contamination with E. coli (Fig. 9), K. pneumoniae (Fig. 10) and C. freundii (Fig. 11) On the other hand P. aeruginosa (Fig. 12) was eliminated only from the water of the copper pipe. All the other pipes and hoses, even those with microbial inert materials like glass and high-grade steel, showed a contamination during the whole time of the experiment (100 days).

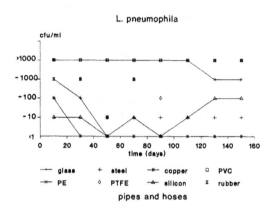

Fig. 8.   Contamination of the water by Legionella pneumophila in different pipes and hoses when the water was exchanged regularly.

E. coli

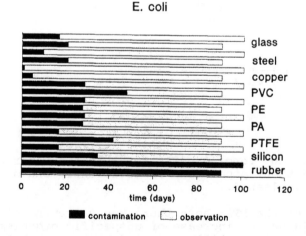

Fig. 9.   Interval of observation and colonization of the water from pipes and hoses by Escherichia coli. There was an initial contamination with E. coli.

K. pneumoniae

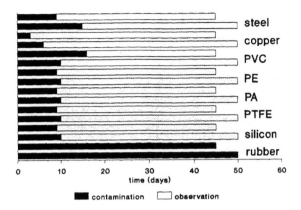

Fig. 10. Interval of observation and colonization of the water from pipes and hoses by Klebsiella pneumoniae. There was an initial contamination with K. pneumoniae.

C. freundii

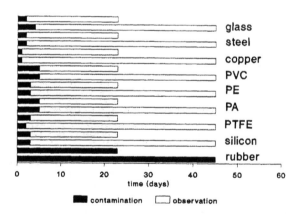

Fig. 11. Interval of observation and colonization of the water from pipes and hoses by Citrobacter freundii. There was an initial contamination with C. freundii.

All these organisms are unsuitable for the testing of the influence of materials on microbial growth, E. coli, K. pneumoniae and C. freundii die off although there is a growth promoting influence of the material e.g. PVC, or there is a growth of P. aeruginosa with inert materials.

P. aeruginosa

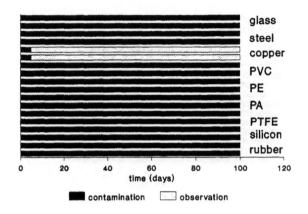

Fig. 12. Interval of observation and colonization of the water from pipes and hoses by Pseudomonas aeruginosa. There was an initial contamination with P. aeruginosa.

7 Conclusions

Materials with which drinking water comes into contact can promote the multiplication of microorganisms. Colony count increases in the water and a microbiological growth on the surfaces of the materials themselves have been found in field observations. Usually the increase in unspecific microorganisms, which are normally present in the water in low numbers, was registered. Occasionally, there was an increase in microorganisms of hygienic relevance, such as Legionella pneumophila, Pseudomonas aeruginosa and coliform organisms.

In comparing investigations with different methods it was possible to prove that the microbial growth is promoted by materials and not by components of the water. Disinfectants can suppress the growth. So far, the observations chiefly concern the public drinking water supply. However, these instances of impairment are not limited to this domain. In home installations and plants that are permanently or temporarily joined to the distribution system, a disadvantageous growth may occur.

To exclude microbial impairment of the water by materials it is necessary to examine the materials. In the U.K. and in West Germany there are acknowledged testing methods to assess materials used in the drinking water sphere. So far it has not been possible to replace microbiological testing by chemical testing. To avoid incidences in the field caused by tested materials, it is important to achieve the same conditions for the examination as prevail in the water supply system.

8 References

Adams, G.O. and Kingsbury, F.H. (1937) Experiences with Chlorinating New Water Mains. J. New England Water Works Assoc., 60-68.

Ahrens, W. and Siegert, C.B.F.W. (1957) Weitere Untersuchungen und Beobachtungen über die Beeinflussung der Trinkwasserkeimzahlen durch Kunststoffrohre. Gas- und Wasserfach (Wasser/Abwasser), **98**, 661-663.

Alben, K. Bruchet, A. Shpirt, E. and Kaczmarczyk, J. (1987) Solvents Leached from Vinyl Coated Storage Tanks Methodology to Estimate Human Exposure. Annual Conference Proceedings, Amer. Water Works Assoc., 1995-2007.

Althaus, H. Barth, H. Hamburger, B. Müller, G. Schubert, R.W. and Selenka, F. (1979) Stand der Entwicklung von mikrobiolgischen Prüfmethoden zum Verhalten von Kunststoffen, die im Trinkwasserbereich eingesetzt werden sollen. Bundesgesundheitsblatt, **22**, 374-377.

Anonym (1984) Vermehrung von Mikroorganismen und Materialien für den Trinkwasserbereich; Prüfung und Bewertung, Technische Regeln, DVGW-Arbeitsblatt W 270, ZfGW-Verlag, Frankfurt/Main.

Anonym (1977) Gesundheitliche Beurteilung von Kunststoffen und anderen nichtmetallischen Werkstoffen im Rahmen des Lebensmittel- und Bedarfsgegenständegesetzes für den Trinkwasserbereich 1. Mitt. Bundesgesundheitsblatt, **20**, 10-13. 2. Mitt. Bundesgesundheitsblatt, **20**, 124-129. 3. Mitt. Bundesgesundheitsblatt, **22**, 213-216. 4. Mitt. Bundesgesundheitsblatt, **22**, 264-265. 5. Mitt. Bundesgesundheitsblatt, **28**, 371-374. 6. Mitt. Bundesgesundheitsblatt, **30**, 178.

Anonym (1983) Rubber Seals - Joint Rings for Water Supply, Drainage and Sewerage Pipelines - Specification for Materials. International Organization for Standardiza-tion,, **First Edition - 1983-06-01, UDC 678.4-447: 621.643.44, Ref.No. ISO 4633-1983 (E).**

Ashworth, J. and Colbourne, J.S. (1981) The Effect of Non-Metallic Materials Employed in Water Supply Distribution and Plumbing Systems, and Consumer Fittings upon Potable Water Quality. International Water Industry Conference, Brighton U.K. Papers and Abstracts, 130-135.

Barth, H. (1972) Kunststoffe zur Oberflächenbehandlung, ein mikrobiologisches Problem. II. Voraussetzungen und Entwicklung von Methoden zur Prüfung der Kunststoffe auf physiologische Unbedenklichkeit und auf Verwertbarkeit für Bakterien und Pilze. Zbl. Bakt. Hyg., I. Abt. Orig.

B, **155**, 360-373.

Behr, H. Schöler, H.F. Schoenen, D. and Sturm, E. (1985) Kunststoffhaltiger Mörtelzusatzstoff als Kohlenstoffquelle für das Wachstum von Mikroorganismen im Trinkwasser. Vom Wasser, **65**, 107-118.

Bell, F.B and Sorg (eds), T.J.(1985) Plumbing Materials and Drinking Water Quality, Proceedings of a Seminar Cincinnati, Ohio May 1984, U.S. Environmental Protection Agency EPA /600/9-85/007, Washington 1985.

Bernhardt, H. and Liesen, H.-U. (1988) Trinkwasserverkeimungen in Verteilungsnetzen durch Korrosionsschutz auf Bitumenbasis. Gas- und Wasserfach (Wasser/Abwasser), **129**, 28-32.

Böing, J. (1957) Über die Erhöhung der Keimzahlen in Polyäthylen-Rohren. Zbl. Bakt. Hyg., I. Abt. Orig., **168**, 324-328.

Burlingame, G.A. and Brock, G.L. (1985) Water Quality Deterioration in Treated-Water Storage Tanks. Annual Conf. Proceedings, Amer. Water Works Assoc., 351-369.

Burman, N.P. and Colbourne, J.S. (1977) Techniques for the Assessment of Growth of Microorganisms on Plumbing Materials Used in Contact with Potable Water Supplies. J. Applied Bacteriology, **43**, 137-144.

Calvert, C.K. (1939) Investigation of Main Sterilisation. J. Amer. Water Works Assoc., **31**, 832-836.

Campbell, M.S. (1940) Gastro-Enteritis Outbreak Traced to Painting of Water Tank. J. Amer. Water Works Assoc., **32**, 1928-1939.

Colbourne, J.S. (1985) Materials Usage and Their Effects on the Microbiological Quality of Water Supplies. J. Applied Bacteriology Symposium Supplement, 47S-59S.

Colbourne, J.S. Pratt, D.J. Smith, M.G. Fisher-Hoch, S.P. and Harper, D. (1984) Water Fittings as Sources of Legionella pneumophila in a Hospital Plumbing System. The Lancet i, 210-213.

Dott, W. and Schoenen, D. (1985) Qualitative und quantitative Bestimmung von Bakterienpopulationen aus aquatischen Biotopen. 7. Mitt.: Entwicklung der Aufwuchsflora auf Werkstoffen im Trinkwasser. Zbl. Bakt. Hyg., I. Abt. Orig. B, **180**, 436-447.

Ellgas, W.M. and Lee, R. (1980) Reservoir Coatings Can Support Bacterial Growth. J. Amer. Water Works Assoc., **72**, 693-695.

Frensch, K. Hahn, J.U. Levsen, K. Niessen, J. Schöler, H.F. and Schoenen, D. (1987) Lösemittel aus dem Anstrichmaterial eines Hochbehälters als Ursache für erhöhte Koloniezahlen in Trinkwasser. Vom Wasser, **68**, 101-109.

Gärtner, Aug.(1915) Die Hygiene des Wassers, Friedr. Vieweg & Sohn, Braunschweig 1915, p. 95.

Gärtner, W. (1912) Über Bakterienwachstum in Wasserreservoiren mit Innenschutzanstrichen. J. Gasbeleuchtung und Wasserversorgung, **55**, 907-908.

Gerstein, H.H. (1928) Bacteria Growing on Leather
Packing Cause Errors in Sampling. Engineering
News-Record, 13, 407.
Gettrust, J.S. (1932) Contamination of Water in a New
Pipe Line by Jute and Hemp Packing. Ohio Conference on
Water Purification, 23-24.
Groth, P. (1975) Aufwuchsbildung auf Kunststoff-Folien
in Reinwasserbehältern. Wasser und Boden, 27, 257-259.
Houston, A.C. (1916) Leather Bacillus. 12th Research
Report Metropolitan Water Board, 9-11.
Jones, F.E. and Greenberg, A.E. (1964) Coliform Bacteria
in Redwood Storage Tanks. J. Amer. Water Works Assoc.,
56, 1489-1493.
de Jong, B. (1973) Veränderungen der
Wasserbeschaffenheit durch Anstrichmittel und Folien.
Schr.Reihe Verein Wasser-, Boden- und Lufthygiene, 40,
89-94.
Karrenbrock, F. and Haberer, K. (1982) Nachweis
organischer Lösungsmittel im Wasser eines
frischgestrichenen Trinkwasserbehälters. Zbl. Bakt. Hyg.,
I. Abt. Orig. B, 176, 519-524.
Klopfer, X. (1976) Anstrichschäden. Bauverlag GmbH,
Wiesbaden, p. 104.
Kooij, D.v.d. Visser, A. and Hijnen, W.A.M. (1982)
Determining the Concentration of Easily Assimilable
Organic Carbon in Drinking Water. J. Amer. Water Works
Assoc., 74, 540-545.
Krasner, S.W. and Means, E.G. (1985) Returning a Newly
Hypalon-Covered Finished-Water Reservoir to Service
Health and Aesthetic Considerations. Annual Conference
Proceedings, Amer. Water Works Assoc., 371-394.
Kreft, P. Trussell, A. Lang, J. Kavanaugh, M. and
Trussell, R. (1981) Notes and Comments Leaching of
Organics from a PVC-Polyethylen Plexiglass Pilot Plant.
J. Amer. Water Works Assoc., 73, 558-560.
Larson, Ch.D. Love, O.Th. and Reynolds, G. (1983)
Tetrachloroethylene Leached from Lined Asbestos-Cement
Pipe into Drinking Water. J. Amer. Water Works Assoc.,
76, 184-190.
Mäckle, H. Mevius, W. Pätsch, B. Sacré, C. Schoenen, D.
and Werner, P. (1988) Koloniezahlerhöhungen sowie
Geruchs- und Geschmacksbeeinträchtigungen des
Trinkwassers durch lösemittelhaltige
Auskleidematerialien. Gas-und Wasserfach
(Wasser/Abwasser), 129, 22-27.
McFarren, E.F. Buelow, R.W. Thurnau, R.C. Gardels, M.
Sorrell, R.K. Snyder, P. and Dressman, R.C. (1977) Water
Quality Deterioration in the Distribution System. Water
Quality and Technology Conference Proceedings, Amer.
Water Works Assoc., 2A-1, 1-10.
Miller, H.C. Barrett, W.J. and James, R.H. (1982)
Investigation Potential Water Contamination by
Petroleum-Asphalt Coatings in Ductile-Iron Pipe. J. Amer.

Water Works Assoc., **74**, 151-156.
Montiel, A. and Rauzy, S. (1983) Dosage des monomères et solvants légers dans les matériaux plastiques en contact avec l'eau et dans l'eau en contact avec ces matériaux. REVUE FRANCAISE DES SCIENCES DE L'EAU, **2**, 255-266.
Niedeveld, C.J. Pet, F.M. and Meenhorst, P.L. (1986) Effect of Rubbers and Their Constituents on Proliferation of Legionella pneumophila in Naturally Contaminated Hot Water. The Lancet ii, 180-183.
Roggenkamp, K.-H. (1982) Bakteriologische Trinkwasserbeeinträchtigung nach Auskleidung und Beschichtung von Behältern. Neue DELIWA-Zeitschrift, 62-64.
Schmidt, B. (1960) Bakteriologische Untersuchungen zur Frage der Verwendbarkeit von Polystyrol Erzeugnissen in Trinkwasserleitungen. Zbl. Bakt. Hyg.,I. Abt. Orig., **178**, 381-392.
Schoenen, D. (1985) TOC-, TIC- und TC-Bestimmungen an Werkstoffen für den Trinkwasserbereich. Z. Wasser Abwasser Forsch., **18**, 254-257.
Schoenen, D. and Colbourne, J.S. (1987) Microbiological Evaluation of Drinking Water Construction Materials - Comparison of Two Test Procedures -. Zbl. Bakt. Hyg. B, **183**, 505-510.
Schoenen, D. and Karrenbrock, F. (1984) Geruchs- und Geschmacksbeeinträchtigungen des Trinkwassers durch Werkstoffe. Gas- und Wasserfach (Wasser/Abwasser), **125**, 142-145.
Schoenen, D. and Schöler, H.F. (1983) Trinkwasser und Werkstoffe. Praxisbeobachtungen und Untersuchungsverfahren. Gustav Fischer Verlag, Stuttgart. (1985) Drinking Water Materials: Field Observations and Methods of Investigations. Ellis Horwood Limited, Chichester GB
Schoenen, D. Schulze-Röbbecke, R. and Schirdewahn, N. (1988) Mikrobielle Kontamination des Wassers durch Rohr- und Schlauchmaterialien. 2. Mitt.: Wachstum von Legionella pneumophila. Zbl. Bakt. Hyg. B, **186**, 326-332.
Schoenen, D. and Tuschewitzki, G.-J. (1982) Mikrobielle Besiedlung benetzter Bitumen- und Edelstahlflächen in rasterelektronenmikroskopischen Aufnahmen. Zbl. Bakt. Hyg., I. Abt. Orig. B, **176**, 116-123.
Schoenen, D. and Wehse, A. (1988) Mikrobielle Kontamination des Wassers durch Rohr- und Schlauchmaterialien. 1. Mitt.: Nachweis von Koloniezahlveränderungen. Zbl. Bakt. Hyg. B, **186**, 108-117.
Schwartz, A. and Schwartz, W. (1962) Verhalten von gechlortem Wasser gegenüber Kunststoff-Rohrverbindungen in Trinkwasserleitungen. Kunststoffe, **52**, 137-138.
Seidler, R.J. Morrow, J.E. and Bagley, S.T. (1977) Klebsielleae in Drinking Water Emanating from Redwood Tanks. Applied and Environmental Microbiology, **33**,

893-900.
Sorg, T.J. and Bell (eds), F.A. (1986) Plumbing Materials and Drinking Water Quality, Noyes Publications, Park Ridge New Jersey (USA) 1986.
Voss, L. Button, K.S. Lorenz, R.C. and Touvinen, O.H. (1986) Legionella Contamination of a Preoperational Treatment Plant. J. Amer. Water Works Assoc., **78**, Jan. 70-75.
Wolff, L. and Heintz, A. (1966) Beobachtungen über Wassergüteveränderungen durch Kunstharzanstriche von Wasserbehältern. Städtehygiene, **17**, 58-59.
Yoo, R.S. Ellgas, W.M. and Lee, R. (1984) Water Quality Problems Associated with Reservoir Coatings and Linings. Annual Conference Proceedings, Amer. Water Works Assoc., 1383-1400.
Zimmermann, W. (1956) Trinkwasserhygiene und Kunst-stoffrohre. Städtehygiene, **7**, 266-268.

12 DEGRADATION OF CONCRETE IN SEWER ENVIRONMENT BY BIOGENIC SULFURIC ACID ATTACK

A.C.A. VAN MECHELEN and R.B. POLDER
Institute for Building Materials and Structures (TNO-IBBC),
Rijswijk, The Netherlands

Abstract
Since the seventies Biogenic Sulfuric Acid Attack (BSA) is
considered a serious problem in concrete sewer pipes and manholes in
countries with a moderate climate. Some unanswered questions concern
the level of aggressivity in a certain section, a method for
measurement of the level of aggressivity, the rate of concrete loss
given a certain level of aggressivity, also in relation with the
type of concrete and the remaining lifetime of a certain section. In
1987 experiments were started in Rotterdam to give answers and clues
to the questions. Concrete prisms of four different types of cement
were exposed in 10 different sewer manholes. Certain parameters of
sewage were measured and the level of aggressivity caused by BSA was
assessed. Also the concrete material loss was examined. Differences
in concrete material loss between the different sewer manholes were
clearly shown. Also there were significant differences in material
loss of the different concrete types.
Keywords: Concrete Sewer Degradation, Biogenic Sulfuric Acid Attack
(BSA), Assessment of Aggressivity, Exposure.

1 Introduction

Degradation of concrete sewer pipes has been recognized as a problem
of large economic consequences in Western Europe in the seventies.
German research efforts have increased our understanding of the
processes involved, after severe damage of Hamburg's large diameter
sewers. The dominant mechanism, Biogenic Sulfuric acid Attack,
abbreviated BSA, is understood at least qualitatively
[Thistlethwayte, 1972; Bielecki and Schremmer, 1987]. It will be
explained very briefly here.
 Bacteriae in the anaerobic slime layer on the under water pipe
surface reduce sulfates and other sulfur compounds in the sewage to
sulfide; the sulfide is liberated into the sewage, and consequently
oxydized to sulfate if the sewage contains sufficient oxygen. If
there is not enough oxygen available, however, which maybe the case
if the sewage residence time is long, or if there is no contact with
air (in highly filled sewers), the sulfide remains in the reduced
form and may be liberated to the sewer atmosfere as hydrogen
sulfide, the well known odourous gas. Here it can be oxydized to
sulfuric acid in the moist layer on the concrete surface by sulfur

oxydizing bacteriae called Thiobacilli. The most acid resistant type, Thiobacillus Thio-oxydans or Concretivorus, can produce sulfuric acid of about 10% w/w having a pH below 1. Concrete is a material consisting of usually inert aggregate particles bound together by an alkaline cement matrix. The acid dissolves the matrix, and gypsum and other degradation products are formed, which may be seen on the concrete surface, or may be washed away, exposing gravel in the surface. Under conditions favourable for the process the degradation rate may be up to 6 millimeters per year. This process is called Biogenic Sulfuric acid Attack because micro-organisms play an essential role in it. It is the most important internal degradation mechanism in concrete sewers.

The consequences of BSA occurring can be rapid loss of wall thickness of concrete pipes, causing structural risks. Dutch sewer authorities have become aware of BSA reducing their sewers' service life, at least in some parts of the sewer networks. A literature survey undertaken in 1986 showed some areas of incomplete knowledge [Polder, 1987]. A series of investigations has been undertaken since then. The most important unanswered questions regarding concrete attack were:

What is the level of aggressivity in a particular section?
How can the aggressivity level be measured?
What is the degradation rate?
What is the remaining service life of this particular section?

An example may illustrate the relevance of these questions. It quite often happens that a remotely controlled video camera inspection of a sewer section shows exposed gravel, so there is some degree of attack, but the penetration depth and the degradation rate are unknown. The sewer management authority wants a reliable prediction of the remaining service life for its maintenance and replacement scheme. It needs to know what sections have to be replaced with high priority. If a section is attacked only superficial, it hardly needs any attention at all. Both sections may look very similar as viewed by video, because of the rather superficial nature of the technique.

In 1987 experiments have been started to answer these questions. In cooperation with the Public Works Department of the Municipality of Rotterdam 200 concrete specimens have been exposed to sewer environment in ten manholes [Polder and van Mechelen, 1989]. When samples were placed and taken after exposure, tests were performed in order to characterize the sewers chemically. After exposure the mass loss of concrete samples was determined. This paper will give the most important results of both tests. It will put together the various pieces of information in order to answer the questions mentioned above, and conclusions will be drawn on testing methods.

2 Exposure of concrete to sewer environment

The concrete used for the exposure experiments was designed to be as similar as possible to the concrete used for normal pipe production, with exception of the cement type, which was varied. The question investigated is what cement type provides best resistance against

BSA. The American approach states that the more lime the concrete contains, the better resistance it has against acid degradation. German experiments seem to support this [Pohl and Bock, 1986; Sand, et al, 1987]. On the other hand, a more dense cement matrix, such as developed from cements blended with pozzolanic additives, has proved to provide superior resistance against penetration of chloride, e.g. in marine environments, and of sulfates in aggressive soil. A certain type of blended cement, blast furnace slag cement, has been used successfully for decades in the Netherlands, a.o. for sewer pipes and manholes. Recently fly ash and condensed silica fume have been added to concrete in order to decrease the permeability to aggresssive substances such as chlorides. Either of these might improve the resistance against BSA as well. Hall [1989] describes exposure experiments of acid-proof concretes and ordinary Portland cement concrete in sewers done in the U.S. to test the resistance against BSA. Our experiments are set up to test three cement types marketed in The Netherlands for their performance under BSA conditions.

The cements used were Ordinary Portland Cement, Blast Furnace Slag Cement (BFSC, containing about 70% blast furnace slag) and Portland Fly Ash Cement (PFAC, containing 25% fly ash). Three concrete mixes were made with concrete sand and gravel (d_{max} = 8 mm). As a reference specimens were included from a normal production sewer pipe, made with BFSC (d_{max} = 12 mm). All surfaces were sawn, to avoid the influence of cement-rich surface layers. Specimen size was about 150 x 45 x 45 mm^3. Five specimens of each mix were mounted in a steel frame, containing 20 specimens, and suspended in each of the 10 manholes at the level of the crown of the pipes, as is shown schematically in figure 1. The specimens will be washed when the water level is high, just as the crown of the pipes.

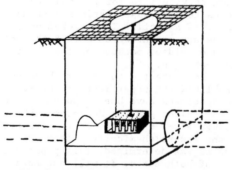

Fig.1. concrete prisms in manhole

The selected exposure locations all concerned gravity sewers, transporting mainly domestic sewage by a mixed system. Pipe sizes ranged from circular 400 mm diameter to egg shaped 800/1200 mm.

Specimens have been placed in September 1987. The first series of samples were taken in January 1988, after 3.5 months of exposure (about 0.3 year). The second series of samples were taken in October 1988, after 13 months or 1.1 year and the third series of samples in August 1989. This paper deals with results up till then. It is planned to take another series of samples in September 1990.

2.1 Results of exposure tests

After 0.3 year exposure some of the prisms show signs of attack under visual examination. Clear distinction can be made between manholes where some level of aggressivity is active, and others where nothing seems to happen. Attacked prisms show discoloration and red staining by dissolution of iron and subsequent redeposition. Some show significant loss of material, at maximum about one millimeter deep around gravel grains.

After 1.1 year exposure the attack was stronger than after 0.3 year. Generally the same manholes showing visual signs of aggressivity before, showed substantial mass loss now. One site, number 7, was particularly aggressive. Here the material loss was equivalent to about 0.5 to 2 mm of concrete per year. Other sites showed somewhat less aggressivity. Still others showed no material loss and no visual attack anyway.

After 2 years of exposure the attack again was stronger than after 1.1 year. Differences between manholes were clearly shown.

In figure 2 an overview is given of the total material loss of concrete prisms per manhole after 1.1 and 2 years of exposure. The material loss after 0.3 year was too small to take into account in this figure.

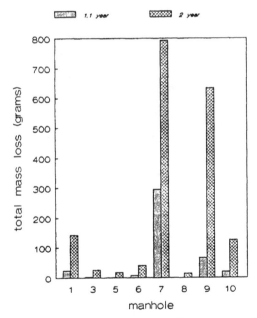

Fig.2. total concrete mass loss per manhole

The concrete prisms, especially those from manhole 7 and 9, showed after 2 years remarkable material loss and gypsum formation at the surface. It occured to us that a gypsum layer is not part anymore of the concrete and does not contribute to its strength. In our view this gypsum layer had to be considered material lost. It showed that the thickness of the gypsum layer was different for the different types of concrete. It was thicker for concrete made of Blast furnace slag cement than it was for concrete made of Portland cement.

An even more striking result was that differences between the concrete types were now significant. Concrete prisms made of Portland cement showed a degradation rate less than those made of Blast furnace slag cement or Portland fly ash cement. The observed mean degradation rates (± S.D) in mm concrete per year were:

Blast furnace slag cement: 1.0 ± 0.7 (n = 9)
Portland cement   : 0.8 ± 0.6 (n = 8)
Portland fly ash cement : 1.3 ± 0.9 (n = 8)

The most severe concrete degradation rate, observed in manhole 7, was between 2 and 2.8 mm of concrete per year.

3 Chemical characterisation of sewer environment

Every time when specimens were installed or samples taken the ten exposure manholes were investigated chemically. In addition some thirty more sewer manholes in Rotterdam were investigated, at least once. The measurements involved oxygen and sulfide content, pH and temperature of the sewage and the pH of the concrete surface. Additionally the level of the sewage in every pipe section was registrated. The oxygen content of the sewage will show its tendency to get anaerobic and produce sulfide; it was measured using an oxygen specific electrode. Sulfide itself is the source of hydrogen sulfide gas and the fuel for BSA; it was measured by a sulfide specific electrode in sewage samples buffered at high pH. Sewage pH and temperature may show a relation to sulfide build up or BSA activity. The sewage level may provide an indication of uneven settlement and consequent stagnation in certain sections. Stagnation of course increases the residence time of the sewage, possibly giving rise to sulfide build up.

The pH of the concrete surface is a direct measure for the actual BSA activity. It was measured by spraying the concrete surface with two different pH indicator solutions and observing the resulting colour by a video camera suspended from a tripod in the manhole. The indicator solutions were chosen to distinguish the most important pH regions: pH=1 and lower; pH between 1 and 3; pH between 3 and 5; pH above 5. These pH levels were expected to correspond approximately to conditions which are very aggressive; moderately; weakly and not aggressive.

3.1 Results: Sewage analysis
In table 1 results are shown of the sewage analysis in January, October, September and August per manhole.

The temperature of the sewage follows the expected trend, being higher in August than in September, which is higher than in October, which is again higher than that in January; averages are 22.0, 18.0, 16.7 and 11.5°C respectively. The oxygen content is generally below the saturation level (which is about 8 mg/l), but mostly above 1 mg/l except for the summer. In summer the sewage is almost completely anaerobic. The warmer water temperature will encourage oxygen depletion by increasing bacteriae metabolism.

Table 1. Results of sewage analysis in January,
October, September and August per manhole

	date	manhole 1	2	3	4	5	6	7	8	9	10
Temp.	Jan.	9	11	13	12	12	11	12	12	12	11
[^{o}C]	Oct.	16	16	17	17	17	17	17	17	18	16
	Sep.	18	18	19	19	18	18	18	18	19	18
	Aug.	22	22	23	22	22	23	23	22	24	23
O_2	Jan.	7.3	3.4	4.9	4.2	2.4	2.3	2.8	5.9	2.2	1.6
[mg/l]	Oct.	4.4	4.7	4.8	3.2	2.7	3.2	4.0	6.1	3.0	3.1
	Sep.	-	-	-	-	-	-	-	-	-	-
	Aug.	1.3	0.0	1.2	0.0	0.0	0.0	0.0	0.0	0.1	0.0
S^{2-}	Jan.	1.0	1.4	1.4	1.0	2.6	3.1	3.4	1.2	2.4	5.6
[mg/l]	Oct.	2.2	1.4	1.5	1.1	3.8	4.1	3.9	1.9	5.0	7.0
	Sep.	3.5	2.3	1.8	1.4	4.1	4.7	5.1	2.4	4.7	8.3
	Aug.	9.0	6.0	6.8	3.5	10.7	10.3	11.3	5.6	18.5	31.0
pH	Jan.	3	5	5	5	5	5	1	5	3	3
	Oct.	1	5	1	5	1	3	1	5	1	1
	Sep.	1	5	1	5	3	1	1	5	1	3
	Aug.	1	3-5	1	3	1	1-3	1	1-3	1	1
Turbulence:		1	0	1	0	0	1	1	1	1	0
0 = no											
1 = yes											

The sulfide content varies stongly, between about 1 and 31 mg/l. The sulfide level certainly is related to temperature, as shown in figure 3, at least when conditions are anaerobic. This is clearly shown in the results of August, which gave very high sulfide levels.

3.2 Results: Concrete surface pH
The concrete surfaces of the manholes show a wide distribution of pH's from as low as 1 to and above 5. It can be seen that there is considerable temperature influence as well. There is also a certain pH distribution in one manhole. Usually a narrow zone some 100 mm high above the low water level the pH is about 7, which is probably caused by regular contact with fresh sewage. From that level to higher regions the lower pH values are found.

3.3 Conclusions from sewage analysis and concrete surface pH
Because several locations have been inspected several times, various types of conclusions can be drawn, including those related to temperature and seasonal changes. Other conclusions concern

interrelations between sewage components and between sewage
components and the resulting BSA strength.

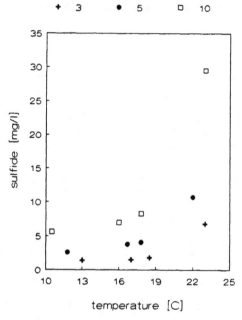

Fig.3 Temperature of sewage of manhole 3, 5 and
        10 versus the sulfide content

3.3.1 Oxygen and sulfide
There is a general relation between sulfide content and oxygen
content, which is however not very clear. We have seen a strong
correlation between a low oxygen content and a high sulfide content,
and vice versa, although there is considerable scatter. It is
somewhat surprising, however, that up to several mg O_2/l is measured
even when sulfide concentrations are high. It may be caused by the
sampling and measurement procedure, which will be modified for the
next series of tests.

3.3.2 Sulfide and concrete pH
As expected there is a strong relation between the sulfide content
and the pH of the concrete surface of the manhole, as shown in
figure 4.
   High sulfide contents above 5 mg/l cause a low pH, indicating
active sulfuric acid production. Very low sulfide, below 1 mg/l,
does not cause BSA activity, indicated by pH's above 5. Sulfide
contents between 1 and 5 mg/l may have different effects: in some
cases the surface pH is above 5, and in others it is as low as 1. A
closer look at the data showed that there is a relation with the
type of flow of the sewage when it enters the manhole. If the sewage
falls turbulently from a certain height into the manhole, even a low
sulfide content causes strong BSA, because apparently the liberation
of hydrogen sulfide gas is promoted. Sewage with the same sulfide

152

content, but flowing gently into the manhole, does not liberate as much hydrogen sulfide, and causes hardly any BSA activity.

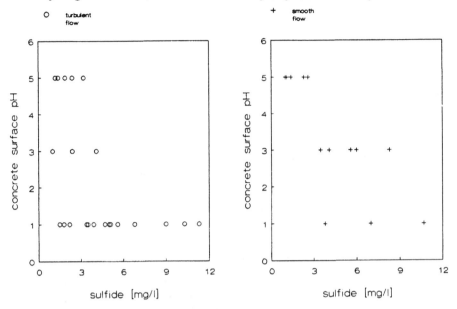

Fig.4. Sulfide content of sewage versus the concrete surface pH of the manholes

3.3.3 Temperature influence

The average sulfide content of the sewage in the exposure locations is influenced significantly by temperature; even a difference of $1.3°C$ (between September 1987 and October 1988) significantly changed the sulfide concentration. The oxygen content of the sewage does not seem to depend on the temperature, at least according to our observations, but at temperatures higher than $20°C$ the oxygen content will fall sharply.

The concrete wall pH depends on temperature, as expected. Higher temperatures promote microbial processes, producing more sulfide and sulfuric acid, so lower pH results than at low temperatures. Comparing the average pH of the ten exposure locations is not very meaningful, because some do not exhibit active BSA. Temperature effect on BSA activity is illustrated in table 2 more clearly by counting the frequency of different pH values. The difference between low and higher temperatures (about $10°C$ between January '88 and the other three) is clear. The effect of $1.3°C$ between October '88 and September '87 is not significant. The effect of $4.0°C$ between September '87 and August '89 however is significant.

Table 2. Frequency of concrete wall pH values in January, October,
September and August

Date and sewage water temperature		pH = 1 frequency	pH = 3 frequency	pH = 5 frequency
January '88	(11.5°C)	1	3	6
October '88	(16.7°C)	6	1	3
September '87	(18.0°C)	5	2	3
August '89	(22.0°C)	8	2	

3.3.4 Various observations

The variation of the sulfide content is relatively small during one
day between 9 am and 6 pm, less than 0.5 mg/l. The pH of the sewage
is almost constant, being 7.2 on the average, and lying almost
always between 6.9 and 7.5. When the pH was outside this region, the
sewage always originated from non-domestic sources, i.e. from
industrial sources such as laundries.

4 Compilation: aggressivity assessment

A practical method for aggressivity assessment of sewer environment
with respect to biogenic sulfuric acid attack of concrete should be
based on relatively simple measurements to be executed in the field,
and/or laboratory testing on samples that can be easily extracted
from the site. Furthermore a method should be available weighting
the input parameters, and leading to an unambiguous result.

From the measurements described we have concluded upon the
important parameters, and we have proposed a weighting method. The
important parameters are:

(Total) sulfide content of the sewage,
Surface pH of the (manhole) concrete,
Turbulence of the sewage flow into the manhole,
The sewage temperature.

Of these, sulfide provides the fuel for BSA; sewage turbulence
promotes the liberation to the atmosphere, and concrete surface pH
shows the strength of the actual acid production process. The
temperature is important, as we have seen that only the most
aggressive spot maintained a high level of BSA activity at lower
temperatures.
There is no or little influence of (or variation in) sewage pH,
sewage level in pipes, or daily fluctuations.
The weighting of these factors can be done in many different
ways. We propose one way of doing this, which gives near optimal
results when used on the exposure locations and the other manholes
investigated. It is possible, however, that a somewhat different way
of weighting develops in the near future. We think the elements will
be the same anyway.

154

The aggressivity A level is given by:

$$A = S + C + T + t, \qquad (1)$$

with: S the sulfide parameter
C the concrete surface pH parameter
T the turbulence parameter, and
t the temperature parameter.

The parameters have the following values:

if total sewage sulfide < 1 mg/l, S = 0,
if between 1 and 5 mg/l, S= 1,
if > 5 mg/l, S = 2;

if concrete surface pH > 5, C = 0,
if pH between 2 and 5, C = 1,
if pH < 2, C = 2;

if sewage flow is smooth, T = 0,
if sewage flow is turbulent, T = 1,

if temperature is below 15°C, t = 1.
if temperature is between 15°C and 20°C, t = 0,
if temperature is above 20°C, t = - 1

Note that the temperature influence is accounted for in a reversed way. As we have seen, a pH of 3 at 11°C may correspond to pH 1 at 17°C. Consequently a certain level of aggressivity is to be taken more serious when it occurs at low temperature. In other words, if a sewer is inspected in cold conditions, the aggressivity is predicted for the warmer part of the year.

The aggressivity of the exposure locations can be analyzed using this approach, and the results can be compared to the mass loss of concrete specimens. In table 3 the A-level for the measurements in January, October, September and August and the concrete degradation rate are given. The manholes with the most severe degradation rate are marked with a (*). As can be seen the mass loss or degradation rate is reflected at best in the months January and October. This means that this method should not be applied in the summer (July, August and September), because BSA and high sulfide content will also detected in manholes which are in other seasons not attacked at all.

As the degradation rate of the exposed prisms cannot yet be determined accurately, the interpretation of the aggressivity level in terms of a degradation rate (millimeters material loss per year) is not yet possible. The order of magnitude for the most aggressive environment in our exposure experiments, location 7, is between 2.0 and 2.8 mm per year. Location 9 with the same A-level (January and October together) however has a degradation rate between 1.3 and 2.8 mm per year. Interpretation gives a rough estimation of the degradation rate, which can be very usefull.

Table 3. Aggressivity level for 10 sewer manholes in January,
          October, September and August and degradation rate

manhole		January	October	September	August	degradation rate (mean) mm per year
1	(*)	4	4	4	4	1.0
2		2	1	1	2	0.6
3		3	4	4	4	0.6
4		2	1	1	1	0.6
5		2	3	2	3	0.6
6		3	3	4	4	0.6
7	(*)	5	4	5	4	2.4
8		3	2	2	4	0.6
9	(*)	4	5	4	4	1.8
10	(*)	4	4	3	4	0.9

5 Conclusions

A few of the questions mentioned in the introduction can be answered
now. Others are waiting until the investigation, which will carry on
for some years, is developed into a further stage.

The aggressivity of a particular sewer section can be
characterized by a few rather simple measurements performed in the
manholes. To this end one needs to measure the sulfide content and
the temperature of the sewage, and the lowest pH occurring on the
manhole surface. The presence of turbulent sewage flow must be
registrated. The observed parameters can be classified and a simple
weighting method generates the level of the aggressivity.

The outcome may be valuable for the sewer authority in many
respects. For instance, when a video inspection of a 30 year old
sewer shows exposed gravel, it is not clear, whether the attack is
still going on, and how severe it is. A relatively simple inspection
using the described chemical techniques shows the aggressivity
level. If it is low, the attack must have happened in the past; as
no BSA is active anymore, no measures are necessary to maintain the
sewer in its present state. It may be useful to have some cores
taken and tested, however. Recently a technique for taking cores
through the road surface has been developed by the Rotterdam Water
Management Department, which is marketed now in The Netherlands.

In another case, the sewer authority may worry about the
durability of relatively new sections. The proposed method can
estimate the BSA activity in sewers operative only a few years,
where no attack is visible yet. If severe attack is to be expected,
measures can be taken before serious amounts of concrete have been
lost.

Manholes showed a different level of aggressivity. The total mass
loss in the manholes was most severe in those manholes with the
greatest level of aggressivity.

The concrete degradation rate cannot be estimated with sufficient
accuracy from the experiments up till now. The most severe

degradation rate of concrete prisms in a manhole was between 2 and 2.8 mm of concrete per year. In another manhole with almost the same level of aggressivity the degradation rate was between 1.3 and 2.8 mm of concrete per year.

After two years of exposure of concrete prisms in sewer manholes, differences showed up between the different concrete types, with concrete made of Portland cement having less material loss. The material loss is related to the aggressivity in the sewer manhole. More measurements at the locations will be done to make clear whether the differences in material loss will remain, show even better or will disappear.

Acknowledgements

The authors wish to acknowledge the creative, technical and financial contributions of the Rotterdam Water Management Department group 'Concrete Degradation'. Mrs. Jacqueline Pluym-Berkhout is gratefully ackowledged for starting the experiments as part of her training as a civil engineer. Part of the work has been financed by the Innovative Research Programme for the Construction Industry (IOP Bouw).

6 References

Thistlethwayte, D.K.B. (1972) Control of Sulfides in Sewage Systems. Butterworths, Sydney.

Bielecki, R. and Schremmer, H. (1987) Biogenic Sulphuric Acid corrosion in gravity sewers. Heft der Mitteilungen des Leichtweiss-Instituts fuer Wasserbau der Technische Universitaet Braunschweig., 94.

Polder, R.B. (1987) Duurzaamheid rioolleidingen (durability of sewer pipes; in dutch with a summary in English and German). TNO-CHO-IBBC report., 17.

Pohl, M. and Bock, E. (1986) Bakterielle Steinschaden. Goldschmidt Informiert., 1, 19-25.

Polder, R.B. and van Mechelen, A.C.A. (1989) Assessment of Biogenic Sulfuric Acid Aggressivity of Sewer Environment. Proceedings 2nd International Conference on Pipeline Construction, Hamburg.

Sand, W. and Bock, E. and White, D.C. (1987) Biotest System for Rapid Evaluation of Concrete Resistance to Sulfur Oxidizing Bacteria. Materials Performance., 26, 14-17.

Hall, G.R. (1989) Control of Microbiologically Induced Corrosion of Concrete in Waste-Water Collection and Treatment Systems. Materials Perfomance., 28, 45-49.

13 MICROMORPHOLOGICAL ASPECTS OF THE MICROBIAL DECAY OF WOOD

R. VENKATASAMY, R. MOUZOURAS, E.B.G. JONES
and S.T. MOSS
School of Biological Sciences, Portsmouth Polytechnic, UK

Abstract
To study the ultrastructure of bacterial wood decay, species of softwoods and
hardwoods of varying lignin to cellulose contents, were exposed, either untreated
or treated with copper chrome arsenate (CCA), in seawater, riverwater and soil.
Examination by scanning and transmission electron microscopy (SEM and TEM)
revealed three distinctive patterns of decay. These are described as: (1) Erosion
(wide and narrow), (2) Cavitation and (3) Tunnelling (types I and II). All timbers
were subjected to the 3 types of bacterial attack, irrespective of their lignin to cellulose contents, the environment in which they were exposed and preservative
treatment.
    Following initial bacterial colonization, fungi are in general terms the second
group of microorganisms to inhabit or invade wood. The four main types of fungal
decay are discussed with respect to their micromorphological effects on the wood.
These comprise the staining, soft rot, white rot and brown rot fungi.
Keywords: Bacteria, Erosion, Cavitation, Tunnelling, Fungal Decay, Preservative
Treatment.

1 Introduction

Wood has become an invaluable natural material in the life of man and its
degradation is of great consequence, as it can be beneficial or disadvantageous.
Organisms which cause the deterioration of wood are widespread in terrestrial and
aquatic systems. They all contribute to the recycling of lignocellulose in the environment. Wood decay becomes a problem when man-made structures are inadequately protected. "To the wood-rotting microorganisms colonizing it, wood consists of a series of conveniently orientated holes surrounded by food" (Levy, 1982).
    Under favourable conditions wood is severely degraded by such microor-
ganisms as fungi, Actinomycetes and bacteria. Within the fungi, species of the
Basidiomycotina are the most important decomposers of wood in terrestrial environments. The Ascomycotina and Deuteromycotina each play a greater role in
the decay of wood in aquatic systems. Actinomycetes have been shown to have a
significant role in metabolising wood components (Crawford and Crawford, 1980;
Crawford, Barder, Pometto and Crawford, 1982; Pettey and Crawford, 1985; Mc-
Carthy, 1987).

1.1 Bacteria
More recently the potentials of bacteria as agents of decay of whole wood have
been recognised and investigated by Holt (1981), Fengel and Wegener (1984),

Higuchi (1985) and Venkatasamy (1988). Earlier work concentrated mainly on the micromorphology of bacterial action on whole wood, and this has been reviewed by Liese and Greaves (1975), Holt (1981), Nilsson (1982), Jeffries (1987) and Venkatasamy (1988). It is now recognised that to have a better understanding of the pathways involved in bacterial wood decay, several parameters must be considered. These include ecological implications, community structure and interactions, enzyme resources, ultrastructure of wood cell wall layers and the chemistry of the various wood components. Consequently, more attention is now being focussed on the degradation of the different cell wall layers, the bacterial enzyme systems involved and their action on the different wood components.

It has been proposed that initial colonisation and degradation of wood by bacteria starts with the ray parenchyma and pits through the action of pectinolytic, hemicellulolytic and cellulolytic organisms (Greaves, 1969; Johnson and Giovik, 1970; Liese and Greaves, 1975; Levy, 1975). Attack then proceeds through the cell walls causing lysis zones at the point of contact between bacterial and wood cell wall layers. These zones eventually grow into erosion troughs and cavities until the entire wood cell wall is destroyed (Holt and Jones, 1978). Bacterial attack has been observed mostly in sapwood rather than heartwood (Fengel and Wegener, 1984). The micromorphology of bacterial decay has, in the past, been variously described as: conical depressions and pitting (Harmsen and Nissen, 1965); erosion zones (Greaves, 1969); erosion troughs and lysis zones (Eaton and Dickinson, 1976); honey-comb attack (Cundell and Mitchell, 1977) and tunnels (Leightley and Eaton, 1977).

This variety of definitive terms has led to some confusion and misunderstanding and a need to define terms has been expressed (Mouzouras, Jones, Venkatasamy and Moss, 1986). Recent work, especially at the ultrastructural level, has led many investigators to limit themselves to a few definitive terms. The most widely accepted ones are erosion, cavitation and tunnelling.

It is generally agreed that bacterial decay of whole wood is restricted to environments of high humidity and particularly in aquatic habitats (Fazzani, Furtado, Eaton and Jones, 1975; Kohlmeyer, 1980; Holt and Jones, 1983; Daniel and Nilsson, 1986). Consequently, most investigations on bacterial wood decay have been confined to the aquatic environment. The major problem in studying bacterial action in wood has been attributed to difficulties in the isolation of pure strains to follow different stages of decay in the laboratory (Holt and Jones, 1978; Schmidt and Liese, 1982; Daniel and Nilsson, 1985a).

1.2 Fungi

The fungi are widely distributed in most environments and can cause major economic losses of wood. Fungal elements in or on the wood can cause its disfigurement and the action of their extracellular enzymes can lead to a loss in strength. The costs of repairing the damage caused by wood-decaying fungi in UK buildings in 1977 was estimated at £3 million per week (Scobie, 1980). In most cases such decay was due to the fact that such timbers were inadequately protected from water. Wood-rotting fungi require a certain moisture content in the wood before they can establish themselves.

Generally, it is accepted that wood decay fungi fall into four taxonomically unrelated categories; brown rot, white rot, soft rot and stainers. The former two types of decay are caused principally by species of the Basidiomycotina, whereas the latter two are caused by members of the Ascomycotina and Deuteromycotina. However, it is popular to refer to decay in buildings as either 'dry' or 'wet rot' (Dickinson, 1982). The term 'dry rot' is nowadays used to describe decay caused by *Serpula lacrymans* which is the true 'dry rot' fungus. In situations where decay occurs under relatively wetter conditions the term 'wet rot' is commonly used. This paper mainly describes the four categories mentioned above on the basis of their growth patterns in and within the wood cell walls.

2 Materials and methods

2.1 Bacteria

Test blocks (25 x 25 x 10 mm) of three hardwoods (*Betula verrucosa* Eh., *Quercus robur* L., *Aesculus hippocastanum* L.) and three softwoods (*Picea abies* Karst., *Pinus sylvestris* L., *Thuja plicata* D. Don) were exposed either untreated or vacuum impregnated (9.6 x 10^5 Nm^{-2}) with 3.5% CCA to a final preservative loading of 25.6 Kg m^{-3}. For exposure in seawater and riverwater test blocks were kept submerged at a depth of 1.5 - 2.0 m. Exposure in soil was by burying horizontally at a depth of 100 mm. Test blocks were removed at regular intervals processed for scanning and transmission electron microscopy following the techniques described by Mouzouras *et al.* (1986; 1988). Coated stubs were examined in either a JEOL T20 SEM operated at 20 kV or a JEOL 35C SEM operated at 20, 25 or 35 kV. Grids were examined in a JEOL 100S transmission electron microscope operated at either 60 kV or 80 kV.

2.2 Fungi

For light microscopy, freehand sections of fungal infected wood were made using stainless steel razor blades. Sections were mounted in lactophenol and examined using the Leitz interference microscope. Material for SEM and TEM was prepared and examined as previously described by Mouzouras *et al.* (1986).

3 Results

3.1 Bacteria

3.1.1 Scanning Electron Microscopy

A. Penetration of cell walls
A large number of bacteria initially attached to the exposed cell wall surfaces either passively or as a first step towards wood decay. More commonly following polar attachment, bacteria penetrated the S3 through a small bore hole, thus enabling them to get into the S2 wall layer (Figure 1). These bore holes were either discrete, round to ovoid, not much larger than the diameter of the causative bacteria or slightly larger and angular as illustrated. In other instances attached bacteria progressively eroded an area of the cell wall layer, sinking deeper as cell wall components were hydrolysed.

B. Erosion
Wide and narrow erosion were caused by morphologically different types of bacteria. Narrow erosion was invariably caused by long rod-shaped bacteria. A large variety of bacteria ranging from long rods to cocci were associated with the wide erosion type of decay.

(a) Wide erosion
Attached bacteria initally caused a small area of pitting within their immediate vicinity. In time these widened forming eroded areas that extended away from the

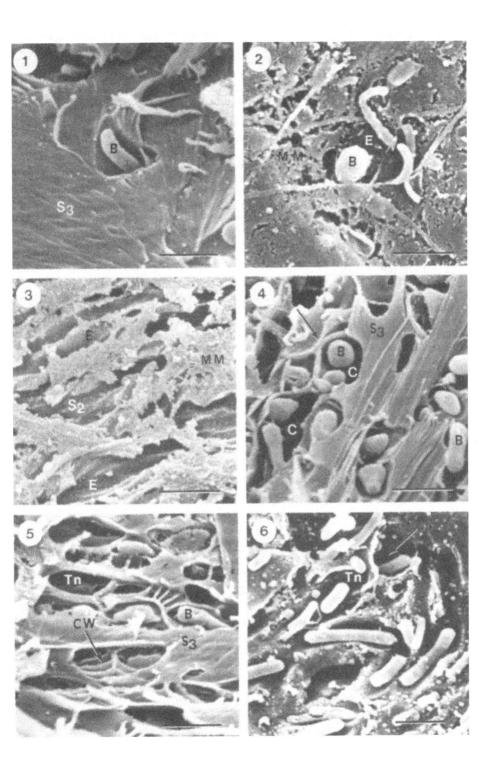

causative bacteria (Figure 2). These were irregular in shape, not reflecting on the morphology of the bacteria which very often were either in close contact to a small uneroded area of the wood cell wall layer or joined to it by a short mucilaginous fibril. This may have been a strategy to keep the bacterium anchored at a specific point owing to the availability of nutrients. Generally, wide erosion bacteria were associated in consortia embedded within thick mucilaginous matrices.

(b) Narrow erosion
Narrow erosion was limited to cellulose-rich timbers. Bacteria were preferentially attached laterally to the exposed S_2 wood cell wall surfaces where they caused shallow erosion channels (Figure 3). They were always associated with thick mucilaginous matrices and the erosion channels formed were not much wider than the diameter of the bacteria. Through cellular division the bacteria were able to cause long narrow channels along the S_2 and these tended to follow the orientation of the cellulose microfibrils. At some stages the bacteria would detach, probably as a result of depletion of nutrients. They were consequently not often observed within the erosion channels formed.

C. Cavitation
Cavitation bacteria were found to initially attach to the exposed wood cell wall layer and cause small depressions at the attachment points. These were eventually enlarged as the bacteria started boring through the cell wall layer (Figure 4). Once within the deeper layers of cell walls, bacterial division caused enlarged cavities. Daughter cells would then penetrate deeper into the cell wall layer through small bore holes and cause fresh cavities to be formed. Cavities were preferentially formed in the S_2 wall layer, but the bacteria were able to hydrolyse and cross the middle lamella to gain access to adjacent cell walls. Cavities varied in shape, depending on the morphology of the causative bacteria. Thick mucilaginous linings were commonly formed on cavity walls, especially in adjoining cavities. Eventually separating walls of adjacent cavities collapsed, leaving the mucilage linings together with partially decayed cell wall layers within large decayed areas.

D. Tunnelling
Wood decay is known to be caused by two types of tunnelling bacteria. The main

Figures 1 - 6. Scanning electron micrographs showing the ultrastructure of bacterial wood decay.

Figure 1. Rod-shaped bacterium (B) penetrating the S_3 into the S_2 cell wall layer of untreated *Pinus sylvestris* exposed in seawater for 6 weeks. Bar = 4.8 μm. Figure 2. Wide erosion zones caused by bacteria (B) attached within a mucilaginous matrix (MM) to the S_3 of CCA treated *P. sylvestris* exposed in seawater for 6 weeks. Bar = 4.0 μm. Figure 3. Narrow erosion channels (E) formed below a mucilaginous matrix (MM) on the S_2 layer of treated *Aesculus hippocastanum* exposed in soil for 9 weeks. Bar = 6.0 μm. Figure 4. Cavities (C) formed by bacteria (B) in the S_3/S_2 layers of *Betula verrucosa* exposed in seawater for 6 weeks. Mucilaginous borders (arrowed) are distinctive. Bar = 2.6 μm. Figure 5. Type I tunnels (Tn) with cross walls (CW) and caused by pleomorphic bacteria (B) in the S_3/S_2 layers of *P. sylvestris* exposed in seawater for 9 weeks. Bar = 2.8 μm. Figure 6. Type II tunnels (Tn) formed along and through (arrowed) S_2 layer of *P. sylvestris* exposed in seawater for 6 weeks. Bar = 3.4 μm.

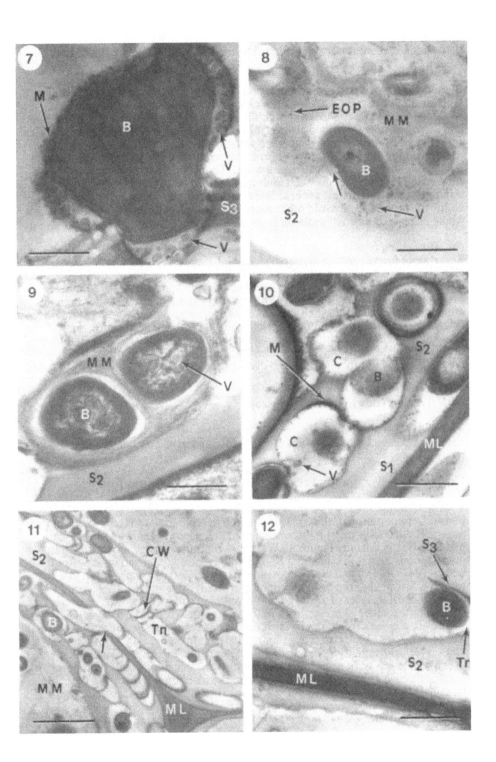

163

difference lies in the fact that type I tunnelling bacteria form cross walls along their tunnels whereas no such walls are formed by tunnelling type II bacteria.

(a) Tunnelling type I
In tunnelling type I, after attachment and penetration of the S3 wood cell wall layer through a narrow bore hole, the causative bacteria turned and tunnelled along the S2 parallel to the S3 and middle lamella. Long tunnels were thus formed and these typically contained cross walls laid at irregular intervals (Figure 5). The causative bacteria were rod-shaped or pleomorphic. Although they showed a preference for the S2 and S1 wall layers, they were also able to hydrolyse the middle lamella to accede to adjacent cell walls. Tunnels were always occupied by single bacteria and whenever cell division occurred, daughter cells would form separate tunnels. Tunnels never crossed each other, the bacteria showing an ability to recognise and avoid other tunnels. Through extensive tunnelling the whole cell wall eventually collapsed.

(b) Tunnelling type II
This type of tunnelling was caused by long rod-shaped bacteria which initially aligned themselves within a shallow channel formed along the exposed cell wall layer. They then penetrated the cell wall at an angle and formed long tunnels preferentially in the S2 (Figure 6). These were in some instances wider than the diameter of the bacteria which showed no tendency to align themselves with the orientation of the cellulose microfibrils, as is the case with narrow erosion bacteria. At some stages during tunnelling bacteria would penetrate deeper into the wood cell wall layer through bore holes and cause new tunnels within undecayed areas. The lack of tunnel cross walls distinguish between type I and type II tunnelling.

3.1.2 Transmission Electron Microscopy

A. Penetration of cell walls
After attachment to the cell walls of exposed timbers, bacteria penetrated the S3

Figures 7 - 12. Transmission electron micrographs of the ultrastructure of bacterial wood decay.

Figure 7. Penetration of the S3 layer of *Quercus robur* exposed for 6 weeks in riverwater. The bacterium (B) is encapsulated in mucilage (M) and vesicles (V) are present. Bar = $0.8\,\mu$m. Figure 8. Wide erosion caused by bacteria (B) within a mucilaginous matrix (MM) in the S2 of *Betula verrucosa* exposed in seawater for 6 weeks. Vesicles (V) and electron opaque particles (EOP) are present and a bacterium is attached to the S2 at one point (arrowed). Bar = $0.6\,\mu$m. Figure 9. Narrow erosion of S2 layer of *Thuja plicata* exposed in soil for 12 weeks. Bacteria (B) containing vesicles (V) are within a mucilaginous matrix (MM). Bar = $0.5\,\mu$m. Figure 10. Cavities (C) in the S2 layer of *Pinus sylvestris* exposed in seawater for 12 weeks. Vesicles (V) and mucilage linings (M) are present. Bar = $5.4\,\mu$m. S1 = inner layer of secondary wall, ML = middle lamella. Figure 11. Tunnelling type I in cell wall layers of *B. verrucosa* exposed in riverwater for 9 weeks. Tunnels (Tn) have distinctive cross walls (CW). Arrow indicates a tunnel traversing the middle lamella (ML). Bar = $5.4\,\mu$m. Figure 12. Narrow tunnels (Tn) formed by type II bacteria (B) in the S2 layer of *P. sylvestris* exposed in seawater for 6 weeks. Bar = $0.5\,\mu$m. S3 = outer secondary wall layer, ML = middle lamella.

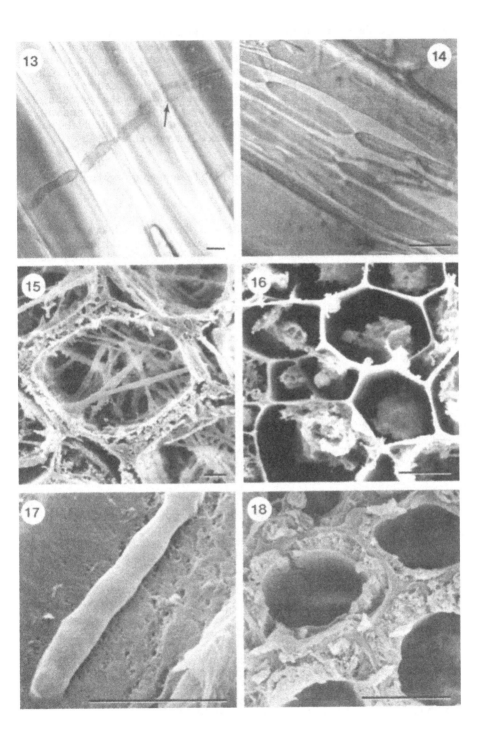

165

layer. Mucilage played an important role in both attachment and the process of cell wall penetration. When associated with single cells mucilage was in the form of a cocoon, a sheath or capsule; mucilaginous matrices were invariably present wherever consortia were involved in the decay process. The attached bacterium normally formed a penetration point in the form of a sharp, rounded or blunt penetration peg (Figure 7). In this instance vesicles released by the bacterium are actively involved in the hydrolysis of the S_3 wall layer. Vesicles were found within the mucilaginous sheath and may be involved in production. Eventually the bacterium enters the S_2 wall layer and causes decay of the wood.

B. Erosion
The difference between wide and narrow erosion became more evident when examined at the TEM level.

(a) Wide erosion
Wide erosion was found to be caused by diffusion of enzymes well away from the causative bacteria (Figure 8). A mucilage matrix enrobes the bacteria involved and large areas of the S_2 have been hydrolysed with little residue left within the decayed areas. This suggests that all components of the cell wall layer are being utilised. In Figure 8 a bacterium is anchored at one point to an undecayed area of the S_2, a typical feature of wide erosion bacteria. Vesicles are produced intracellularly and released into areas being hydrolysed. The presence of electron-opaque particles in decayed areas of treated timbers are presumed to be immobilised preservative elements.

(b) Narrow erosion
Narrow erosion was caused by rod-shaped bacteria closely associated with the wood cell wall layers undergoing decay. Figure 9 illustrates a cross section of bacteria. The bacterial cells are embedded in a thick fibrillar mucilaginous matrix, causing decay in the S_2 layer. Decay is limited to areas within close proximity of the bacteria and appears as narrow eroded channels. Vesicles were commonly observed within the bacterial cell walls but these were rarely released into areas being decayed. Narrow erosion bacteria were never observed causing decay of the S_3 wall layer or the middle lamella; they were restricted to the S_2 of cellulose-rich timbers.

Figures 13, 14. Light micrographs of decayed balsa wood.

Figure 13. A longitudinal section of showing the typical growth pattern of staining hypha. Note fine penetration peg (arrowed) in the wood cell wall. Figure 14. Polarized light micrograph of a longitudinal section illustrating typical soft rot cavities. Bars = 10 μm.

Figures 15 - 18. Scanning electron micrographs of decayed pine.

Figure 15. Transverse section of a tracheid with numerous soft rot cavities within the wood cell wall and fungal hyphae in the lumen. Figure 16. Transverse section through fibre cells showing severe decay with only the middle lamella and possibly primary wall remaining. Figure 17. Longitudinal section showing a hypha of the brown rot fungus *Poria placenta*. Note the absence of an erosion trough. Figure 18. Transverse section through wood infected with *P. placenta*, showing the typical amorphous pattern of brown rot decay. Bars = 10 μm.

C. Cavitation

In the cavitation type of decay, bacteria attached to the exposed S3 wall layer formed small cavities or bore holes which enlarged once the S2 was reached. Several such cavities in an area of a cell wall layer (Figure 10) illustrates this typical form of wood decay. The cavities are round, ovoid or rectangular with mucilaginous borders appearing as rims of variable thickness. Vesicles released by the causative bacteria were commonly found associated with mucilage in areas undergoing decay.

D. Tunnelling

The two types of tunnelling were observed as being distinctively different at the TEM level. Whereas type I produced tunnel cross walls, tunnels of type II bacteria were devoid of these walls.

(a) Tunnelling type I

Pleomorphic bacteria were found to be involved in type I tunnelling and these usually attached to the wood cell wall within a mucilaginous cocoon. They then penetrated the S3 wall layer, turned and tunnelled along the S2 producing distinctive tunnel cross walls (Figure 11). These cross walls are laid at irregular intervals, being thick or thin and fibrillar in nature. Although tunnels were mainly restricted to the S2 wall layer, crossing of the middle lamella to get into adjacent wall layers was commonly observed. Vesicles were abundantly produced at stages during tunnelling. Vesicles were either associated with hydrolysis of wood cell wall components or formation of the cross walls. Tunnels did not contain any form of unutilised material suggesting that the bacteria were able to hydrolyse and assimilate both lignin and cellulose. Whenever cell walls were heavily tunnelled and nutrient sources became scarce, tunnelling type I bacteria formed resting spores. It is assumed that these spores would eventually attach to fresh wood cell wall layers where the process of cell wall penetration and decay would recommence.

(b) Tunnelling type II

Type II bacteria were found to be rod-shaped which attached to and penetrated the S3 wall to get into the S2 layer where they formed long narrow tunnels. Figure 12 illustrates a cross section of type II tunnels. These conform to the contour of the causative bacteria and are formed mainly in the S2. Vesicles were often observed in areas being hyrolysed and little mucilage was found associated with the bacteria. Extensive tunnelling of the S2 eventually caused collapse of the unattacked but weakened S3 layer. Unlike type I tunnelling bacteria, type II bacteria often crossed into adjacent tunnels resulting in the formation of enlarged tunnels. They rarely decayed the middle lamella which, however, they were able to cross to reach undecayed wall layers. By-products of cell wall metabolism in the form of granular material were commonly found in tunnels and these were assumed to be wood cell wall components which the bacteria were unable to utilise.

3.2 Fungi

Figure 13 illustrates the typical growth pattern of the hyphae of a stainer fungus. Such hyphae are capable of passing from cell to cell through pits, or by direct penetration of the wood cell wall. When hyphae come into contact with the wood cell wall, a fine penetration peg is formed (Figure 13) which remains constricted while in the cell wall and regains normal size on emerging into the lumen of the next cell. This can repeat itself across a large number of wood cells.

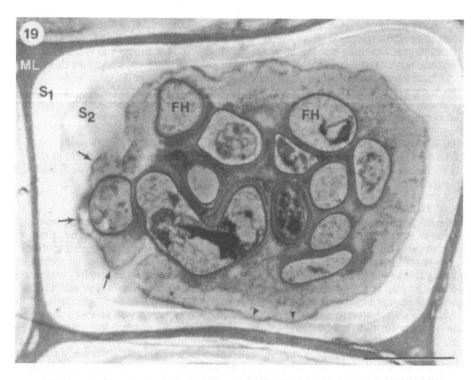

Figure 19. Transmission electron micrograph of a transverse section through balsa wood illustrating white rot decay. Erosion troughs (arrowed) are evident in the S_2 layer of the wood cell wall. The coalescence of erosion troughs has resulted in the progressive thinning of the wood cell wall, down to the S_1 layer (arrowheads) in some places. ML = middle lamella, FH = fungal hypha. Bar = 2 μm.

A similar pattern of hyphal growth is also observed for those fungi causing soft rot. The penetration hyphae of soft rot fungi commonly form branches within the S_2 layer of the wood cell wall. Such fine branches may be bidirectional (T-branch) or unidirectional (L-bend) and each branch grows parallel to the direction of the cellulose microfibrils. Enzymatic hydrolysis of the S_2 layer ensues leading to the formation of cylindrical cavities with conical ends or biconical (diamond) shaped cavities. Continued growth of the hypha within the cavity results in the formation of chains of cavities such as those illustrated in Figure 14. Cavities also arise by the formation of lateral branches from existing or mature cavities. In advanced stages of decay such cavities will eventually coalesce leading to the destruction of all the S_2 layer of the wood cell wall. Figure 15 illustrates such coalescence of numerous cavities within the wood cell walls. Also visible are a large number of hyphae running across the lumen and penetrating from cell to cell. The severity of attack illustrated by Figure 15, will eventually lead to the collapse of the S_3 layer and reach a situation similar to that illustrated in Figure 16.

Members of the wood-rotting Basidiomycotina (both white and brown rot) enter the wood cells via ray cells and spread by penetrating through pits or through the wood cell wall by bore hole formation. The typical micromorphology of brown rot decay is illustrated by Figures 17 and 18. Brown rot fungi produce enzymes which diffuse considerable distances from the hyphae leading to an amorphous or porous pattern of degradation. Thus in longitudinal sections (Figure 17) there is no distinct erosion of the cell wall around the hypha (unlike white rot decay). The amorphous nature of brown rot decay is illustated in Figure 18.

Figure 19 illustrates white rot decay at the transmission electron microscope level. The wood cell lumen contains several fungal hyphae which have formed erosion troughs from the S$_3$ layer inwards. Decay in Figure 19 is considerably advanced in that the S$_3$ layer is absent and a substantial area of the S$_2$ layer has been eroded. Dissolution of the S$_1$ layer has also occurred in some areas. This gradual thinning of the wood cell wall will proceed towards the middle lamella and the eventual collapse of the cell.

4 Discussion

4.1 Bacteria

4.1.1 Erosion

The terms used to describe erosion of wood cell walls by bacteria have been reviewed by Holt (1981), Nilsson (1982) and Venkatasamy (1988). Two patterns of erosion attack, as observed by the LM, SEM and TEM, have been described and these cause either wide or narrow erosion zones. Wide erosion zones are caused by diffusion of enzymes away from the causative bacteria (Furtado, 1978; Holt, Jones and Furtado, 1979; Nilsson, 1982; Butcher, Nilsson and Singh, 1984; Singh and Butcher, 1985). These authors examined bacterial erosion with the LM and described it as irregular in morphology, extending from the S$_3$ through the S$_2$ cell wall layers to the middle lamella which is not decayed.

Holt (1983) examined erosion decay with the SEM and described narrow erosion troughs which conformed to the dimensions of the causative bacteria. These are caused by rod-shaped bacteria which align themselves with the cellulose microfibrils where they cause long, narrow erosion zones. Bacteria causing these erosion troughs eventually penetrated into the S$_2$ cell wall layer. More recent investigations at the TEM level have shown that this type of decay is caused by localised enzyme activity (Schmidt, Nagashima, Liese and Schmitt, 1987; Singh, Nilsson and Daniel, 1987b). Some cellulolytic bacteria are known to associate closely with, and form narrow erosion zones on cellulose microfibrils, owing to the cell-bound nature of their celluloytic enzymes (Lamed, Setter and Bayer, 1983; Lamed, Setter and Kenig, 1983; Ljungdhal, Petterson, Eriksson and Wiegel, 1983; Murray, Snowden and Colvin, 1986; Kudo, Cheng and Costerton, 1987).

4.1.2 Cavitation

Cavitation, one of the typical forms of bacterial wood decay, has been observed and recognised for some time. The micromorphology of bacterial attack on wooden pilings described by Boutelje and Bravery (1968) were, in fact, typical cavitation forms. At the LM level, cavities formed in wood cell walls by bacteria have been described by Eaton and Dickinson (1976), Leightley and Eaton (1977) and Holt and Jones (1978). Other investigators have used different terms to describe this pattern of bacterial wood decay. The 'cheese cake' pattern observed with the SEM on pine discs exposed in the sea (Cundell and Mitchell, 1977) conforms to a cavitation form of decay. Holt (1981) examined bacterial wood decay with the LM and SEM and described and illustrated a 'honeycomb' type of attack which, again, was typical of the patterns of decay associated with cavitation bacteria.

Owing to the confusion caused by the various definitive terms used by several investigators, Nilsson and Singh (1984) examined this pattern of wood decay with the TEM and coined the term 'cavitation' to describe it. At the LM level it is not possible to distinguish between bacterial cavitation and erosion, both appear as pitted areas in wood cell wall layers. However, observations at the SEM and TEM levels have enabled the characterisation of this type of decay.

4.1.3 Tunnelling

The existence, in natural environments, of a group of bacteria that cause wood decay by tunnelling through the cell wall layers has been observed with the LM and reported by several investigators (Courtois, 1966; Eaton and Dickinson, 1976; Leightley and Eaton, 1977). The mechanism by which these bacteria penetrate and tunnel through wood cell wall layers was, however, not fully understood. Holt (1981) proposed and illustrated several methods by which bacteria may penetrate and tunnel through wood, causing a typical form of decay.

Two types of tunnelling were recognised and proposed by Mouzouras *et al.* (1986) and these are referred to in this investigation as tunnelling type I and tunnelling type II. More commonly, tunnels are formed by a group of pleomorphic bacteria, preferentially in the S2 cell wall layer, with distinctive chambers and chamber cross-walls. This type (type I) has been described by Nilsson and Daniel (1983). A second type (type II), caused by rod-shaped bacteria, forms long, narrow tunnels which lack chambers and cross-walls and were first described by Mouzouras *et al.* (1986).

A. Tunnelling type I

The inability of earlier investigators to demonstrate the finer details of tunnelling type I by bacteria was mainly due to the limitations of light microscopy and scanning electron microscopy. Investigations at the TEM level enabled Nilsson and Daniel (1983), Daniel and Nilsson (1985a,b), Daniel, Nilsson and Singh (1987) and Venkatasamy (1988) to illustrate and describe the micromorphology of tunnelling type I wood decay. A group of pleomorphic bacteria, not yet identified, was found to attack and penetrate the S3 layer of timbers exposed in natural environments. These bacteria subsequently tunnel preferentially along the S2 cell wall layer forming tunnels with distinct chambers and chamber cross walls.

Tunnelling type I by bacteria has been reported in a range of naturally durable and non-durable softwoods and hardwoods in various environments. The causative bacteria are known to decay wood in freshwater (Holt, 1981; Venkatasamy, 1988), seawater (Nilsson and Holt, 1983; Nilsson and Daniel, 1983; Singh, Daniel and Nilsson, 1987a; Mouzouras *et al.*, 1986; Mouzouras, Jones, Venkatasamy and Holt, 1988) and in soil (Nilsson, 1982; Venkatasamy, 1988). Tunnelling type I has also been reported on chemically modified wood (Nilsson and Rowell, 1982; Nilsson and Daniel, 1983).

The fact that tunnelling type I bacteria cause decay of wood of high lignin to cellulose content has been assumed to be an indication of utilisation of the lignin component (Singh *et al.*, 1987a; Venkatasamy, 1988). Tunnel chambers are usually clear of deposited, unutilised metabolites. In some other forms of wood decay a granular, amorphous layer of cell wall components, presumed to be mostly lignin, is left in decayed regions of wood cell walls. Evidence of lignin utilisation by tunnelling bacteria has been shown by Daniel and Nilsson (1986) using ^{13}C-NMR and klason lignin. Biochemical experiments using ^{14}C-labelled natural and synthetic lignins have also confirmed the ability of type I bacteria to break down and utilise the lignin component of wood cell walls (Singh *et al.*, 1987a).

B. Tunnelling type II

Wood decay, caused by tunnelling type II bacteria, was first reported by Mouzouras *et al.* (1986). These authors demonstrated at both the SEM and TEM levels how a group of rod-shaped bacteria burrowed through wood cell wall layers and caused long tunnels without the formation of chambers and chamber cross walls. The causative bacteria attached to exposed wood cell surfaces and initially burrowed into the S_3 wood cell wall layer. Cell division and elongation of daughter cells caused the formation of short tunnels in the S_3 and S_2 layers, not much wider than the diameter of the bacteria involved. Mucilage or exopolysaccharides were generally not found associated with this pattern of wood decay and the causative bacteria were rarely in contact with the lateral walls of the tunnels.

Tunnelling type II can be easily confused with the erosion type of decay. The basic differences between erosion and tunnelling type II only become obvious at the TEM level. Erosion bacteria either cause hydrolysis of cell wall components from a distance or are closely associated with the cell wall layers. Tunnelling involves bacteria that penetrate cell walls and hydrolyse cell wall components forming narrow tunnels. The causative bacteria do not show a tendency to align with cellulose microfibrils and are not restricted to cellulose-rich timbers, delignified or otherwise chemically modified wood.

4.1.4 Effects of environmental conditions

There were indications that decay proceeded faster in soils than in seawater or riverwater. This was attributed to the high number of bacterial species and complex communities associated with soils (Cromach, 1981; Kaiser and Hanselmann, 1982a, b; Burns, 1982, 1983). There is also a higher input of lignocellulosic materials in soils (Nioh, 1977; Cromach, 1981; Heal and Ineson, 1984) and it is to be expected that bacteria adapted for the breakdown and utilisation of lignocellulose are higher in numbers and more active in the soil environment. The stability of the soil environment has also been associated with enhanced bacterial degradation of a wide variety of materials (Stotsky and Krasovsky, 1981).

Bacterial decay was found to be more active in seawater than in riverwater and patterns of decay were similar to those produced in soil. There have been several recent reports on bacterial activity on timbers exposed in marine environments (Mouzouras *et al.*, 1986, 1988; Daniel *et al.*, 1987). The most diverse patterns of decay were observed in timbers exposed in seawater. This could have been the consequence of less competition from other wood decaying microorganisms, allowing bacteria to develop and proliferate. It has been shown that bacterial activity in seawater is higher than fungal activity (Hoppe, 1986; Robra, 1986).

In riverwater decay was slower than in the other two environments and restricted to localised areas of wood cell walls. However, lignocellulolytic bacteria are known to be abundant in riverwater (Suberkropp and Klug, 1976; Eslyn and Moore, 1984; Meyer-Reil, 1987). Several patterns of wood decay have ben observed and described on timbers exposed in freshwater (Schmidt *et al.*, 1987).

4.1.5 Effects of wood components

It is clear from the results presented that bacteria showed no specific preference for any of the timber species used. Neither were decay patterns related to timber species. The lignin to cellulose contents of the timbers did not affect the decay patterns although there was evidence that high lignin and wood extractive contents slowed down bacterial decay. Timbers with high cellulose contents were attacked and decayed faster. Bacteria causing narrow erosion were limited to cellulose-rich timbers. It has been demonstrated that narrow erosion bacteria are more active on delignified wood (Holt, 1981).

No relationships were found that correlated decay with the syringyl and guaiacyl lignin types of the timbers. Both types were equally attacked and decayed. Tunnelling types I and II bacteria showed a preference for the cellulose rich S_2 wall layers, although they were able to break down other wall layers, including the lignin rich middle lamella. Bacteria responsible for wide erosion decayed all cell wall layers including the middle lamella.

4.1.6 Effects of preservative treatment

The ability of bacteria to cause decay in timbers treated with high loadings of CCA has been shown (Daniel and Nilsson, 1985a) and is of concern where wood is used in wet environments. The preservative treatment did not inhibit any of the bacteria involved in producing the various patterns of decay. A number of bacteria, singly, as colonies or communities, are believed to have an ability to immobilise the elements of CCA. The way in which preservatives are immobilised and detoxified has not been fully elucidated. Binding to exopolymers as well as internally by the bacteria has been demonstrated by Mittelman and Geesey (1985) and Ross (1986). It has also been advanced that plasmids may play an important role in the binding of metal ions by bacteria (Rouch, Camakaris, Lee and Luke, 1985). An adaptation to quarternary ammonium compounds by bacteria has also been demonstrated by Ventullo and Larson (1986). The complex way in which preservatives are inactivated by bacteria poses serious problems in finding methods to counteract such abilities. In exposed treated timbers electron-opaque particles believed to be immobilised CCA elements were consistently observed to be associated with mucilage. Similar particles analysed by Daniel and Nilsson (1985a) were shown to be metal ions of the wood preservatives used.

4.2 Fungi

Staining fungi are amongst the early inhabitants of wood following initial bacterial colonisation (Clubbe, 1980). Stains occur as a result of the growth of pigmented hyphae in or on the wood surface, mainly by members of the Ascomycotina and Deuteromycotina. Stain fungi utilise the food reserves in the ray parenchyma of the sapwood, or residual sugars and simple carbohydrates elsewhere in the wood (Levy, 1982). Loss in strength properties of the timber as a result of growth by such fungi does not generally occur (Findlay and Pettifor, 1937; Findlay, 1967). However, it has been shown that several stain fungi can cause soft rot (Krapivina, 1960).

In ground contact situations the first wood-rotting fungi to colonise wood are the soft rot fungi whereas the Basidiomycotina form the climax of the succession (Levy, 1982). Soft rot fungi become more active if for any reason the Basidiomycotina are eliminated from the mycota (Levy, 1982), such as the presence of preservative on the lumen of hardwood fibres. With the suppression of the Basidiomycotina, soft rot fungi penetrate the S_2 layer, especially if it is poorly treated with preservatives (Dickinson, 1982).

The three distinct micromorphological types of fungal decay can be explained on a biochemical basis. The subject has been extensively studied and has been reviewed by Eriksson (1981), Kirk (1983), Kirk and Cowling (1984) and Jeffries (1987). For example, Fukazawa, Imagana and Doi (1976) have shown histochemically that the region of decay in brown rot fungi spreads very rapidly through the entire S_2 layer of the wood cell wall, which is in contrast to the localised forms of attack exhibited by the other two forms of decay. Green (1980) postulated that the three different micromorphologies can be explained by the degree of localisation of their cellulases and hemicellulases. Brown rot organisms retain less of their enzymes than the white and soft rot fungi.

White rot fungi are capable of extensively degrading the lignin component of wood (Kirk, 1983; Kirk and Cowling, 1984) and some preferentially delignify wood (Blanchette, 1984). White-rotted wood becomes bleached in appearance, with a fibrous nature along the grain and generally softened. Brown rot fungi can only partially degrade or modify lignin (Kirk, 1983; Kirk and Cowling, 1984) and the residual lignin maintains the cell shape. It is only in the late stages of decay that the remaining cell wall material collapses. The wood rapidly becomes brittle, later becomes discoloured and finally brown and soft. On drying extensive checking occurs across the grain giving a cubical pattern and finally the wood crumbles; the characteristic macroscopic feature of brown rot.

Soft rot fungi also degrade lignin to a limited extent (Kirk, 1983; Kirk and Cowling, 1984). The mechansims of soft rot cavity formation have been a subject of study for a number of years and to mention all the literature is beyond the scope of this paper. However, a series of papers by Hale and Eaton (1983, 1985a, b, c, 1986) describe the mechanisms of cavity formation. Soft-rotted wood, especially in wet situations, takes on a darkened appearance and becomes softened, although below certain depths wood may still be sound. Some soft rot fungi have also been observed to form the erosion type of decay, similar to that observed for white rot fungi. This type of decay is commonly referred to as Type 2 (T2) soft rot and the cavity form is referred to as Type 1 (T1) soft rot (Corbett, 1963).

5 Conclusion

It is generally agreed that bacterial decay on wood is restricted to environments of high humidity, particularly in aquatic habitats (Fazzani, Furtado and Jones, 1975; Kohlmeyer, 1980; Holt and Jones, 1983; Daniel and Nilsson, 1986). Consequently most investigations on bacterial wood decay have been confined to the aquatic environment, cooling towers and in soils of high humidity. The major problem in studying bacterial action on wood has been attributed to difficulties in isolating pure strains to monitor different stages of decay in the laboratory (Holt and Jones, 1978; Schmidt and Liese, 1982; Daniel and Nilsson, 1986).

Although a large spectrum of bacteria may settle and attach to wood in the natural environment, only a few have the ability to metabolise certain wood components and still fewer to cause decay in whole wood (Schmidt and Dietrichs, 1976; Nilsson, 1982). Lignin, known for its resistance to bacterial attack, forms a protective barrier around easily degraded cellulose and hemicelluloses. It is now recognised that for any appreciable cellulolytic degradation to occur, the lignin barrier must be at least partially removed, either chemically or through the action of lignolytic microorganisms (Schmidt, 1980; Holt, 1981). Other workers are in agreement that whole wood and synthetic lignins are metabolised by certain bacterial strains (Kawakami, 1976; Haider, Trojanowski and Sundman, 1978; Haider and Trojanowski, 1980; Crawford and Crawford, 1980).

Preservative treated wood is also vulnerable to bacterial decay, preservative concentrations showing little degree of toxicity to wood degrading strains. The ability of bacteria to cause decay in preservative treated timbers in a range of habitats, marine, freshwater, soils and cooling towers, has been observed and investigated (Eaton and Dickinson, 1976; Holt, Jones and Furtado, 1979; Nilsson, 1982). Copper chrome arsenate treated poles in ground contact have suffered decay through the action of cavitation and tunnelling bacteria (Drysdale and Hedley, 1984; Willoughby and Leightley, 1984). Treated wood exposed in the sea and under laboratory conditions have been found severely decayed by tunnelling bacteria, (Nilsson and Daniel, 1983).

Wood degrading bacteria are known to have an ability to survive in the presence of heavy metals which appear to have little or no effects on their metabolic processes (Robra, 1986). They have been demonstrated to have intracellular and extracellular mechanisms to bind and detoxify heavy metals (Murray, 1987). The immobilisation of copper by bacteria has been demonstrated (Daniel and Nilsson, 1986; Blunn and Jones, 1987). Research on wood degradation by microorganisms has, in recent years, concentrated on the decay caused by microorganisms with a view to improve in-service performance of timbers through the formulation of better and more efficient broad spectrum wood preservatives.

It is evident that the fungi will always be present on wood if given the right conditions (i.e. water). The only time that fungi become totally eliminated is under conditions of anaerobiosis, where anaerobic bacteria will continue the process of wood decay at a slower rate. The type of decay fungi prevailing at any time will very much depend on the environmental conditions. For example, it has been shown (Mouzouras, 1989) that cavity-forming soft rot species are the main agents of fungal decay of wood in the sea. As discussed earlier in this paper, bacteria are capable of immobilising the heavy metal components of wood preservatives. Similarly some wood-rotting fungi are tolerant of heavy metals, such as copper tolerance in *Poria* and *Phialophora* species (Sutter, Jones and Wälchli, 1983, 1984; Sutter and Jones, 1985). Without doubt, both bacteria and fungi are major agents in the recycling of wood in the environment and adequate measures should be taken to protect structures of value to man.

5 Acknowledgements

The authors would like to thank Mr. Colin Derrick for photographic reproduction and Mrs. T. Elliott for typing the manuscript.

6 References

Blanchette, R.A. (1984) Screening wood decayed by white rot fungi for preferential lignin degradation. **Appl. Environ. Microbiol.**, 48, 647-653.

Blunn, G.W. and Jones, E.B.G. (1988) Immobilization of copper and copper-nickel alloys, in **Marine Biodeterioration, Advanced Techniques Applicable to the Indian Ocean** (eds M.F. Thompson, R. Sarojini and R. Nagabhushanam), Oxford and IBH Publishing Co. PVT Ltd., New Delhi, pp. 747-756.

Boutelje, J.B. and Bravery, A.F. (1968) Observations on the bacterial attack of poles supporting a Stockholm building. **J. Inst. Wood Sci.**, 20, 47-57.

Burns, R.G. (1982) Carbon mineralisation by mixed cultures, in **Microbial Interactions and Communities**, Vol. I (eds A.T. Bull and J.H. Slater), Academic Press, London, pp. 475-543.

Burns, R.G. (1983) Extracellular enzyme-substrate interactions in soil, in **Microbes in their Natural Environments** (eds J.H. Slater, R. Whittenburg and J.W.T. Wimpenny), Cambridge University Press, London, pp. 249-298.

Butcher, J.A., Nilsson, T. and Singh, A.P. (1984) The bacterial attack of CCA-treated horticultural posts. **Proc. New Zealand Wood Pres. Assoc.**, 24, 34-49.

Clubbe, C.P. (1980) The colonization and succession of fungi in wood. **IRG/WP/1107.**

Corbett, N.H. (1963) Anatomical, ecological and physiological studies on microfungi associated with decayed wood. **Ph.D. Thesis**, University of London, U.K.

Courtois, H. (1966) Über den Zellwandabbau durch Bakterien im Nadelholz. **Hozforschung.**, 20, 148-154.

Crawford D.L. and Crawford, R.L. (1980) Microbial degradation of lignin. **Enzyme Microb. Technol.**, 2, 11-22.

Crawford, D.L., Barder, M.J., Pometto, A.L. and Crawford, R.L. (1982) Chemistry of softwood lignin degradation by *Streptomyces viridosporus*. **Arch. Microbiol.**, 13, 140-145.

Cromach, K. (1981) Below-ground processes in forest succession, in **Forest Succession - Concepts and Application** (eds D.C. West, H.H. Shugart and D.B. Botkin), Springer-Verlag, New York, pp. 361-373.

Cundell, A.M. and Mitchell, R. (1977) Microbial succession on a wood surface exposed in the sea. **Int. Biodeter. Bull.**, 13, 67-73.

Daniel, G. and Nilsson, T. (1985a) Ultrastructural and T.E.M.-Edax studies on the degradation of CCA treated radiata pine by tunnelling bacteria. **IRG/WP/126.**

Daniel, G. and Nilsson, T. (1985b) EM studies on the bacterial degradation of wood. **J. Ultra. Res.**, 91, 261-262.

Daniel, G. and Nilsson, T. (1986) Ultrastructural observations on wood-degrading erosion bacteria. **IRG/WP/1283.**

Daniel, G., Nilsson, T. and Singh, A.P. (1987) Degradation of lignocellulosics by unique tunnel-forming bacteria. **Can. J. Microbiol.**, 33, 943-948.

Dickinson, D.J. (1982) The decay of commercial timbers, in **Decomposer Basidiomycetes: their Biology and Ecology** (eds J.C. Frankland, J.N. Hedger and M.J. Swift), Cambridge University Press, Cambridge, pp. 179-190.

Drysdale, J.A. and Hedley, M.E. (1984) Types of decay observed in CCA-treated pine posts in horticultural situations in New Zealand. **IRG/WP/1226.**

Eaton, R.A. and Dickinson, D.J. (1976) The performance of copper-chrome-arsenic treated wood in the marine environment. **Mat. Org.**, 3, 521-529.

Eriksson, K.-E. (1981) Fungal degradation of wood components. **Pure and Applied Chemistry**, 53, 33-43.

Eslyn, W.E. and Moore, W.G. (1984) Bacteria and accompanying deterioration in river pilings. **Mat. Org.**, 19, 264-282.

Fazzani, K., Furtado, S.E.J., Eaton, R.A. and Jones, E.B.G. (1975) Biodeterioration of timber in aquatic environments, in **Microbial Aspects of the Deterioration of Materials** (eds D.W. Lovelock and R.J. Gilbert), pp. 39-58.

Fengel, D. and Wegener, G. (1984) **Wood, Chemistry, Ultrastructure, Reactions.** Walter de Gruyter, Berlin.

Findlay, W.P.K. (1967) **Timber Pests and Diseases.** Pergamon Press, Oxford.

Findlay, W.P.K. and Pettifor, C.B. (1937) Effect of sapstain on the properties of timber. I. Effect of sapstain on the strength properties of Scots pine sapwood. **Forestry,** 11, 40-52.

Fukazawa, K., Imagana, H. and Doi, S. (1976) Histochemical observation of decayed cell wall using UV and fluorescence microscopy. **Res. Bull. Coll. Exp. For.,** 33, 101-114.

Furtado, S.E.J. (1978) The interaction of organisms in the decay of timber in aquatic habitats. **Ph.D. Thesis,** C.N.A.A., Portsmouth Polytechnic, U.K.

Greaves, H. (1969) Micromorphology of the bacterial attack of wood. **Wood Sci. Technol.,** 3, 150-166.

Green, N.B. (1980) The biochemical basis of wood decay micromorphology. **J. Inst. Wood Sci.,** 8, 221-228.

Haider, K. and Trojanowski, J. (1980) A comparison of the degradation of ^{14}C-labelled lignins, phenols and phenolic polymers in relation to soil humus formation, in **Lignin Biodegradation, Microbiology, Chemistry and Potential Applications** (eds T.K. Kirk and H.M. Chang), CRS Press Inc., Bow Raton, Florida, pp. 111-134.

Haider, K., Trojanowski, J. and Sundman, V. (1978) Screening for lignin degrading bacteria by means of ^{14}C-labelled lignins. **Arch. Microbiol.,** 119, 103-106.

Hale, M.D.C. and Eaton, R.A. (1983) Soft rot decay of wood: the infection and cavity-forming processes of *Phialophora hoffmannii* (Van Beyma) Schol-Schwarz, in **Biodeterioration 5** (eds T.A. Oxley and S. Barry), Wiley, London, pp. 54-63.

Hale, M.D.C. and Eaton, R.A. (1985a) The ultrastructure of soft rot fungi. I. Fine hyphae in wood cell walls. **Mycologia,** 77, 447-463.

Hale, M.D.C. and Eaton, R.A. (1985b) The ultrastructure of soft rot fungi. II. Cavity-forming hyphae in wood cell walls. **Mycologia,** 77, 594-605.

Hale, M.D.C. and Eaton, R.A. (1985c) Oscillatory growth of fungal hyphae in wood cell walls. **Trans. Br. mycol. Soc.,** 84, 277-288.

Hale, M.D.C. and Eaton, R.A. (1986) Soft rot cavity widening: a kinetic approach. **Proc. R. Soc., Lond.,** B 227, 217-226.

Harmsen, L. and Nissen, T.V. (1965) Timber decay caused by bacteria. **Nature,** 266, 319.

Heal, O.W. and Ineson, P. (1984) Carbon and energy flow in terrestrial ecocsystems: Relevance to microflora, in **Current Perspectives in Microbial Ecology** (eds M.J. Klugg and C.A. Reddy), Amer. Soc. Microbiol., Washington D.C., pp. 394-404.

Higuchi, T. (1985) Degradative pathways of lignin model compounds, in **Biosynthesis and Biodegradation of Wood Components** (ed T. Higuchi), Academic Press, Orlando, pp. 557-578.

Holt, D.M. (1981) Bacterial breakdown of timber in aquatic habitats and the relationship with wood degrading fungi. **Ph.D. Thesis,** C.N.A.A., Portsmouth Polytechnic, U.K.

Holt, D.M. (1983) Bacterial degradation of lignified wood cell walls in aerobic aquatic habitats: decay patterns and mechanisms proposed to account for their formation. **J. Inst. Wood Sci.,** 9, 212-223.

Holt, D.M. and Jones, E.B.G. (1978) Bacterial cavity formation in delignified wood. **Mat. Org.,** 13, 13-30.

Holt, D.M. and Jones, E.B.G. (1983) Bacterial degradation of lignified wood cell walls in anaerobic aquatic habitats. **Appl. Environ. Microbiol.,** 46, 722-727.

Holt, D.M., Jones, E.B.G. and Furtado, S.E.J. (1979) Bacterial breakdown of wood in aquatic habitats. **Rec. Ann. Conv. BWPA,** 13-24.

Hoppe, H.G. (1986) Degradation in seawater: microbial degradation, in **Biotechnology,** Vol. 8 (eds H.J. Rhem and G. Reed), VCH, Weinheim, pp. 453-474.

Jeffries, T.W. (1987) Physical, chemical and biochemical considerations in the biological degradation of wood, in **Wood and Cellulosics: Industrial Utilization, Biotechnology, Structure and Properties** (eds J.F. Kennedy, G .O. Phillips and P.A. Williams), Ellis Horwood Ltd, Chichester, pp. 213-230.

Johnson, B.R. and Giovik, L.R. (1970) Effect of *Trichoderma viride* and a contaminating bacterium on microstructure and permeability of loblolly pine and Douglas fir. **Proc. Amer. Wood Preserv. Assoc.,** 66, 234-242.

Kaiser, J.P. and Hanselmann, K.W. (1982a) Fermentation metabolism of substituted monoaromatic compounds by a bacterial community from anaerobic sediments. **Arch. Microbiol.,** 133, 185-194.

Kaiser, J.P. and Hanselmann, K.W. (1982b) Aromatic chemicals through anaerobic conversion of lignin monomers. **Experientia,** 38, 167-176.

Kawakami, H. (1976) Isolation and identification of the aquatic bacteria, utilising biphenyl-type lignin molecules. **Mokuzai Gakkaishi,** 22, 537-538.

Kirk, T.K. (1983) Degradation and conversion of lignocelluloses, in **The Filamentous Fungi, Vol. 4, Fungal Technology** (eds J.E. Smith, D.R. Berry and B. Kristiansen), Edward Arnold, London, pp. 266-295.

Kirk, T.K. and Cowling, E.B. (1984) Biological decomposition of solid wood, in **The Chemistry of Solid Wood** (ed R.M. Rowell), American Chemical Society, Washington D.C., pp. 455-487.

Kohlmeyer, J. (1980) Bacterial attack on wood and cellophane in the deep sea, in **Biodeterioration 4** (eds T.A. Oxley, G. Becker and D. Allsopp), Pitman Publishing Co. Ltd., London, pp. 187-192.

Krapivina, I.G. (1960) Razrusenie gribami sinevy vtoricnogo dloga kletocnoj stenki. **Lesnoi Z. Archangel'sk,** 3, 130-133.

Kudo, H., Cheng, K.J. and Costerton, J.W. (1987) Electron microscopic study of the methylcellulose-mediated detachment of cellulolytic rumen bacteria from cellulose fibres. **Can. J. Microbiol.,** 33, 267-272.

Lamed, R., Setter, E. and Bayer, E.A. (1983) Characterisation of a cellulose binding, cellulose-containing complex in *Clostridium thermocellum.* **J. Bacteriol.,** 156, 828-836.

Lamed, R., Setter, E. and Kenig, R. (1983) The cellulosome: A discrete cell surface organelle of *Clostridium thermocellum* which exhibits separate antigenic cellulose-binding and various cellulolytic activities. **Biotechnology for Fuels and Chemicals Symposium,** Gotlinburg, Tennessee.

Leightley, L.E. and Eaton, R.A. (1977) Mechanism of decay of timber by aquatic micro-organisms. **Rev. Ann. Conv. BWPA,** 1-26.

Levy, J.F. (1975) Bacteria associated with wood in ground contact, in **Biological**

Transformation of Wood by Microorganisms (ed W. Liese), Springer-Verlag, Berlin, pp. 64-73.

Levy, J.F. (1982) The place of basidiomycetes in the decay of wood in contact with the ground, in **Decomposer Basidiomycetes: their Biology and Ecology** (eds J.C. Frankland, J.N. Hedger and M.J. Swift), Cambridge University Press, Cambridge, pp. 161-178.

Liese, W. and Greaves, H. (1975) Micromorphology of bacterial attack, in **Biological Transformation of Wood by Microorganisms** (ed W. Liese), Springer-Verlag, Berlin, pp. 74-78.

Ljungdhal, L.G., Petterson, B., Eriksson, K.E. and Wiegel, J. (1983) A yellow affinity substance involved in the cellulolytic system of *Clostridium thermocellum*. **Curr. Microbiol.**, 9, 195-200.

McCarthy, A.J. (1987) Lignocellulose degrading actinomycetes. **FEMS Microbiol. Rev.**, 46, 145-163.

Meyer-Reil, L.A. (1987) Seasonal and spacial distribution of extracellular enzymatic activities and microbial incorporation of dissolved organic substrates in marine sediments. **Appl. Environ. Microbiol.**, 53, 1748-1755.

Mittelman, M.W. and Geesey, G.G. (1985) Copper-binding characteristics of exopolymers from a freshwater-sediment bacterium. **Appl. Environ. Microbiol.**, 49, 846-851.

Mouzouras, R. (1989) Soft rot decay of wood by marine microfungi. **J. Inst. Wood Sci.**, 11, 193-201.

Mouzouras, R., Jones, E.B.G., Venkatasamy, R. and Moss, S.T. (1986) Decay of wood by microorganisms in marine environments. **Rec. Ann. Conv. BWPA**, pp. 27-44.

Mouzouras, R., Jones, E.B.G., Venkatasamy, R. and Holt, D.M. (1988) Microbial decay of lignocellulose in the marine environment, in **Marine Biodeterioration, Advanced Techniques Applicable to the Indian Ocean** (eds M.F. Thompson, R. Sarojini and R. Nagabhushanam), Oxford & IBH Publishing, New Delhi, pp. 329-354.

Murray, W.D. (1987) Metabolism of cellobiose and cellulose by *Bacterioides cellulosolvens*. **J. Ind. Microbiol.**, 1, 393-398.

Murray, W.D., Snowden, L.C. and Colvin, J.R. (1986) Localisation of the cellulose activity of *Bacteroides cellulosolvens*. **Letters Appl. Microbiol.**, 3, 69-72.

Nilsson, T. (1982) Bacterial degradation of untreated and preservative treated wood. **Proc. 16th Conv. Deutsche Gesselschaff für Holzforschung**, 35, 217-222.

Nilsson, T. and Rowell, R.M. (1982) Decay patterns observed in butylene oxide modified ponderoza pine after exposure in unsterile soil. **IRG/WP/3211**.

Nilsson, T. and Daniel, G. (1983) Tunnelling bacteria. **IRG/WP/1186**.

Nilsson, T. and Holt, D. (1983) Bacterial attack occcurring in the S2 layer of wood fibres. **Holzforschung.**, 37, 107-108.

Nilsson, T. and Singh, A.P. (1984) Cavitation bacteria. **IRG/WP/1235**.

Nioh, I. (1977) Characteristics of bacteria in the forest soils under natural vegetation. **Soil Sci. Plant Nutri. (Tokyo)**, 23, 523-529.

Pettey, T.M. and Crawford, D.L. (1985) Characterization of acid-precipitable lignin (APPL) produced by *Streptomyces viridosporus* and protoplast fusion recombinant *Streptomyces* strains. **Biotechnol. Bioeng. Symp.**, 15, 179-190.

Robra, K.H. (1986) Special methods in microbial degradation, in **Biotechnology - A Comprehensive Treatise, Vol. 8** (eds H.J. Rehm and G. Reed), VCH, Weinheim, pp. 651-670.

Ross, I.S. (1986) Uptake of heavy metals by microorganisms. **Internatl. Ind. Biotechnol.**, 6, 184-188.

Rouch, D., Camakaris, J., Lee, T.O. and Luke, R.K.J. (1985) Inducable plasmid-mediated copper resistance in *Eschericia coli*. **J. Can. Microbiol.**, 131, 939-943.

Schmidt, O. (1980) Laboratory experiments on the bacterial activity towards the woody cell wall, in **Biodeterioration, Proc. 4th Int. Symp.** (eds T.A. Oxley, G. Becker and D. Allsopp), Pitman Publishing Ltd., London, pp. 63-66.

Schmidt, O. and Dietrichs, H.H. (1976) Zur Activitat von Bacterien gegenuber Holzkomponenten. **Mat. Org., Beih. No. 3**, 91-102.

Schmidt, O. and Liese, W. (1982) Bacterial decomposition of woody cell walls. **Internatl. J. Wood Preserv.**, 2, 13-19.

Schmidt, O., Nagashima, Y., Liese, W. and Schmitt, U. (1987) Bacterial wood degradation studies under laboratory conditions and in lakes. **Holzforschung**, 41, 137-140.

Scobie, D. (1980) **Timber Trades Journal**, 312, 21.

Singh, A.P. and Butcher, S.A. (1985) Degradation of CCA-treated *Pinus radiata* posts by erosion bacteria. **J. Inst. Wood Sci.**, 10, 140-144.

Singh, A.P., Nilsson, T. and Daniel, G.F. (1987a) Ultrastructure of the attack of the wood of two high-lignin tropical hardwood species (*Alstonia scholaris* and *Homalium foetidum*) by tunnelling bacteria. **J. Inst. Wood Sci.**, 11, 26-42.

Singh, A.P., Nilsson, T. and Daniel, G.F. (1987b) Attack of pulped fibres by erosion bacteria. **Mat. Org.**, 22, 257-269.

Stotsky, G. and Krasovsky, V.N. (1981) Ecological factors that affect the survival, establishment, growth and genetic recombination of microbes in natural habitats, in **Molecular Biology, Pathogenicity and Ecology of Bacterial Plasmids** (eds S.B. Levy, R.C. Clowes and E.L. Koenig), Plenum Press, New York, pp. 31-42.

Suberkropp, K. and Klug, M.J. (1976) Fungi and bacteria associated with leaves during processing in a woodland stream. **Ecology**, 57, 707-719.

Sutter, H.P. and Jones, E.B.G. (1985) Interactions between copper and wood degrading fungi. **Rec. Ann. Conv. BWPA**, pp. 29-41.

Sutter, H.P., Jones, E.B.G. and Wälchli, O. (1983) The mechanisms of copper tolerance in *Poria placenta* (Fr.) Cke. and *Poria vaillantii* (Pers.) Fr. **Mat. Org.**, 18, 241-262.

Sutter, H.P., Jones, E.B.G. and Wälchli, O. (1984) Occurrence of crystalline sheaths in *Poria placenta* (Fr.) Cke. **J. Inst. Wood Sci.**, 10, 19-23.

Venkatasamy, R. (1988) Mechanisms of bacterial deterioration of wood. **Ph.D. Thesis**, C.N.A.A., Portsmouth Polytechnic, U.K.

Ventullo, R.M. and Larson, R.S. (1986) Adaptation of aquatic microbial communities to quarternary ammonium compounds. **Appl. Environ. Microbiol.**, 51, 356-361.

Willoughby, G.A. and Leightley, L.E. (1984) Patterns of bacterial decay in preservative treated eucalypt power transmission poles. **IRG/WP/1223**.

GROUNDWATER ENGINEERING

PART FIVE

GROUNDWATER
ENGINEERING

14 BIOFOULING IN SIERRA COLORADO WATER SUPPLY: A CASE STUDY

R.E. ALCALDE
Departamento Provincial de Aguas, Rio Negro, Argentina
M.A. GARIBOGLIO
Universidad Nacional de La Plata, Buenos Aires,
Argentina

Abstract
Sierra Colorada located in the south of Rio Negro Province (Argentina)
has suffered several troubles in two wells of its water supply, which
have been imputed to a biofouling process. Bacteriological examination
of water and material from encrustations has revealed the presence of
sheathed iron precipitating bacteria and sulphate reducing bacteria
(SRB). Both wells clogged completely, and one of them was successfuly
treated by using hydrochloric acid and calcium hypochlorite. The
development of events through time that led to clogging of these wells
is described.
Keywords: Sierra Colorada, Biofouling, Water Wells, Iron Bacteria,
Sulphate Reducing Bacteria, Clogging.

1 Introduction

Sierra Colorada is a rural town located in the south of Rio Negro
Province, an arid region of Patagonia (Argentina) where the water is
scarce and of mediocre quality for use as drinking water source. Almost
all the water available is ground water and this has a fluoride content
exceeding the allowable level. In the past decade the drilling of new
wells has increased in response to a major demand in water for public
supply and irrigation. This, as elsewhere in the province, has
brought about   certain phenomena that have an important impact on
production and quality of water and maintenance costs.

This paper deals with a case study of biofouling in two wells of the
Sierra Colorada water supply which clogged completely interrupting
potable water production. Following a rehabilitation treatment, one of
the clogged wells was successfully returned to its original specific
capacity, allowing renewal of the service. The other, older, well was
kept for study purposes and its casing and screen pulled out.

2 Description of Sierra Colorada water supply

Originally the source of the water supply was a well drilled in 1970,
well nº 1, whose characteristics are summarized in Table 1. This well
gave water with excesive fluoride content, therefore a defluoridation
plant using bone char was installed.

In 1983 with the purpose of increasing the water production a second well was drilled, well nº 2, at a distance of 30 m from well nº 1 (see Table 1). Both wells were connected to a common pipeline and functioned alternatively.

Table 1.   Characteristics of wells of Sierra Colorada

Well nº	Total depth (m)	Static level (m)	Casing (in.)	Screen (in.)
1	36.9	15.4	6 (0.1524 m)	4 (0.1016 m)
2	30.5	15.75	8 (0.2032 m)	6 (0.1524 m)

In 1987 as result of the impairment of wells nº 1 and 2 a third well was drilled. A diagrammatic representation of Sierra Colorada water supply is showed in Fig. 1.

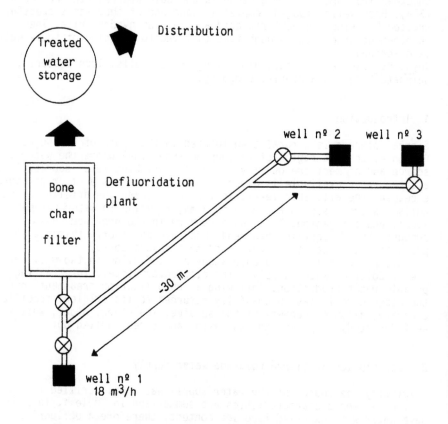

Fig. 1.   Diagrammatic representation of Sierra Colorada water supply.

3   Events, Symptoms and Diagnosis in the Biofouling Process

For over fifteen years well nº 1 gave water without any problem and
the maintenance of its pumping equipment was related to mechanical
problems. During this time there were no reports of corrosion or
encrustation. In January 1983 the drilling of well nº 2 finished and
in May, when the pumping equipment was pulled out for repairs, it was
evident there had been intense phenomenon of corrosion and encrustation
in the pump and pipes. The bowls and impellers of the submersible pump
showed severe pitting, while its casing was heavily encrusted with
rusty material. The water contained in the pump's pipes had a green
colour and a septic odour (Alcalde and Castronovo de Knott, 1986). All
these symptoms suggested a possible biological cause of the problem,
instead of a process of purely chemical nature. Bacteriological
examination of the water of the well nº 2 was carried out to look for
iron bacteria using the technique outlined in Fig. 2. This technique
(Alcalde and Castronovo de Knott, 1986) uses the bacterial adherence
to surfaces as medium of collecting specimens for microscopical
identification and was complemented with enrichment cultures in alfalfa
straw medium (Stokes, 1954; Alcalde and Castronovo de Knott, 1986).
By mean of both methods it was possible to detect the presence of
bacteria of the Sphaerotilus-Leptothrix group in all samples tested.

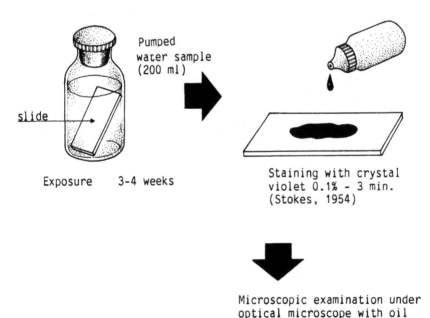

Pumped
water sample
(200 ml)

slide

Exposure   3-4 weeks

Staining with crystal
violet 0.1% - 3 min.
(Stokes, 1954)

Microscopic examination under
optical microscope with oil
immersion objective

Fig. 2. Observation of sheathed iron precipitating bacteria attached
to a glass slide.

Later, it was reported that well nº 1, which had been out of use for some time for repairs to its pumping equipment, gave green coloured water when it was put in service again. To check the possibility that this well was also infected, bacteriological examinations were performed which corroborated this.

During 1987 both wells showed different problems. In February when a new pump was to be installed in well nº 1 at a depth of 23 m there were difficulties in lowering it below 18 m. At this depth the casing had reduced from 6 in. to 4 in. In May the pump was pulled out and it showed intense pitting. Examinations for iron bacteria and sulphate reducing bacteria (SRB) were positive.

The investigation of SRB in water and tubercles or encrustations was carried out using the methodology proposed by Gariboglio (1986). These involve culturing the water sample or homogenized solid material in sterile distilled water in Postgate's C medium contained in a vial with atmosphere of CO_2 and H_2 at specific conditions of redox potential and with the addition of iron filings (Fig. 3).

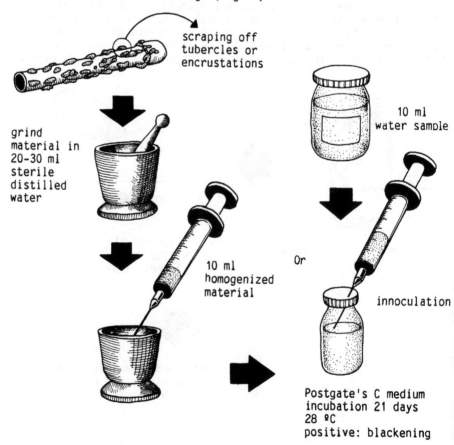

scraping off tubercles or encrustations

grind material in 20-30 ml sterile distilled water

10 ml water sample

10 ml homogenized material

Or

innoculation

Postgate's C medium incubation 21 days 28 ºC positive: blackening

Fig. 3. Investigation of SRB under tubercles or water samples.

The submersible pump in well nº 2 then developed some operating problems, indicated by abnormal noise and air aspiration. In October, well nº 2 suddenly stopped giving water.

During December as a part of cooperative program between Departamento Provincial de Aguas (DPA) and Consejo Federal de Inversiones (CFI), the casing and screen of well nº 1 were pulled out in order to study the phenomenon in more detail. The casing showed extensive tuberculation and the screen was totally clogged. Samples of tubercles were cultured for SRB by means of the method outlined in Fig. 3 giving positive results. On the other hand, the tubercles treated with hydrochloric acid dissolved in it giving off a strong smell of hydrogen sulphide.

At the same time the rehabilitation treatment described below was carried out in the well nº 2.

4   Treatment for rehabilitation

The rehabilitation treatment was designed so as to recover well nº 2. Taking into account the local conditions, acid treatment and shock chlorination were chosen (Hackett and Lehr, 1985; Hackett, 1987). The same was realized by adding 160 l of 28% hydrochloric (muriatic) acid by means of the pipe of an air-lift pumping system. This volume was twice the capacity of the screen and was added with the objective of forcing the acid into the formation. To this end an equal volume of water was added afterwards. The contact time of acid into the well was 24 h during which the liquid column was agitated periodically by mean of air injection, and samples were taken for pH control. The pH values ranged between 1.3 and 2.8 at which iron is soluble. The samples were mainly greenish to rusty in colour and of high turbidity. Acid treatment was performed twice in identical conditions. When acid was added perceptible emanations of hydrogen sulphide were produced.

In succession shock chlorination was carried out using calcium hypochlorite (30-35% available chlorine) which was fed continuosly to mantain a residual chlorine level above 1000 mg/l during 24 h of contact time. Residual chlorine was monitored by mean of the iodometric method in samples taken periodically, from which it could be seen that there was a high chlorine demand. This treatment was also performed twice. After shock chlorination, as well as acid treatment, the chemical products and plugging material were removed by means of air lift pumping. The shock chlorination using calcium hypochlorite gave some problems such as blocking the inlet pipe.

Fig. 4 shows the characteristics of well nº 2 and Fig. 5 summarizes the rehabilitation treatment.

5   Effectiveness of rehabilitation treatment

5.1   Well performance after treatment

In January 1988 a pumping test was carried out in well nº 2 to determine the degree of effectiveness of treatment. Table 2 shows the data recorded in that test and in the original one.

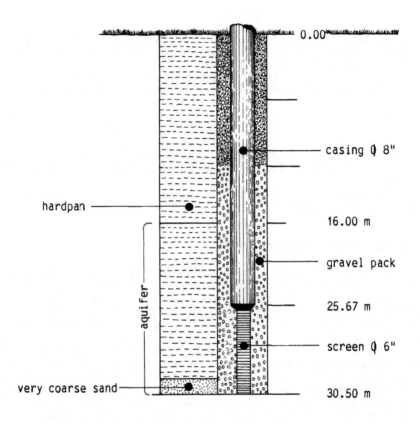

Fig. 4. Characteristics of well nº 2 of Sierra Colorada water supply.

Table 2. Pumping test data of well nº 2 - (A) original and (B) after rehabilitation treatment.

Pumping test	Static level (m)	Yield (m3/h)	Dynamic level (m)	Drawdown (m)
A	15.75	25.0	23.10	7.35
B	15.18	24.8	22.57	7.39

From these data the respective specific capacities are:

A: 3.40 m³/hm

B: 3.36 m³/hm

that is a 99% return to original specific capacity.

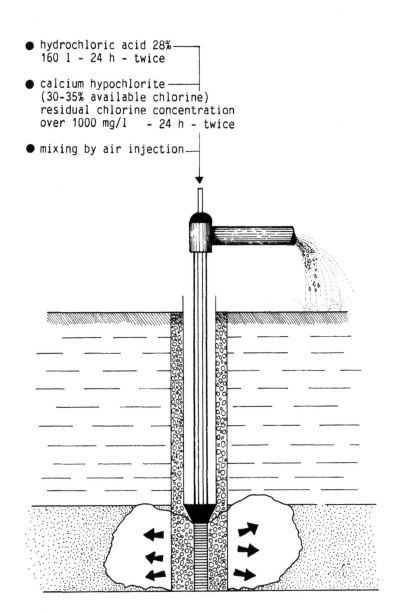

- hydrochloric acid 28%——
  160 l - 24 h - twice

- calcium hypochlorite———
  (30-35% available chlorine)
  residual chlorine concentration
  over 1000 mg/l   - 24 h - twice

- mixing by air injection—

Fig. 5. Chemical treatment of well nº 2.

5.2 Bacteriological monitoring after treatment

Immediately after treatment, and in the subsequent months, bacteriological checks were made in well nº 2 for iron bacteria and SRB. These were negative for the former but positive for SRB in spite of subsequent disinfection.

6 Monitoring and maintenance program

Although bacteriological monitoring could not be continued to date because of the distance of the site from the laboratory, the pumping equipment has been inspected. In fact, in November 1989 the pump and pipes of well nº 2 were pulled out verifying that they remained unaltered since installation in January 1988.

A pumping test of this well is planned for the near future. During this year, as part of DPA-CFI program of cooperation, a plan for more complete monitoring will be established.

7 Concluding remarks

Biofouling in water wells is a recently observed problem in Rio Negro province. It appeared in a noticeable form ten years ago and ever since it has increased and extended (Alcalde and Castronovo de Knott, 1986).

Sierra Colorada's case is the first one in which the clogging of biofouled wells has completely interrupted the water supply. The rehabilitation treatment has been successful, although laboratory studies have shown that this kind of treatment is not always completely effective (Cullimore, 1988, personal communication) for this reason we intend to continue to work for improvements.

This problem is now of such magnitude that we must understand it better and avoid having to deal with it in the future.

8 References

Alcalde, R.E. and Castronovo de Knott, E. (1986) Occurrence of iron bacteria in wells in Rio Negro (Argentina), in Proc. **International Symposium on Biofouled Aquifers: Prevention and Restoration** (ed. D.R. Cullimore), American Water Resources Association, Bethesda, Maryland, U.S.A., pp. 127-136.

Cullimore, D.R. (1988) Personal communication.

Gariboglio, M.A. (1986) Ensuciamiento biológico y corrosión inducida microbiológicamente en sistemas de captación y distribución de agua de Caleta Olivia, Santa Cruz, **Informe Técnico Convenio UNLP-CFI** para el Estudio hidrogeológico en Caleta Olivia Santa Cruz - Universidad Nacional de La Plata, Argentina.

Hackett, G. (1987) A review of chemical treatment strategies for iron bacteria in wells. Water Well J. 41, 37-42.

Hackett, G. and Lehr, J.H. (1985) Iron bacteria occurrence, problems and control methods in water wells, in report prepared for **U.S. Army**

Corp of Engineers Waterways Experiment Station, National Water Well
Association, Ohio, U.S.A.

Stokes, J.L. (1954) Studies on the filamentous sheathed iron bacterium
Sphaerotilus natans, J. Bacteriol., 67, 278-291.

Drawing: Marcelo Ferreyra (DPA)

15 IRON BIOFOULING IN GROUNDWATER ABSTRACTION SYSTEMS: WHY AND HOW?

P. HOWSAM and S.F. TYRREL
Department of Agricultural Water Management, Silsoe
College, Cranfield Institute of Technology, Cranfield, UK

Abstract
A study into the diagnosis and treatment of iron biofouling in the UK, has revealed that not only can all borehole/aquifer systems be expected to support biofilm development, but that a number of engineering /operational factors play a significant part in whether a biofouling problem will occur. This paper summarises the findings of the study and highlights those processes involved of which awareness amongst engineers may be low. It also discusses other water flow systems where similar biofouling processes should be expected.
Keywords: Iron Biofouling, Environmental/Engineering Factors, Diagnosis, Prediction, Water Flow Systems.

1 Introduction

There is a growing awareness in the UK and elsewhere of the need to properly monitor and maintain groundwater abstraction systems. One of the problems which many well owners/operators are becoming increasingly aware of is biofouling (Howsam 1988). The consequences of biofouling in a groundwater abstraction system can be many-fold but all are usually negative.

Biofouling refers to the deleterious effect of the development of a biofilm on a surface within an engineered environment. The problems caused include:-
DETERIORATION of hydraulic efficiency by clogging of screen slots, formation and gravel pack, pump inlets/bowls and by increasing frictional resistance to flow in pipes;
DETERIORATION of material properties (Eg steel, concrete, rubber);
DETERIORATION of water quality.
A biofilm develops due to the attachment and growth of bacteria on a surface. Most of us think of bacteria in relation to personal/public health. In this respect the bacteria are perceived to be swimming around in the media involved; these are planktonic bacteria. The bacteria with which we are concerned are ones which attach themselves to surfaces and these are called sessile bacteria. Sessile bacteria within a biofilm have several significant advantages: (i) they are protected to a large degree from anti-microbial agents; (ii) they experience better nutrient uptake conditions; (iii) they benefit from inter-species cooperation (Costerton et al 1981).

A biofilm usually consists not just of bacterial cells but also proportionately large volumes of extra-cellular slime. This slime has the ability to trap particulate material carried by the water which flows past and also provides an environment where chemical and microbially enhanced precipitation of metal oxides (particularly iron oxyhydroxides) occur. Thus the appearance of a biofilm will vary depending on the proportion of these various elements, ie from slimy and pastey, to hard brittle encrustations, in texture; and from whitish, grey, black to buff, brown, orange-brown in colour (Kay et al 1989). Certainly, in many cases, the appearance of biofouling deposits will not readily convey to an engineer that microbiology is involved.

2 Biofouling: environmental factors

The obvious environmental factors are the sources of bacteria and the nutrients they require to grow.

2.1 Where do the microorganisms come from?
All natural waters will contain bacteria and soil is absolutely teeming with them. Therefore it is not difficult to imagine that in drilling a well that there is plenty of scope for inoculation of any part of the hole from the upper soil/water horizons. These we might regard as immigrants, but there is also growing evidence of migrant and resident bacteria at depth in soil/rock/water environments. There is a common misconception that groundwater, having been 'filtered' through the ground is bacteriologically pure. Yet whilst groundwater generally does contain far fewer microorganisms (especially pathogenic bacteria) than surface water, it should not be regarded as sterile. In the UK study of over 40 boreholes, involving all the major aquifer systems, 'iron' bacteria, as well as other microorganisms were found to be present in all cases (Tyrrel 1990).

2.2 What nutrients do they require?
The main nutrients which bacteria in general need to grow are carbon, nitrogen, phosphorus and sulphur. The carbon source can be carbon dioxide, which is normally present in groundwater, or organic carbon, often in the form of organic acids which are produced during the decomposition of plant/animal matter in soils. Nitrogen can be derived from ammonia, nitrite or nitrate. The latter has been found to be increasing in concentration in groundwaters in the UK. Sulphate too is relatively abundant in groundwaters. Phosphorus in the form of inorganic phosphate is less common and therefore may be a limiting nutrient in some cases. In the UK borehole survey, it was found that a significant biofilm-based population could be sustained by the nutrients dissolved in what would normally be classed as unpolluted groundwater (Tyrrel 1990).
    Other important factors relating especially to iron biofouling are the existence of an aerobic/anaerobic interface, the presence of ferrous iron and water flow.

2.3 The presence of iron

The bacteria involved with iron biofouling are of many varieties. Of these some need the presence of ferrous iron because the oxidation of ferrous to ferric iron provides energy for their metabolism. This transformation results in the precipitation of iron as the hydroxide, imparting the familiar orange/brown colour to the biofilm. Others utilise iron-containing organic compounds with the same result (McCrae et al 1973). It should be noted that ferrous iron may also be oxidised and precipitated by a chemical process which is initiated at an aerobic/anaerobic interface.

2.4 The presence of an aerobic/anaerobic interface

An aerobic/anaerobic interface can develop in a borehole due to the effects of pumping in conjunction with the construction of the borehole and the hydrogeology of the aquifer. As anaerobic groundwater nears a pumping borehole it is liable to encounter oxygenated water from the near-surface aquifer zone. The interface is important as it is the point at which aerobic bacterial activity and chemical iron oxidation is initiated.

2.5 Water flow

The flow of water is required to transport nutrients to the biofilm. It is logical and has been demonstrated that for low nutrient waters (which most un-polluted groundwaters are) the faster the rate of flow the faster the rate of microbial/biofilm growth; ie the more food is supplied the more growth will occur (Caldwell 1986). It is also possible to imagine that for a biofilm on a surface with water flowing past, better nutrient uptake conditions will exist with turbulent flow rather than with laminar flow (Cullimore 1986).

3 Biofouling: engineering/operational factors

The key engineering/operational factors are those which introduce or enhance the key elements in iron biofouling, ie the existence of an aerobic/anaerobic interface; improved nutrient supply/uptake; iron concentration.

The latter is fairly obvious, in that iron availability can be increased, over and above the background level from the aquifer if iron/mild steel are used in any part of the groundwater system.

The introduction of oxygen and the introduction/movement of an aerobic/anaerobic interface will be enhanced by intermittent pumping, or by allowing the pumping water level to fall below the top of the screen.

The UK study demonstrated that intermittent pumping and water cascading down the screen were features commonly found in iron-biofouled boreholes. In addition, the mixing of incompatible oxygenated and anaerobic, iron-containing waters due to the presence of a pumping borehole, was shown to be a way in which design and operation could upset the balance of a natural system.

The flow velocities in different parts of a borehole are variable. Key areas of localised higher velocity are: at the screen slots (especially if the screen open area is too small in relation to the abstraction rate because the slots are too small, they are inadequate in number, or they become blocked); at the pump intake and within the pump itself and at pipework restrictions such as valves. In practice it has been found that these are exactly the points in the borehole where clogging is at its worst. Clogging of the upper section of the screen (whilst the lower section remains clean) is a common phenomenon and may relate to flow velocities – entrance velocities and screen velocities are highest in the upper screen sections – and/or to increased oxygenation of the near-surface groundwater.

Enhanced biofouling was demonstrated in a groundwater dewatering system where the flow rates in the ejector pipes ranged from 2-4m/s. In this case biofouling was occurring throughout the system but was worst in the areas of highest flow. It caused some surprise at the time that deposition rather than erosion was occurring under these conditions and provides a good example of the tenacity of iron biofouling deposits.

Frequency of submersible pump-failure caused by the build-up of iron-based deposits in impeller bowls and pump intakes was shown to be increased by the change from low-speed pumps to high speed pumps (which, in general, generate higher pump velocities). The remedy in some cases has been to grind down the impellers to give greater clearance, which, however, leads to decreased pumping head.

4 Iron biofouling in other water flow systems

During the course of the UK borehole study, a number of cases of iron biofouling came to light in other flowing systems.

4.1 Raw groundwater mains/distribution systems
Although the combined processes of biofilm formation and iron precipitation may have been initiated within the borehole, they have been found to continue to cause problems in the raw water supply pipework and out into distribution. The build up of iron biofouling deposits increases head losses and also can clog up in-line monitoring devices. Any ferrous iron that has not been deposited in the borehole may subsequently precipitate out, provided that sufficient oxygen is still available. Problems in supply may be caused by inadequate aeration and filtration at the treatment works.

Non-iron biofilms have been reported in treated water supply systems. In addition to potential public health problems associated with biofilm-based bacteria such as Legionella (Colbourne et al 1989), it is possible that these growths may result in subtle but significant head-losses in a distribution system over a long period of time. Work on biofouling of ships hulls has shown that a biofilm of only 10um thickness can cause a 10% increase in drag (Cooksey 1989).

4.2 Irrigation systems
There is growing awareness of the potential for iron biofouling in drip/trickle irrigation systems (Kay et al 1989). Drip/trickle irrigation is a method of applying small but frequent applications of water to crops through small emitters. The emitters may become clogged with microbial slimes, the products of their metabolism and particulate matter which becomes entrapped within the biofilm (Bucks et al 1979).

Good emitter design and good water quality are important in preventing these problems.

4.3 Pressure relief/drainage systems in embankment dams

Biofouling in pressure relief wells has recently been investigated in the USA by the US Corp of Army Engineers. Although there has been little coverage in the civil engineering literature of this problem, dam engineers will admit, when quizzed, to seeing orangey-brown deposits in the end of outflow pipes of pressure relief systems. Analysis of these deposits has shown the composition to be similar to that of borehole biofouling deposits. Whilst there are no proven cases of dam failure due to pore pressure build-up, there is certainly concern about the potential risk.

5 Conclusions

Iron biofouling describes a complex set of reactions which result in the deterioration of groundwater abstraction system performance. In UK boreholes, the process appears to be initiated by the introduction of oxygen into iron-containing groundwater. Biofouling can be enhanced, in certain circumstances at flow velocities which might have been expected to cause scour. Organisms capable of biofilm formation given the right circumstances and suitable nutrient concentrations appear to be omnipresent, even in groundwaters considered to be unpolluted. Interaction of the environmental and engineering aspects of the aquifer and borehole are considered to govern the process.

Biofouling affects a variety of flowing systems. It is an often unseen, yet insidious process which may have subtle or disastrous effects on engineered systems. The development of biofilms should come as no surprise to the operator of any flowing system and as such, importance should be attached to an awareness of available monitoring and prevention technology .

6 References

Bucks, D.A., Nakayama, F.S. and Gilbert, R.G. (1979) Trickle irrigation water quality and preventive maintenance. **Agricultural Water Management**, 2, 149-162.

Caldwell, D.E. (1986) Microbial colonisation of surfaces. **Int. Symp. on Biofouled Aquifers: Prevention and Restoration.** American Water Works Association., pp. 7-9.

Colbourne, J.S., Dennis, P.J., Trew, R.M., Berry, C. and Vesey, G. (1989) Legionella and public water supplies. **Water Science Technology.**, 20, 5-10.

Cooksey K.E.(1989) Personal communication

Costerton, J.W., Irvin, R.T. and Cheng, K-J. (1981) The bacterial glycocalyx in nature and disease. **Annual Review of Microbiology.**, 35, 299-324.

Cullimore, D.R. (1986) Physiochemical factors in influencing the biofouling of aquifers. **Int. Symp. on Biofouled Aquifers: Prevention and Restoration.** American Water Works Association., pp. 23-36.

Howsam, P. (1988). Biofouling in wells and aquifers. **J.Institution of Water and Environmental Management**, 2.2, 209–215.

Kay, M.G., Tyrrel, S.F. and Howsam, P. (1989) Biofouling in drip /trickle irrigation systems. **Irrigation: Theory and Practice**. (Eds Rydzewski and Ward) Pentech Press, pp. 654–660

McCrae, I.C., Edwards, J.F. and Davis, N. (1973) Utilisation of iron gallate and other iron complexes by bacteria from water supplies. **J.Applied Microbiology.**, 25, 991–995.

Tyrrel, S.F. (1990) Unpublished MPhil Thesis. Silsoe College. Cranfield Institute of Technology.

16 COMPLEXITY OF CAUSES OF WELL YIELD DECREASE

F. BARBIC and O. KRAJCIC
Institute for Technology of Nuclear and Other Mineral Raw
Materials, Belgrade, Yugoslavia
I. SAVIC
Faculty of Science, University of Belgrade,
Belgrade, Yugoslavia

Abstract
Physio-ecological characteristics of iron and manganese
bacteria populations were studied in the phreatic and
subartesian springs of Serbia together with the ecology of
the spring itself. Hydrogeological, physico-chemical, biological and technological parameters of the springs, as
well as the exploitation of the wells, were interpreted
from the abstraction system yield decrease standpoint. A
decrease in the specific well yield associated with clogging of the zones surrounding the filter, is a complex and
interrelated process. Reactions that produce in ochre
formation, depend on ecological factors that are mutually
dependent and interlinked. The potable groundwater spring
is a specific ecosystem, this fact being reflected in it's
relative simplicity. Ochre depositing dynamics and the
increase in iron bacteria population growth associated with
the more intense water movement in the spring, are treated
separately.
Keywords: Groundwater, Iron and Manganese Bacteria, Ecological Factors, Abstraction Systems, Ochre Deposits, Well
Yield Decrease.

1 Introduction

A detailed research study of iron and manganese bacteria
populations in the wells of potable groundwater in Serbia
and also in Yugoslavia, was initiated in the early
seventies by our Institute; also by other organizations
even earlier than that, when the first Ranney well to
exploit a Belgrade spring was consructed. In these investigations, the yield decrease was considered to be mainly
due to hydrogeological characteristics and changes in well
characteristics. The "biological factors" were only
marginally mentioned.
   Investigations associated with the presence of iron and
manganese bacteria population in wells and its contribution
to the formation of ochre were first carried out on
Belgrade spring and later extended to the other springs of

Serbia. The need for these more comprehensive investigation was forced by the fact that in the case of almost every well, irrespective of its type, intense yield decrease, the so called "well ageing", was detected.

Nowadays, a great number of wells in Serbia, varying in type and way of exploitation, have already been investigated. Due to the results obtained, a number of water supply organizations are investing in these studies, including them in their development programs. With this paper, we want to contribute to a better and more comprehensive understanding of the well yield decrease, first of all by pointing out its complexity and also by emphasizing the need for a multidisciplinary approach to its solution.

2 Materials and methods

To be able to determine the interdependences and changes of ecological factors and their influence onto specific yield decrease of water abstraction systems, we monitored: the changes qualitative and quantitative composition of iron bacteria population, the quality and quantity of deposited ochre and the dynamics of its deposition, the quality of water and the changes in specific yield.

Depending on the technical possibilities, various methods and equipment were used for monitoring the yield decrease of the wells, i.e. one that is associated with clogging of the filter and surrounding zone. Determination of the qualitative and quantitative composition of iron and manganese bacteria populations and dynamics of ochre deposition were based upon microbiological analyses of water and ochre samples and upon simulators plates.

The method of simulators proved to be the most practical the most reliable and the fastest way of determining changes in qualitative and quantitative composition of the iron and manganese bacteria populations. Depending on whether simulators were intended for re-use and on their location (wells or piezometers), the simulators were made out of various materials, in various sizes and positioned in various ways. As in our previous papers, we expressed the degree of presence of bacteria and the amount of deposited ochre with an index value, ranging from 1-10. In this paper, if not emphasized otherwise, an index value from 1-10 represents the degree of bacterial occurrence and the amount of deposited ochre on simulators placed in wells or piezometers for the period of 25 days.

In an attempt to standardize the quantitative presence of bacteria and deposited ochre on the simulators, we established a reference criterion to enable the comparisons of various wells, or springs as a whole. We obtained these comparative values after years long investigation of a great number of analysed simulators. At the Obrenovac spring more then 250 simulators were placed, while at the Belgrade spring this number went up to 4000.

More detailed taxonomic order of iron bacteria present in the wells was determined at the begining of our 15-years long research study. Determination of bacteria was carried out according to Bergey's system. Later on, in this paper too, all present species of iron bacteria were mostly expressed cumulatively, as relative values. The investigations showed that in wells of one spring, nearly the same species and genera (3-5 of the latter) are present. These bacteria usually form over 90% of the total number of present iron bacteria.

The use of simulators (the way of placing them in and taking them out) and their further treatment was reported in detail by Bracilovic,et al.(1975), Barbic,et al. (1987), Hässelbarth and Lüdemann (1967). For deposit collecting, some specific simulators, i.e. larger parts of well filters submerged for more than three months, were used too. From these deposit samples, physico-chemical, granulometric and mineralogical analysis were carried out. The physico-chemical analysis of water were performed according to the Yugoslavian standards, Standard Method for the Physical, Chemical and Bacteriological Examination of Water (1961), mostly during the regular water monitoring.

3 Results and discussion

The spring area of Obrenovac makes a part of a broad area of springs along the right riverbank of the Sava river, 30 km upstream from its mouth (Fig. 1.1). These springs are bounded by the Sava river along three sides. The abstraction of the groundwater is carried out by vertical tube wells (12 wells), with diameters of 800 mm, their initial yield being about 12-20 l/s. The total yield of these wells is nowadays about 110 l/s. The depth of the wells is in the range of 17-20 m. We included in our investigations a great number of piezometer boreholes, located 5-200 m away from the wells. The wells are located between 150 and 250 m from the Sava river.

The Obrenovac spring area, like the whole lower Sava spring area, consists of sediments of Tertiary and Quarternary periods. A Tertiary clays aquiclude lies about 11-21 m below the ground surface. Above this occurs extensive Quarternary pebble and sand aquifers of varying thickness. The lower layer of pebble-sandy sediments is highly permeable in comparison to the upper layer of clayer sands, which has the low permeability, but acts as a barrier against the surface contamination. The lower layer is characterized by an intergranular porosity of a super-capillary type, which satisfies the conditions of the dense spring formation. The permeability of the lower, coarser porous zone is in the range of $2.5-8.5 \times 10^{-4}$ m/s, which means that. in hydrogeological sense. these sediments are the typical aquifers. This spring is fed in three ways:

From the surface waters depositing from the atmosphere.
From the main underground water flow, and
Mostly from the riverbed water of the Sava river; the river also represent the main regulator of the groundwater level — when its level goes up, the river feeds the springs and vice versa.

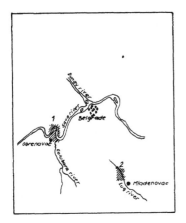
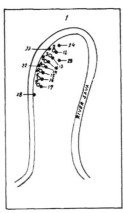
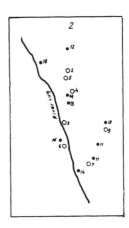

Fig.1. Location of the Obrenovac and the Mladenovac springs with investigated wells and piezometers

Another investigated spring with its wells and piezometer tubes (research boreholes) is located about 30 km southeast from Belgrade. near the town of Mladenovac (Fig. 1.2). Based on hydrogeological researches carried out so far, one can say that the aquifers are found in a series of Neocene sediments where impermeable clays, interstratified with thin irregular layers of coal and fine to medium of sand. Nowadays, about 140 l/s are abstracted from the Mladenovac spring by vertical tube wells. The depth of these is between 28 and 200 m. with diameters between 320 and 650 mm. The specific yield of wells at this locality is very low and within the range of 0,05-0,5 l/s/m. The filtration effects of sand layers in this Neocene series vary (filtration coefficient K is in the range of 10^{-4}-10^{-7} m/s), as well as granulometric composition. The Neocene series of sediments at the Mladenovac spring are mainly subartesian.

The choice of these two particular springs was determined for several reasons. Both show evidence of decreasing well. The wells of various age and yield have been constructed on these springs, some of them are not exploited, either permanently or temporarily. A large number of piezometers also exists. Some of the wells are chemically and mechanically regenerated. There are however some differences between these two springs. The Obrenovac spring is of a phreatic type , fed mainly from the Sava river. In the ecological sense, this spring is more open, the influence of external factors is much faster and more obvious then in the case of Mladenovac spring.

Within the investigation of the cause – effect relationship between presence and activity of iron bacteria and ochre sedimentation in the researched springs, the quality of their groundwater was constantly followed. On the bases of numerous chemical analyses of groundwater, the average values of some characteristic parameters are presented in Table 1. For the purpose of comparison the same table contains results of the chemical analyses of the Sava river water, that feeds the Obrenovac spring. From the Table 1 it is obvious that all water quality parameters (pH, Eh, iron and manganese concentration, etc.) are favourable for iron and manganese bacteria growth.

Table 1. Comparative survey of physico-chemical characteristics of the Sava river water and the groundwater from the Obrenovac and Mladenovac springs

Physico-chemical parameters	Sava river	Obrenovac	Mladenovac
pH	7.92	8.15	7.42
Eh (mV)		135	96
dry residue (mg/l)	329	355	
Cl^- (mg/l)	13.27	13.2	51.3
NH_4^+ (as N,mg/l)	0.4	0.95	0.8
Total hardness (°dH)	11.33	16.4	17.8
HCO_3^- (mg/l)	211.2	396	
CO_2 (mg/l)		27.5	16.4
$KMnO_4$ consump.(mg/l)	18.6	5.5	4.05
SO_4^{2-} (mg/l)	30.29	24.1	
Ca (mg/l)	62.49	71.9	66.0
Mg (mg/l)	13.61	26.7	
Fe^{2+} (mg/l)		2.5	0.98
Total Fe (mg/l)	0.243	2.7	1.07
Mn (mg/l)	<0.05	0.5	0.06

Relative occurrence of iron and manganese bacteria population and ochre sedimentation during normal exploitation in wells and piezometers boreholes of Obrenovac and Mladenovac water supply systems is presented in Table 2. Results of bacterial occurrence are the average values obtained for a greater number of simulators and from analyses of water samples during the two year research period. All investigated wells (marked 2–8) are eight to fourteen years old, the well number 2 being the oldest one and the well number 8 the youngest.

Table 2. Occurrence of iron and manganese bacteria and ochre sedimentation at the Obrenovac and Mladenovac springs

Object mark	Leptothirix				Gallionella				Siderocapsa				Sedimentation index			
	1	2	3	4	1	2	3	4	1	2	3	4	1	2	3	4
B-2	8	8	7		3	3	2		9	8	9		8	8	7	
B-3	6	8	7	7	4	5	4	4	8	8	9	8	9	8	8	8
B-4	4	4	4	3	2	2	1	2	7	7	8	7	6	6	7	6
B-5	5	3	3		4	2	3		7	6	6		5	6	6	
B-6	7	5	6		1	1	1		6	6	6		5	5	5	
B-7	3	2	2	2	1	–	1	1	2	4	4	4	2	2	3	2
B-8	5	5	4	6	2	2	2	2	7	8	8	7	8	8	9	8
p-12	1	–	1		–	–	1		2	2	2		2	2	2	
p-13	–	x	x		1	1	1		1	2	2		1	1	1	
p-14	–	–	–		–	x	x		1	2	x		1	2	1	
p-15	–	–	x		x	–	x		1	x	2		1	1	1	
p-16	1	1	2		–	–	–		1	2	3		2	1	1	
p-17	x	x	x	x	–	–	–	–	x	1	1	1	x	x	x	x
p-18	1	1	x	x	2	–	–	x	4	x	1	3	2	1	2	2
p-28	–	–	–	–	–	x	x	–	1	x	x	1	x	–	x	–
p-22	–	–	–	x	–	–	–	–	–	x	x	–	–	–	x	x
p-23	–	–	–	x	–	–	–	–	1	x	–	–	–	–	x	x
p-20	2	–	x	x	–	–	–	–	x	2	3	1	1	1	2	1
p-24	x	–	–	–	x	x	x	x	1	1	1	1	2	1	1	1
PV-2	1	1	3	2	4	4	3	4	7	6	7	7	6	5	5	5
PV-4	7	7			2	1			8	8			7	7		
PV-5	4				2				2				4			
PV-6	6				–				9				6			
PV-7	5	5			2	3			–	2			5	5		
PV-9	1	1			2	2			2	2			3	2		
PV-13	x	1	1		1	1	1		3	3	4		1	3	2	
pp-2	1	1	x		2	1	x		2	2	2		1	1	1	
pp-4	x	2			x	x			2	1			1	1		
pp-6	1				–				2				2			
pp-7	x	x			–	–			–	x			x	x		
pp-9	–	–			x	–			x	x			–	x		

Legend: B –wells on the Obrenovac spring
p –piezometers on the Obrenovac spring
PV–wells on the Mladenovac spring
pp–piezometers on the Mladenovac spring
x –index of bacterioflora occurrence and ochre sedimentation bellow 1

203

It is evident from Table 2 that iron and manganese bacteria are present in all investigated wells and piezometers of the Obrenovac spring. Occurrence of bacteria ranged between 1 and 9. On the average, bacteria numbers and ochre deposition are five times greater in wells than in their piezometers. It has to be emphasized that wells number 3 and 6 were chemicaly-mechanically regenerated two years before this investigation. In every well of the Obrenovac spring at least two bacterial genera are identified, while in their piezometers only the genus Siderocapsa is present (hardly noticeable, with index 0-1). A great amount of deposited ochre is associated with the presence of the genera Leptothrix and Gallionella . Piezometers immediately beside the wells (marked with number 12 to 18) show the same bacteria occurrence and sedimentation of ochre as in their well, but with several times smaller indices. Piezometer boreholes immediately beside the Sava river, by the bank of the river or several meters away (marked 22,23 and 28) are essentially different in bacterial composition and ochre deposition from piezometers located deeply in the spring or by the well itself.

At the Mladenovac spring, seven wells and five piezometers were investigated: in all cases the presence of iron bacteria and ochre deposition was registered and without exception, three genera (Siderocapsa, Leptothrix and Gallionella) were determined. More or less unchanged state of bacteria and ochre depositing in wells for a certain period of time during normal exploitation (Table 2), lead to the conclusion that every well or piezometer has a characteristic index of deposition. Also every change that happens in the spring ecosystem causes changes to the characteristic index of deposition. One of the factors, that cause qualitative and quantitative modification of the iron bacteria population and the change of dynamics of ochre deposition, is the regime of the groundwater movement (velocity) within the zone of wells and piezometers. This is established by: 1) comparing the state of bacteria and ochre deposition in a well during its normal exploitation and that after the pumping has stopped; 2) comparing the state in old and in new wells; 3) comparing the state in old and in new piezometers; 4) comparing the state in the well with that in the piezometers placed by the river.

The qualitative and quantitative composition of iron and manganese bacteria populations and the dynamics of ochre deposition which depend on the chemical composition of the groundwater, were monitored in one well and in a large number of piezometers located in a straight line, perpendicular to the Sava river (Fig. 2A). The results of these investigations are shown in Table 3. As can be seen from the table the most intense ochre deposition and occurrence of bacteria was in the well itself. This apparent difference in the qualitative composition bacteria and ochre deposit in piezometers is explained as due to changes

in the water quality, its mineralization and enriching in
iron on the way from the river to the well. The water from
the piezometers placed by the river (pa,pb) has a similar
quality to that of the river water rather then the well
water (pH, Eh, total salts), while the water from the
piezometer next the well (pd) has similar quality to the
well water (Fig. 2A).

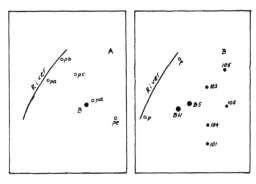

Fig.2.   The scheme of arrangement of wells and
          piezometers upon the Obrenovac spring
          A-Well and piezometers towards Sava river course
          B-The arrangement of new and old piezometers

Table 3. The effect of river interference on the occurrence
          of bacteria activity and ochre deposition in the
          well and in the piezometers

| Object | sampling | | | |
mark	1	2	3	4
B	6	6	7	6
pa	x	x	x	x
pb	x	x	x	x
pc	1	x	1	1
pd	2	3	2	2
pe	1	1	1	1

Since the wells were of an aproximatelly same age,it was
not possible to monitor the well yield decrease in relation
to the well age. We therefore used an old well ready to be
stopped, and a new well located right next to the old one,
including the newly installed piezometers. The location of
these piezometers is presented in Fig. 2B, and the obtained
results in Table 4. These results point out that the new
well (BN) and piezometers (pn) still do not show the
characteristic values of bacteria occurrence and the steady
dynamics of ochre deposition. The same was confirmed in our

previous investigations of the Belgrade spring, where quite
new wells and piezometers were available. Only in the third
placement of the new piezometers, eight months later, was
iron bacteria presence measurable, with the occurrence
index bellow one. In old piezometers, the ochre deposition
and bacterial occurrence almost immediately reached characteristic values. The comparison of old and new wells is
somewhat different. These comparative investigations of the
old and the new well could not be completed as one year
later the old well was completely closed.

Table 4. Comparative survey of ochre and bacterioflora
         sedimentation on simulators in old and in recently
         built piezometers and in the new and the old well

Object mark	after 2 month	after 4 month	after 6 month	after 8 month
pn 101	−	−	x	1
pn 102	−	−	1	2
pn 103	−	−	x	x
ps 104	2	2	2	2
ps 105	2	3	2	3
BN	4	5	5	7
BS	7	7	4	4

The obtained results show that it requires a certain
period of time to establish conditions in the systems which
are optimal for the growth of iron and manganese bacteria
and for ochre depositon. In our case, this period is from
4 to 8 months (Table 4). The change of conditions in the
spring brings about more intense biochemical reactions, and
aerobic reactions replace of anaerobic ones.
    That changes in qualitative and quantitative composition
of iron and manganese bacterioflora population and ochre
deposition, are associated with the change of the water
quality caused by the slowed water movement in the direction spring − well, was confirmed on wells which were out
of operation, either temporarilly or permanently. As can be
noticed from Table 5, an ochre depositing and bacteria
occurrence is expressed with an index value of 6-9 during
the normal well exploitation ("characteristic value"). One
month after the wells were stopped, this index was significantly lower and two months later it came down to the value
of four, where it remained. After the wells were set in
operation again (Table 5, wells B-A and B-B), the previous
state of bacterial occurrance and ochre deposition was
established again. In case of the well B-C, the decrease in
value of the deposition index was recorded after 8 months.
As the well was closed, we could no longer observe it.
These investigations also show that the velocity of
groundwater movement in the spring (that increases with

the   pumping of water) is one of the decisive factors   that
directly   influence the decrease of the specific   yield   of
the well, through the changes of other ecological factors.

Table 5. Bacteriflora and ochre deposition changes in
the well after stopping the exploitation

Well mark	Deposition index								
	Before stopping	After stopping (months)					After restarting		
		1	2	4	6	8	1	2	4
B—A	7	6	4	4	4		5	5	6
B—B	6	6	5	4	4	4	4	6	6
B—C	9	8	8	6	6	5			

The   change   of   ecological   factors   in   the   well   and
broader,   in the spring,   i.e.   the change of a qualitative
and   quantitative   composition   of bacteria,   can   also   be
caused   by   various   substances used   during   the   chemical
restoration   of   the well.   The intensity and   duration   of
their   effect would depend on the type of substances   used,
chemical concentration in the moment of injection into   the
well,   contact time with the environment   and more or   less
intense   penetration into surrounding area.   All   of   these
substances   used   for   restoration,   regardless   of   their
chemical   composition   (strong   acids,   strong   oxidizing
chemical),   act in two ways. First,   the components of the
substance have a bactericidal effect;   the bactericidal and
lethal   effect of the sum of components is   also   expressed
according to a certain rules,   Barbic (1977).   Second,   the
chemical substance and it's components inhibit the   current
biochemical reactions, change their course and, most often,
cause complex changes of the ecological system. The desired
consequences   of   prevention of bacteria growth   and   ochre
deposition can be prolonged and maintained by a   programmed
chemical restoration.

4 Conclusions

Obtained results show that the aquifer ought to be regarded
as   a   specific   ecological   system   with   all   its
characteristics. In case of the potable water body, whether
exploited   or not,   the general laws   valid for   the   other
ecosystem types are applicable here,   too. The aquifer is a
functional unity where a continuous exchange of energy   and
matter takes place.   It is also a dynamic unity, based upon
mutual interactions of living and nonliving components. The

specifity of the water body, considered as an ecological
system, is reflected above all, in the relative simplicity
and poverty of its community composition. First of all, an
energy is produced in the system, the effect of ecological
factors is limited to the chemistry of the environment and
interspecific relationships are negligible. The basic population atributes are fully present, although rather simplified and specific. Stankovic (1968). The study and the
interpretation of aquifer as a system where the laws of
ecology reign and the effects of ecological factors,
i.e.ecological valences, are considered enable it to be
better and more comprehensively understood therefore making
it possible to take the necessary actions to prevent yield
decrease and to obtain more quantities of water of better
quality. Starting with the fact that iron and manganese
bacteria population is among the major causes of yield
decrease (one that is also associated with clogging of
drainage galeries and drains themselves) and knowing their
eco-physiology and the ecology of the aquifer, the possibilities of prevention of their effects are increasing. For
this purpose, it would be sufficient to change some of the
ecological factors or to stop biochemical reaction in the
aquifer.

4 References

Barbic, F. (1977) Effects of different compounds of metals
    and their mixtures on the growth and survival of Thio-
    bacillus ferrooxidans. Zeitschrift f. Allg. Micribiolo-
    gie, 17(4), 227-281.
Barbic, F. Savic, I. Krajcic. O. Vukovic-Pal, M. Babic. M.
    (1987) Benefits of upholding of iron bacteria for trans-
    formation of iron and manganese in groundwater. Mikro-
    biologija, 24(2), 129-138, Belgrade.
Bracilovic, D. Barbic, F. Djindjic. M. (1975) Chlorination
    for iron removal in well water. Water and Sewage Works,
    1, 40-42.
Hässelbarth, U. Lüdemann, D. (1967) Die biologische veroc-
    kerung von brunnen durch massenentwicklung of eisen und
    manganbacterien. Bohrtechik-Brunnenbau-Rohrleitungsbau,
    10, 363-368 und 401-406.
Kuznetsov, I.S. Ivanov, V.M. Lyalikova, N.N. (1962) Intro-
    duction to Geological Microbiology. Acad.Sci., USSR.
Staley, J.T. (1989) Manual of Systematic Bacteriology.
    Williams and Wilkins, Baltimore.
Standard Methods for the Physical, Chemical and Bacteriolo-
    gical Examination of Water (1961) Fed. Inst. for Prof.
    Health, Belgrade.
Stankovic.S. (1968) Animal Ecology. University of Belgrade,
    Belgrade.

17 THEORETICAL EVLUATION OF PRODUCTION LOSSES FROM A WATER WELL WHICH INVOLVES BIOFOULING

D.R. CULLIMORE
Regina Water Research Institute, University of Regina, Canada

Abstract
Biofouling in a water well is difficult to recognize since all of the evidence is indirect. Biofouling involves a series of events involving primary colonization, biofilm formation, interlocking occlusion (which restricts flow) and stratified biofilm structures generate a corrosivity potential due to the production of hydrogen sulphide and organic acids. Early diagnosis of biofouling involves a direct determination of the production capacity of the well, upward shifts in the suspended particulate mass in the water and increase in the incumbent planktonic bacterial population. By an ongoing monitoring of post-diluvial water being pumped from a well which had been quiescent for a period of time, the level of biofouling can be projected based upon the sheared material observed in the subsequent pumped water. A series of zones have been projected to occur around a water well where different form of fouling could occur. These include aerobic occlusive and anaerobic corrosive. Various environmental factors which can influence the rate of biofouling are discussed.
Keywords: Water Well, Plugging, Corrosion, Iron Related Bacteria, Particulates, Biofouling, Bioaccumulation.

1 Introduction

The life expectancy of a water well often cannot be projected because as the well is developed and put on line to produce water various factors can influence the ability of the well to continue to produce a consistent quality of water in constant quantities (Walton, 1987). These production losses can be related on occasion to mechanical failure, the loss in the ability of the aquifer to deliver an adequate supply of water to the well, but on occasion, the water well can be recognized to be suffering from a biofouling event.

2 Definition of biofouling

Biofouling refers to events in which a biological component will cause a system or process to fail to achieve its expected performance criteria. Such biofouling can be related to a direct impediment of

the process or a biologically driven degeneration of one or more
mechanical or chemical events considered essential to that process.
Classically, biofouling has been linked to the many corrosive
processes. The most common examples of this are the activities of
sulphate-reducing bacteria generating hydrogen sulphide which in turn
leads to electro-chemical corrosivity of steels and concretes. Less
acknowledged is the role of micro-organisms in the direct physical
impediment of a process by the generation of a biological mass which
attachs to surface areas and changes various production and exchange
capacities within the process. Such a biofouling involves the attachment of the micro-organisms to the surface areas where colonization
occurs with the subsequent formation of a biofilm (Lynch and Hobbie,
1979; Costerton and Lappin-Scott, 1989) which may gradually thicken to
form a copious polymeric rich mass commonly referred to as slimes.
These slimes may grow to such an extent sufficient to totally occlude
hydraulic flows through such porous materials as gravel pack around a
well screen and so prevent water entering the well (i.e., plugging).

3 Methodologies to observe biofouling

There is inadequate field data to determine the percentage of water
wells which do suffer from biological impairment to flows and lead to
total plugging events. Indeed, the methodologies to predict these
occlusive phenomena is only now being developed. A sequence of events
has been associated with the biofouling of a water well, but clearly
one major component is the generation of an active biofilm within some
region of the well which can subsequently cause a bioimpairment to
flow, a degeneration in water quality and, corrosion and plugging.
During the formation of such a biofilm, a number of events can be
recorded in the product water from the well. Critical amongst these
changes is the concentration of particulates within the water which
may be associated with the biofouling event itself.
    As the biofilm grows within a groundwater system, it will entrap
iron and manganese within the polymeric structures to give different
colours to the biofilm. If the biofilm were to shear including this
pigmented material, then the sheared biofilm will add pigmented particulates (pp) to the water. The common colour for these pp range
from orange, red, brown through to black. Achromogenic particulates
(non-pigmented, ap) may also shear from time to time from the biofilm
during the natural processes of maturation. Such sheared particles
(sp) can be recovered from product water in amounts which will vary
with the degree of shearing which is occurring from the biofilm into
the water at that particular time. In addition, water will normally
carry a background of particulate material (bp) in relatively low
numbers. It can therefore be extrapolated that there are three major
biological components which could be present in the water as pigmented
particulates (pp), achromogenic particulates (ap) which would form a
variable part of the shearing particulates (sp), and background
particulates (bp).

4 Pump testing to determine biofouling

When a well has been kept quiescent for a period of time (e.g., seven days) and is then activated, the pumping process will magnify any biofouling event which has been occurring. The types of particulates which will be released from the well during the pumping will reflect the shearing of pp and ap to form the sp particulates which would over-layer the background particulates (i.e. bp) which naturally occur in the water on all occasions.

There are three major events that have been observed when wells are pumped for the first time after a period of quiescence. In the first phase, where biofouling has occurred, all three types of particulates can be observed. The water may have a distinct colour resulting from the presence of the pp group and a cloudiness as a result of the ap group. The pigmentation in the pp group can originate from intense aerobic activity often associated with the well water column and the gravel pack around the well screen being exposed to significant levels of oxygen. Oxidative processes particularly affect iron and manganese to shift to the oxidative (ferric and manganic) states from the reduced (ferrous and manganous) states respectively. Consequently, the pp around the well may be associated with these aerobic activities occurring closest to the well screen so that upon the initiation of pumping, any sheared pigmented particles will tend to enter the water during the earlier stages to become recordable in the product water. Non-pigmented biofilm may occur deeper out into the groundwater system beyond the direct influence of any oxygen associable with the well or in unsaturated zones above the aquifer. Here, there would not be the bioentrapment of the iron and manganous salts and the slimes and the biofilm may therefore tend to be achromogenic. As pumping is initiated, the greatest turbulence would occur closest to the well itself which would cause massive shearings at this point with gradually reducing effects as the distance the influence of pumping increases. It may therefore be expect that pp would be observed in the product water for extended periods of time reflecting the ongoing shearing of the biofilm further out from the well itself. These shearing effects may become more infrequent as the patterns of hydraulic flow stabilize throughout the wells sphere of influence. In general, shearing will cause increases in the total suspended solids which may range from 0.5 to >30 ppm TSS.

Once the pumping has been extended sufficiently to stabilize the biofilm, the product water will contain a background of particulates which will usually be less than 0.25 ppm TSS.

5 Determination of Particulate in Water

In the determination of a biofouling event, it is important to be able to differentiate and quantify the sizes of the pp, ap, sp and bp factions within the water. The pigmented particulates can be measured spectrophotometrically since the ferric and manganic salts will absorb light. Additionally, the water can have distinct colorations from yellow through orange to brown where very high concentrations of pp are observed. A simpler method to measure the pp is to entrap the particulates on a membrane filter (MF) with a porosity of 0.45 microns. Here, the pigmented particulates will be entrapped on the surface of the MF which can be air dried (to remove the obstructive water film) and the pigments recorded by inverse reflectivity using a reflectant spectrophotometer. Light is shone down onto the surface and the inverse of the reflected light represents the absorbed light entrapped by the pp. The sp and bp particulates can be crudely measured using turbidometry or more precisely using a lazer driven particle counter. These particle counters are able to size each in- dividual particle scanned to determined composite volumes and numbers. From this data, the total volume of the suspended particles can be computed and registered as total suspended solids (TSS). During the phase of pumping when the water is now dominated by a vacillating load of sp, the TSS values observed will fluctuate simultaneously. When the particulate loading stabilizes (usually at less than 0.25 ppm TSS) only the bp are being observed.

6 Biological activity in biofouling

As water is drawn into the well from further out in the groundwater systems, the particulates recovered represent fractions from the incumbent biofilms overlayered with contaminations of particulates from closer to the well screen. Confirmation of a biofilm event therefore requires that the particles recovered need to be shown to have a biological incumbency rather than a simple inorganic or organic content. Confirmation of the biological incumbency requires some level of biological testing at one of three potential levels: (1) presence and absence (p/a) to confirm whether organisms are indeed present in the particulate matrices, (2) at a semi-quantitative level (s/q) to determine the population of micro-organisms occupying the particulate volume observed, or (3) fully quantitative.

A number of cylindrical zones of biological activity can be postu- lated to occur around a water well, each of which would contain a different and distinct form of biological activity. These zones include: (1) the water column itself, (ii) the gravel pack or media immediately surrounding the well screen, (iii) a zone where there may be some extended but casual oxygen intrusion and (iv) the saturated aquifer forming the source of the groundwater being drawn into the well by the effect of the pumping.

In zone i formed by the water column itself, there are some unique characteristics which will affect the microbial activity. These include a dominance of liquid water with a low surface area to volume ratio (for example, less than 0.1:1, Cm^2:ml). During the passive phase, a considerable stratification is likely to occur amongst the planktonic micro-organisms present within the column due to a sedimen- tation of some of these viable particles towards the bottom of the well. The shearing potential from the limited biofilm occurring on the surface areas would be less likely to impart high levels of pp, ap and sp into the water during the initial pumping than resuspension when the turbulence would be very high. If the well is not sealed to prevent oxygen entering, there would be significant diffusivity of oxygen into this zone which would stimulate considerable levels of aerobic activity.

Zone ii formed by the well screen and the material immediately packed beyond the screen forms a very different eco-niche to zone i. Here, the surface area to volume ratio is very high (>5:1) with 20 to 50% of the total volume being interstitial and vulnerable to hydraulic flow. There would be a high potential for oxygen intrusion, both from the well water column and from the movement of water down the outside of the casing during pumping. The limited variety of types of surface area and the relative availability of oxygen would suggest a high likelihood that any biofilm formation would be aerobic and often dominated by the iron-related bacteria. At the initiation of pumping, the turbulence in this zone would be very considerable leading to a high risk of shearing with the input of pp and ap as sp into the water. Much of the biological activity associated with water wells is thought to occur in this zone.

Zone iii forms a secondary eco-niche around the water well where the surface area to volume ratio would still be high allowing biological attachment and growth. There would be a reduced level of turbulence and the oxygen intrusion may be expected to be much more restricted. Much of the oxygen may be vertically migrating downward from the higher unsaturated zones than from the well itself. There may be a higher proportion of micro-organisms able to function both aerobically and anaerobically and hence able to continue to metabolize when this eco-niche enters a reductive phase (with the oxygen excluded) or oxidative phase. It can be therefore projected that there would be a decline in the size of the microbial population in this zone restricted by the available nutrients. The outer terminals of zone iii would be where the groundwater within the aquifer was now outside the direct and indirect influences of the well.

7 Microbiology of plugging

Plugging in a water well is caused by the biofilm occupying a sufficient volume within the interstitial spaces to cause a restriction in flow to water into the well proper (Mansuy and Cullimore, 1989). Such restrictions can function in three ways: (1) the biofilm formation within the interstitial spaces can directly restrict flow; (2) the shearing polymerics from the biofilm can cause an increase in resistance generated by the biofilm's surfaces to hydraulic flow that it reduces the velocity; (3) gas may be evolved or additional water accumulated and entrapped within the biofilm to cause radical increases in the volume of the biofilm and initiate occlusion of the interstitial spaces. Combinations of these three mechanisms capable of restricting hydraulic flow can cause the well to become partially or totally plugged.

In laboratory experiments, where the development of iron-related bacterial biofilms have been monitored, a number of phases have been observed. These phases reflect some of the potential activities which can occur within a biofouled water well. These can be categorized into six phases. Phase One, radical increase in volume often reaching 50% or more of the total interstitial volume. Parallel and equal increasing resistance to flow through the media; Phase Two, compression in the biofilm volume which can reach as much as a 95% reduction. In parallel, the resistance to flow becomes diminished and on some occasions, the flow may be facilitated to speeds greater than that achieved in the control (unbiofouled) media. The size of the incumbent biofilm is now relatively small but the impact upon hydraulic flows can become exaggerated; Phase Three, a gradual but slow increase in the volume of the incumbent biofilm. Elimination of any facilitative flow; Phase Four, gradual increase in biofilm volume and resistance to flow; Phase Five, continued increase in biofilm volume which fluctuates in harmony with changes in the resistance to flow; Phase Six, after the biofilm has reached 60 to 80% of the interstitial space, interconnection of the various biofilm masses cause a total plugging due to an extreme resistance to hydraulic flow.

These events can be summarized as (i) a radical primary volume expansion; (ii) primary compression; (iii) secondary volume expansion; (iv) vacillative pre-occlusive state; and (v) total occlusion. The length of time which each of these phases occupy cannot be projected. In the laboratory, total plugging can be achieved in times which can be varied from two days to two years depending upon the environmental conditions used.

8 Environmental factors influencing biofouling

8.1 Transitional biological activities

Plug formation is influenced by many physico-chemical factors (Cullimore, 1987). During the process of biofouling a saturated porous medium, there are transient biological events that can be observed. For example, the causal water (cw) entering the zone of influence of the biological events occurring around and within the water well will be subjected to the activities of the incumbent biological interface (b_i). Water being pumped from the well has been subjected to the influences of this b_i and can therefore be considered to be postdiluvial water (pw) and to have been amended in its quality by that effect. The concentration of a given affected compound in the postdiluvial water (i.e., pw_c) could therefore be different from the concentration found in the causal water (cw_c) as a result of the uptake and/or degradation of the compound in the biological interface ($bi_{u/d}$). The amount of this biological activity could be theoretically determined as:

$$bi_{ud} = cw_c - pw_c$$

There are a number of mechanisms involved in the biological interface leading to the removal of chemicals from the cw. The size of the $bi_{u/d}$ reflects a number of potential events which could occur within the interface zone. These events include a bioaccumulation of the chemical substance to within the polymeric matrices of the biofilm itself. Once absorbed, the chemical may become subject to passive entrapment or may become potentially assimilable through biosynthesis and/or biodegradation. Much of the iron and manganese so accumulated appears to be subjected to a passive entrapment while organic molecules are more likely to be subjected to a more complete assimilation through biodegradation or synthesis.

8.2 Bioaccumulation events

Iron and manganese are two major components in biofouling which have been subjected to considerable attention, partially because their presence causes slimes and the sheared particulates to generate obvious pigments which can be readily seen. From field experiments to date, it would appear that iron tends to accumulate in zones i and ii around a well whilst manganese which may be more mobile and assimilated less completely tends to be found in the eco-niches of zones i, ii, and iii.

The site of iron and manganese absorption within bioaccumulation in and around microbial cells is relatively defined. The site of accumulation can be: (1) extruded in an extra-cellular polymeric sheath as a ribbon from the cell (i.e, <u>Gallionella</u>); (2) accumulated inside a sheath (tube of slime) formed around group of microbial cells; (3)   accumulated on the outside of the sheath which forms a tube around a number of microbial cells; (4) accumulated randomly in and around the polymeric slime which encompasses a number of microbial cells; or (5) is assimilated into the microbial cells directly.

The postdiluvial waters may therefore contain some biological markers of a biofouling event, but in addition, the water can contain indicators of having passed through the "biological filter" formed by the biofouling event around the well screens. A potential method for the determination of the extent of a biological fouling around the well would be to monitor the ratio of sodium to potassium in the postdiluvial water and compare that to the ratios observed in the causal water obtained from sources known not to have been effected by a biofouling event. Sodium is biologically neutral and tends not to be assimilated by biological systems whereas potassium is biologically very active as a major nutrient. It can therefore be postulated that as causal water passes through the biologically active zone, potassium is likely to be accumulated within the biomass while the sodium would move passively through in solution. The sodium:potassium (Na:K) would therefore shift in favour of sodium in the dissolved phase as a result of the bioaccumulation of potassium. For example, if a causal water contained Na:K ratio of 2:1 and was found to have shifted to 10:1 in the postdiluvial water, this would indicate that there had been a preferential bioaccumulation of the potassium as opposed to sodium and that an active biological filtering process had been occurring.

8.3 Major controlling factors

Groundwater temperatures may also have an influence on the biofouling, however, the variance in groundwater temperature is at a much lower order of magnitude than for surface-waters. While surface-waters can experience very significant diurnal ranging in temperature (e.g., $>+1/-2^{\circ}C$ per 24 hour period) and radical seasonal variations particularly in the temperate zones, ground water may exhibit seasonal variations which may range over as little as a 0.5 to 1°C range. The type of microbial colonization likely to occur within a groundwater system is therefore likely to be dominated by micro-organisms that are able to grow efficiently at ambient temperatures without experiencing a diurnal fluctuation. The most likely cause of temperature shifts would be the act of pumping which would possibly move water towards the well from a different depth where the temperature gradient may cause an elevation of the water temperature arriving at the well.

The narrow band of operating temperatures occurring in ground water would allow the continuous support for either psychrotrophic (bacteria able to grow at below 15°C) or mesotrophic bacteria (able to grow within the temperature range of 15 to 45°C) through the year. The seasonal fluctuations so commonly reported for surface-waters would therefore not be so applicable to any microbial activities occurring in ground water. Organic materials arriving at a biofouled site around a water well may be in one of two forms. These are: (1) dissolved; and (2) suspended in a particulate mass moving through the water system. As the ground water moves closer to the zone of influence that the well is generating in the surrounding ground water, there may be some shifting of the dissolved organics into the mobile particulate phase as assimilation occurs. Once the organic material arrives in the biofouling zone of influence surrounding the well, a number of interactions will occur. These could relate to the absorption of the materials into the biofilm (i.e, bioaccumulation) or the degradation of the compounds from the aqueous phase (biodegradation).

Organic materials entering the biologically active zone resulting from the influences from the water well can be projected to enter the biological systems in a number of ways. This would include passive accumulation within the polymeric matrices of the biofilm, active accumulation within the viable cells and subsequent degradation and utilization for synthetic and energy-generating functions. As the water flows over the biofilm during these stages, it may be expected that some shearing of the biofilm may occur causing the releases of biofilm derived particulates to the water phase. Thus, the passage of organics over an active biofilm may be expected to include the removal of these organics by the biofilm and subsequent releases of amended and non-amended organics to the water in a dissolved or sheared particulate state. As the water approaches the well with an increasing level of turbulence, there can concurrently be a shifting in the reduction-oxidation state (redox) potential from a relatively reduced to a relatively oxidized state. The shifting in the redox potential will result in adjustments within the incumbent consortia in the biofilms. In the reduced state, the biofilm would tend to be dominated by the anaerobic organisms which do not necessarily require oxygen for growth. In oxidative conditions, at least the more exposed microbial components in the upper strata of the biofilm (nearer to the polymeric-free water interface) may be expected to be dominated by aerobes. In consequence, there would be a movement of organics into and out of the polymeric structures as the environmental conditions change. When the water arrives in the well water column through the screen, it could contain only some of the organic materials that were present in the causal water. These would appear in the dissolved form and as components within the sheared biofilm particulates. The level of organics therefore found in the postdiluvial water pumped from the well may be expected, in a well which has been subjected to significant biofouling, to be amended both in concentration and molecular forms with a mean reduction in the total concentration of organic carbon delivered.

8.4 Impact of shearing on microbial loadings

From laboratory experiments and field observations, it would appear that most of the microbial activities within a biofouling water well are, in fact, attached to the surfaces presented by the well screen, the gravel pack and the natural media occurring in the aquifer. Relatively little active microbial activity appears to occur in the planktonic (freely suspended) state. Most microbiological techniques for determining the occurrence of biofouling rely on an evaluation of the postdiluvial water for a determination of biological activity. However, if no shearing is occurring from the biofilm at the time of sampling, then it can be extrapolated that a much lower population of micro-organisms has been recovered. These "background" micro-organisms may be the result of planktonic activities within the well water column itself along with some resuspension of sedimented particulate masses from the base of the well.

Because the biofilm may shear only periodically to release sessile bacteria from the biofilm to the planktonic state, the absence of microbial indicators from the postdiluvial water does not necessarily indicate that the well is still biologically pristine and has not suffered from biofouling. The ideal time to draw samples from a well in order to more correctly evaluate biofouling would be after a prolonged period of quiescence when the well was not pumped. In these events with transient flow, being compatible to that of the groundwater movement itself, is less likely to cause a shearing in the biofilm due to the stabilized hydraulic conditions. During this period, therefore, the biofilm may grow steadily in the absence of erratic intense turbulences. When the well pumping is again initiated, the turbulence generated will cause a magnifying effect on shearing and greater particulate loadings may be expected to occur from a well suffering from a significant biofouling. A monitoring of the well over this period of exaggerated shearing can indicate not only whether biofouling is indeed occurring but also the extent to which the biofouling has infested a zone around the well.

9 Acknowledgement

The author wishes to acknowledge the financial support of the Natural Science and Engineering Research Council of Canada, Marina Mnushkin for her dedicated assistance in the development of the laboratory model wells and Natalie Ostryzniuk for the preparation of the manuscript.

10 References

Costerton, J.W. and Lappin-Scott, H.M. (1989) Behavior of bacteria in biofilms. ASM News 55 (12), 650-654.

Cullimore, D.R. (1987) Physico-chemical factors in influencing the biofouling of groundwater, in International Symposium on Biofouled Aquifer: Prevention and Restoration (ed D.R. Cullimore), American Water Resources Association, Bethesda, pp. 23-36.

Lynch, J.M. and Hobbie, J.E. (1979) Micro-organisms in Action: Concepts and Application in Microbial Ecology. Blackwell Scientific Publications, Oxford, U.K.

Mansuy, N and Cullimore, D.R. (1989) Treating the source, not just the problem. Can. Water Well, 15(3), 8-9.

Walton, W.C. (1987) Groundwater Pumping Tests. Lewis Publishers Inc., Michigan.

18 OCCURRENCE AND DERIVATION OF IRON-BINDING BACTERIA IN IRON-BEARING GROUNDWATER

K. OLANCZUK/NEYMAN
Technical University of Gdansk, Poland

Abstract
Iron-binding bacteria occurring in iron bearing groundwater of Quaternary formations are the subject of this work. The new definition "iron-binding bacteria" is proposed for very diversified names of this bacteria group ie. iron bacteria, iron oxidizing bacteria, iron depositing bacteria. The new definition stresses the role of bacteria in the iron cycle and includes the sum of two bacteria groups ie. the FE (II) bacteria connected with divalent iron and the Fe (III) bacteria connected with trivalent iron. A quantitative, microscopic method for divalent and trivalent iron binding bacteria is described. The biggest water intake in North of Poland called "Letniki" was the object of research. Physico-chemical and bacteriological water analyses, as well as hydrogeological determinations were carried out. The main statements arising from the obtained results are (1) the organic sediments covering the aquifer layer are the main source of iron-binding bacteria and their nutrients, (2) themoorganotrophic Siderocapsa is the dominant iron-binding bacteria genus and covers over 95% of total count of bacteria on membrane filters, (3) statistical analysis of results show good correlation (R=0.88) between iron-binding bacteria number and ammonium nitrogen concentration and oxygen consumption of the water. Continuous supply of nutrients to bacteria, which can adhere and detach to/from the sand grains in the aquifer, affords chemical possibilities for the development of these organisms.
Keywords: Groundwater, Iron-binding bacteria, Methods, Source, Nutritional requirements

1 Introduction

The quality of groundwater as a source for drinking purposes is controlled by over fifty physico-chemical and only four bacteriological analyses. The bacteriological

analyses are based on cultural methods and some water
quality impairing microorganisms, which are not able to
grow on standarized media, are not detected.   The iron-
binding bacteria group is an example of such organisms.
This group belongs to characteristic groundwater
organisms very often affecting the water quality and
causing biofouling of filters screen of water wells.   As
a rule they are not detected in a routine analysis.

This paper is intended to determine the origin and the
quantitative occurrence of iron-binding bacteria in the
Quaternary formations aquifer in comparison with the
hydrogeochemical conditions.

2 Experimental section

Complex Physico-chemical and bacteriological analyses and
hydrological observations of the largest underground
water intake in Gdansk region, known as "Letniki" have
been carried out.

2.1 Sampling area

The "Letniki" water intake exploits a Pleistocene and
Holocene aquiferous layer, which is made up of coarse,
medium and fine-grained sand.   The aquifer occurring
along the intake line has a total thickness of 20 to 30m.
The aquiferous layer of the main level is covered by
mixed clay.   A series of mixed clays of variable
thickness in the range of 6 to 16m are interbedded with
sand and peat formations (Fig.1).

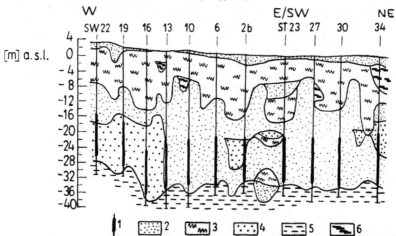

Fig. 1   Hydrological cross-section of "Letniki" water
intake 1-water well, 2-sands, 3-clays, 4-
sand-gravels, 5-loams, 6-peats

The underground water of the "Letniki" intake region are confined in character. The reason is a continuous occurrence of the mixed clay layer within the major floor of the aquiferous level. The initiation of the exploitation of the wells at the "Letniki" intake has resulted in variations of the natural supply conditions of the aquiferous layer. A reduction in pressure has initiated the filtering off processes of the alluvial deposit water which led to disadvantageous qualitative changes of the abstracted water.

The groundwater intake consists of 33 wells bored down to the depths from 32 to 41m, arranged along the 6km length of the Nogat River. The distance between respective wells is from 100 to 150m, and the mean distance of the wells from the riverbed is 200m. The age of the wells varies but is usually from 5 to 20 years.

2.2 Materials and investigation methods

Underground water samples were collected directly from wells at their intake. A sterile hose was fastened to the valve at the head of the well. The other end of the hose was put to the bottom of a bottle. Through the hose the water was flowing slowly to appropriate glass bottles which were corked with ground-in stoppers. Such a procedure made it possible to collect water samples representing natural composition and limited the risk of water aeration.

The physico-chemical analyses of the water were carried out in compliance with the Polish Standard requirements. However, while determining the permanganate value, a slight modification was invoked ie. the determination following immediately after the oxidation of iron compounds.

The samples designed for bacteriological testing were fixed at once with formic acid. The iron-binding bacteria were detected by way of direct counting on the membrane filters of Synpor 7 type. To stain the iron-binding bacteria a new technique was applied based on the paper of Meyers (1958), Rodina (1965) and Spandowska et al. (1979). The iron-binding bacteria placed on membrane filters were stained by a contrast method: ferrous and ferric compounds surrounding the cells - blue (by ferricyanide, or ferrocyanide) while the bacteria cells - pink (by safranines). To stain the iron-binding bacteria two parts of the filter were cut out which were independently coloured: one with respect to the presence of iron-binding bacteria FE (II) and the other Fe (III) (Olanczuk-Neyman, 1989). As a result of the staining procedure the Fe (II) compounds surrounding the bacteria cells in reaction with the ferricyanide produced a dark-blue deposit - the Turnbull's blue, while in reaction of

Fe (III) with potassium ferrocyanide the deposit was blue
- Prussian blue.

The microscopic observations were conducted with a
magnification of 1500 times, and the bacteria cells
surrounded by a blue deposit of iron compounds were
counted in one hundred vision fields. The final result
of determination, defined as the iron-binding bacteria
number, is the sum of the number of the iron-binding
bacteria Fe (II) and Fe (III) contained in one cm^3 of
water.

3 Results

The physico-chemical properties of the groundwater from
33 wells at the intake have been determined on the basis
of the investigation results performed between the period
from February 1987 to October 1988. Simultaneously
investigations were conducted on the iron-binding
bacteria. Calculations were made of the mean
concentration values of the tested water components
obtained from 10-15 measurements.

3.1 Physico-chemical properties of water
The water is of calcium bicarbonate type of total
hardness between 5.91 and 7.85 mval/l and low non
carbonate hardness. The general mineralization does not
exceed 500 mg/l. The content of dissolved oxygen in the
water ranges from 0.00 to 1.20 mgO_2/l. Concentration of
total iron is within a wide range of 3.91 to 10.73 mg/l,
with a domination of divalent iron (72-97% of total Fe).
The content of iron in the water significantly exceeds
the standard requirements relating to drinking water
(from 8 to 21.5 times). The concentration of manganese
ranges from 0.88 to 1.39 mg/l and is also higher (from 8
to 14 times) than the permissible level. In relatively
low concentrations are the oxidixed mineral compounds:
nitrates 0.04 mgN_{NO3}/l, nitrites 0.001-0.006 mgN_{No2}/l,
and sulphates 7.10 - 42.8 mg/l. However the content of
ammonium nitrogen is definitely high and ranges from 0.80
to 2.72 mgN_{NH4}/l. The oxygen consumption ranges between
4.44 and 7.85 mgO_2/l whereas the organic carbon contents
varies from 3.6 to 7.8 mg/l. The colour of the water
depends on a higher concentration of iron and manganese
and the presence of the organic substance decomposition
products not exceeding 50 mg Pt/l. The water turbidity
in some wells is larger than 50 mg/l.

3.2 Type and number of the iron-binding bacteria at the
background of the hydrogeochemical conditions
The most abundant group of organisms represent the iron-
binding bacteria. They include 95 to 97% of the total

number of bacteria on the membrane filters (FM). Among them bacteria of the Siderocapsa type constitute the majority. Other types of bacteria, as for instance, Leptothrix and Gallionella seldom occurred. The mean values of the bacteria number of the Siderocapsa type in water at the intake range from 0.20×10^6 to 1.57×10^6 cells in 1 ml of water. The bacteria of the Siderocapsa type are characterised by the property of liberating iron from ferro-organic compounds (Lundgren and Dean, 1979). This phenomenen takes place in the zone of mixed clays, and the filtrated water from this area carries with it the decomposition products to the aquiferous layer and together with the dissolved and suspended chemical susbtances are transferred bacteria.

Thus, the contents of bacteria in the aquiferous layer to a great extent depends on the composition and volume of water which is filtered from the alluvial deposit zone. Consequently the effect of the filtered water from the alluvial deposit zone upon the quality of water taken from a well depends, among other things, on the hydrogeochemical conditions, location of the filter, and the technique of exploiting the well.

Taking into account the thickness of the alluvial deposits and their variable influence upon the groundwater quality, three basic types of hydrogeochemical conditons (Wijura, 1983) were distinguished:
-   in the first type the aquiferous layer is covered by alluvial deposits of thickness amounting to approximately 10m and more, and in the upper part of the covering layer minor sand and peat interbeddings occur,
-   in the second type, the total thickness of the alluvial deposits with the sand and peat interbeddings as well as the alluvial sand deposits which covers the aquiferous layer is less than 10m,
-   in the third type, the alluvial deposits with sand and peat inserts covering the aquiferous layer have thickness up to 5m.
It has been proved that the mean quantities of the iron-binding bacteria in the water of wells situated in various hydrogeochemical conditons are included in the following ranges:
from 0.31×10^6 to 1.57×10^6 cells/ml - in wells of the first type,
from 0.69×10^6 to 1.26×10^6 cells/ml - in wells of the second type,
from 0.20×10^6 to 0.83×10^6 cells/ml - in wells of the third type.
The least quantitative variations of bacteria in water are observed in the group of wells of the third type, whereas the largest ones - in water from a well of the

first type. The effect of the distance between the upper
edge of the well screen and the floor of the mud series
upon the bacteria number was analysed. From this it was
possible to conclude that the number of bacteria in the
water of various wells decreases with distance (d) of the
filter from the floor of the mud series formations. A
statistical analysis of the mean investigation results
has proved that there exist significant correlations
between the two variables for wells of the first and the
third type of hydrogeochemical conditions. The
coefficient of simple correlation for n=27 wells of the
first and the third type is r = -0.73 (the critical value
of the correlation coefficient for significance level
0.001 amounts to 0.597). (Fig.2).

The analysed values of the bacteria number are
included in the following ranges:
- 0.20×10^6 - 0.60×10^6 cells/ml of water at filtration
  depth (d) between 12.8 - 19.5m
- 0.44×10^6 - 1.57×10^6 cells/ml of water at filtration
  depth (d) between 2.8 - 10.3m

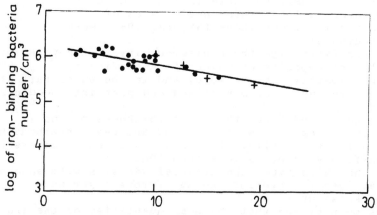

distance (d) between the well screen and the floor
of mud. [m]

Fig.2   The effect of distance between the well
         screen and the base of the mud on iron-
         binding bacteria number (r = -0.7282, p =
         0.001)

3.3 Chemical composition of water and the amount of iron-binding bacteria

Investigations were carried out which were aimed at the determination of the quantity of the iron-binding bacteria in view of the physico-chemical properties of the water. In order to avoid the effect of a non-uniform exploitation of the well upon the water quality, the samples were taken during a continuous, at least a 24-hour, operation of the well. The investigation of water covered an area of 29 different wells. Regression relationships were determined between the logarithm of the iron-binding bacteria number and the ammonia nitrogen concentration and oxygen consumption. The dependences were as follows:

$$y = 7.2453 + 0.6974 \ x_1 \tag{1}$$

$$y = 6.6508 + 0.3564 \ x_2 \tag{2}$$

where: y - logarithm of the number of the iron-binding bacteria in one l of water
           x_1 - concentration of ammonia nitrogen in mgN_{NH4}/l
           x_2 - oxygen consumption in $mg \ O_2/l$

The calculated simple correlation coefficients are: $r_{y,x1} = 0.84$ and $r_{y,x2} = 0.58$ for n=29 measurements and are significant at level p = 0.001
     The relationship of the logarithm of the iron-binding bacteria number and the concentration of both the mentioned water components is expressed by equation (3).

$$y = 3.4761 + 2.4601 \ x_1 + 0.6859 \ x_2 - 0.31257 x_1 x_2 \tag{3}$$

The multiple correlation coefficient with respect to the above relationship amounts to R = 0.88 where the error of the regression model is equal to $B_{mr} = 22.07\%$.
To make sure that the oxygen consumption values in the investigated groundwater reflected the organic carbon contents, comparable analyses were made of both the water indices which involved an analysis of the water samples from 20 water wells. The relationship of the organic carbon concentration and the oxygen consumption ($KMnO_4$) is described by the straight line equation (4)

$$y = 2.1658 + 1.2883 \ x \tag{4}$$

where: y - concentration C $_{org}$ in $mg/dm^{x=3}$
        x - oxygen consumption in mgO_2/l
($r_{y,x} = 0.8609$, p= 0.001, n = 20)

Thus, in the range of both the components concentrations which occur in the investigated water, the oxygen consumption is a good indicator of the C $_{org}$

concentration. Variations between the results of the indication of the oxygen consumption and the organic carbon for respective water samples do not exceed 1 mg/l and are included within the range of 0.65 to 1.1 mg/l given by Alekin (1970).

The above determined regression relationships (1), (2) and (3) shows that an interdependence exists in the investigated water between the contents of organic carbon compounds, as well as the ammonia nitrogen and the number of the iron-binding bacteria. It follows that under the existing environmental conditions at the intake an increase of the above water components concentration is accompanied by a quantitative growth of the iron-binding bacteria.

4 Conclusions

The dominant bacterial flora of the investigated groundwater is made up by the iron-binding chemoorganotrophic bacteria type Siderocapsa. Their presence may have caused biofouling in the aquiferous layer around the well, and the fouling of the well filters. It is open to question and requires further investigation of the groundwater from newly bored wells.

The numbers of iron-binding bacteria found in these studies are similar to those quoted by other authors (Geletin et al. 1981, Lueschow and MacKenthun, 1962, Cullimore and McCann, 1978).

The main source of the iron-binding bacteria of Siderocapsa type in the aquiferous layer is the water that is filtered through the organic deposits. These deposits can be assumed as natural rich medium for the bacteria. The bacteria which develop in them, together with dissolved and suspended in water products of biochemical processes are washed out into the groundwater that occurs below the organic formations covering the aquiferous layer. Filtration of the water to the aquiferous causes deterioration of its quality. This is revealed mainly by an increase of concentration of such components as: total iron, oxygen consumption and increase of number of iron-binding bacteria. This phenomenon occurs at the investigated intake due to transgression of the natural water system resulting from an extensive exploitation, and is intensified during non-uniform consumption which causes vertical movement of the water level.

The dislocation of the bacteria from shallow groundwater to the aquiferous layer is proved by the investigation results of the groundwater at various depths, beginning with the mud series floor. The studies

have indicated that the number of bacteria decreases
depending on that distance.

The growth capability of the iron-binding bacteria in
groundwater is determined both by the chemical and
physical conditions of the environment. The peculiarity
of the environment that is the underground water,
possesses a definite effect on the supply of the bacteria
with nutrient susbtances, and thus also upon their
number.

The groundwater is relatively poor in nutrient
susbtances for the bacteria(mainly organic substances)
and in spite of this it creates an environment for the
bacteria to develop. An ease of the bacteria that are in
water to attach and also to leave stable surfaces, eg.
grains of the aquiferous layer makes it possible to
continuous supply the organisms with substances occurring
in small concentrations. It may be assumed that in the
analysed groundwater the organic substance and the
ammonia nitrogen are a source of two essential biogenic
elements for the iron-binding bacteria. This statement
finds a partial proof in work of Starkey (1945) according
to whom ammonia nitrogen is an indispensable source of
nitrogen for these bacteria. Other authors, like Sepanen
(1987) and Beger (1966) point out an insignificant demand
of these bacteria for organic substances. The
investigation results presented in the paper are only
confined to an assessment of the impact of the chemical
components of water upon the quantity of bacteria.

The paper does not include the investigation results
related to the oxidation-reduction capacity (rH_2) on the
development of bacteria and their precipitation of iron
from the water.

5 References

Alekin,D.A. (1979). Osnowy gidrochimi.
    Gidrometeorologiczeskije Izd. Leningrad
Beger, H. (1966). Leitfader der Trink- und
    Brauchwasserbiologie. **Gustav Fischer Verlag,
    Stuttgart**, 360p.
Cullimore, D.R. and McCann, A.E. (1978). The
    identification, cultivation and control of iron
    bacteria in groundwater in Aquatic Microbiology (eds.
    F.A. Skinner and J.M. Shewan), **Academic Press**, pp.219-
    260.
Geletin, Yu., Gorianowa, G.S., Korobeinikowa, L.I.and
    Rusanova, H.A. (1981). Razwitije zelezobacterij w
    sistemach kommunalnogo wodosnabzenija s podziemnymi
    wodoistocznikami. **Gig. i Sanit.**,1,88-90.

Lueschow, L.A. and Mackenthun, K.M. (1962). Detection and enumeration of iron bacteria in municipal water supplies. **J.Am.Water Works Assoc.**,54,751-756.
Lundgren, D.G. and Dean, W. (1979). Biochemistry of iron. In Biochemical cycling of mineral forming elements. **Studies in Environmental Science 3** (eds. P.A. Trudinger and D.J. Swaine) Amsterdam, pp.211-251.
Meyers, G.E. (1958). Staining iron bacteria. **Staining Techniques 33**, 283-285.
Olanczuk-Neyman, K. (1989). Iron-binding bacteria in the groundwater at the water purification plant. Research of hydraulic engineering, in **Proceedings of the 3rd Polish-Yugoslav Symposium**, held in Gdansk, September 18-20, 1989, Gradjevinski Inst., Zagreb, pp.103-115
Rodina, A.G. (1965). Metody wodnoj mikrobiologii. **Mauka, Moskwa**, pp.353-354.
Spandowska, S. Danielak, K. Ziemkowski, A. (1979). Metodyka bakteriologicznego badania wod podziemnych i gruntow. **Wydawnictwa Geologiczne**, Warszawa, pp.148-151.
Sepänen, H. (1987). Biological treatment of groundwater in basins with floating filters. II. The role of microbes in floating filters. **International Symposium on Groundwater Microbiology: problems and biological treatment**. Kuopio (Finland) 4-6 August 1987.
Starkey, R.L. (1945). Transformation of iron by bacteria in water. **J.Am. Water Works Assoc.**, 37, 963-984.
Wijura, A. (1983). Optymalizacja warunkow eksploatacji Centralnego Ujecia wod podziemnych Wodociagu Zulawskiego. **Instytut Ksztaltowania Srodowiska.**

PART SIX

LAND DRAINAGE AND RECLAMATION AND WASTE DISPOSAL

19   BIOTREATMENT OF CONTAMINATED LAND

B. ELLIS and R.J.F. BEWLEY
BioTreatment Limited, Cardiff, UK

Abstract
This paper examines a range of investigations which focus upon the decontamination of land both in-situ and after excavation. Treatment regimes involve enhancement of natural processes, particularly biological degradation, in order to reclaim contaminated sites economically. The use of laboratory tests to evaluate treatment strategies are discussed with reference to organic contaminants such as phenols, oils and polyaromatic hydrocarbons. Such systems are further tested in field trials which may then lead to full scale remedial programmes. Examples of laboratory tests, field trials and full site reclamation work are given to illustrate the use of biological systems as one of the many techniques available for reclaiming contaminated land.
Keywords: Contaminated Land, Biotreatment, In-Situ Treatment, Organic Pollutants, Reclamation, Microbial Inoculants.

1 Introduction

Contaminated land is a problem that is becoming increasingly important particularly in Europe and North America where sites associated with older industries, such as coal gasification plants, have become redundant leaving behind a legacy of chemical pollution. A polluted site may pose a risk to the environment depending upon its specific circumstances such as the type of contaminants present and their concentration, the geology and proximity of the site and its proposed future use.

Because of its heterogeneity soil often warrants a thorough investigation in order to determine such aspects as geotechnical suitability for a proposed use. When land is chemically polluted problems are often compounded because such contamination is so variable depending upon the site history. Industries known to have caused such pollution include gasworks, wood preservation plants, chemical manufacturers, oil refineries, paint makers, railway yards, steelworks and landfills.

The broad range of sites, the contamination present and proposed future usage, inevitably requires a range of expertise including scientists and engineers, in order to fully understand and evaluate the risk of hazards present and to interpret such information for assisting in the decisions as to the most appropriate remedial strategy and options for future development.

The simplest form of remediation is to move the contaminants elsewhere (e.g. to landfill or toxic tip) but this is becoming increasingly expensive and unacceptable. With greater public concern and impending legislation, the philosophy of shifting contamination from place to place is becoming less attractive (Petts, 1990). However, there are a host of alternative remedial options including incineration, encapsulation, stabilization, thermal fusion and vitrification, chemical oxidation, thermal treatment such as steam stripping, and microbial treatments, (Smith, 1985). The efficacy of a particular technique is limited to specific applications and no single technology can be used for all circumstances of site remediation. Moreover the remedial options available for a particular site will often be limited, not by their potential for reducing environmental risk, but by economics - the time and money available for clean-up.

Usually, several different processes can be incorporated into a site remediation programme and often microbiological techniques are used in order to achieve significant reductions in pollution. One of the major benefits of treating the contamination on-site is that both export and import of material is minimized. Although microbial treatment has economic advantages over alternative remedial measures, there are certain conditions which need to be fulfilled before such technology is applied:

o   the contaminant must be degraded into inert material.

o   the rate of decontamination must fit the remedial programme.

o   treatment should be cheaper than alternatives.

Due to the considerable variability of contaminated sites, a certain amount of investigative work is always necessary before the appropriate remedial approach is chosen. This preliminary study may simply incorporate a site investigation of the geotechnical and chemical properties relevant to future use, but sometimes a more intensive "treatability study" is required in order to develop and test a particular remedial strategy. The intensity of such treatability studies depends upon site conditions, the nature of the contaminants and the degree of clean-up necessary prior to development. Thus, such studies can range from several weeks in the case of well documented contaminants such as those found on gasworks sites, to several months or years if the contamination is due to unusual compounds such as certain pesticides. The present paper illustrates how laboratory and fieldwork investigations have been used to develop full-scale clean-up treatments for contaminated land. Examples have been chosen in order to focus upon the contemporary issue of in situ treatment versus the treatment of excavated soil which is discussed in the recent House of Commons Report on Contaminated Land (1990).

2 Laboratory Programme

The initial scientific programme evaluates the components of contamination and if performed thoroughly will enable delineation and quantification of the contaminants. Apart from these analyses, other factors may require investigation in order to evaluate possible mechanisms of treatment.

Liquid culture experiments using microorganisms normally isolated from the particular site can be used to establish that a contaminant is biodegradable. A vast library of references has now accumulated concerning the microbial degradation kenetics of compounds and therefore accurate predictions of treatability can usually be made. However, the prime aims of the treatability study should be:

o to determine the rate limiting factors affecting biodegradation of contaminants within a particular soil.

o to optimise these rate limiting factors in order to maximise treatment, minimise cost and confer reliability.

2.1 Soil Pot Experiments

The major factors affecting biological treatment of soil include temperature, pH, moisture (water availability), oxygen availability, particle size distribution, soil type, nutrient concentrations, the presence of inhibitory compounds and the availability of the contaminant(s) to microorganisms (this is often related to solubility). Each of these can be altered to improve microbial activity but because of the interrelationship of these factors and the complex nature of soil, it is often best to test such adjustments within soil pot experiments.

Soil collected from a contaminated site is prepared for experiments by sieving (usually 0-6mm fraction) and mixing in order to improve its homogeneity. Normally the capacity of the pot is 0.5 - 50 kg and each treatment should be replicated at least five times to enable statistical evaluation of the final data. Rate limiting factors such as moisture content and pH can be adjusted as separate treatments or in combinations and compared to untreated controls in order to varify the efficacy of a particular treatment regime. A pragmatic approach to experimental design is often required since such tests must allow maximum information directly relevant to the particular project within the constraints of time available.

Figure 1 compares data from several pot experiments using soil from various contaminated sites and illustrates the different rates of loss of different groups of pollutants due both to a combination of factors each limiting the rate of degradation to various degrees, and also to the soil type itself. In each case the optimised, microbially-based treatment resulted in rapid degradation compared to the controls (water only). The data shown in Figure 1 indicate total pollutant concentrations, but more useful information can be derived if the individual components are monitored. Examination of the polyaromatic hydrocarbons of creosote-contaminated soil A, for example, indicated that after the 60 day period of treatment each component compound was reduced to a different extent in comparison with the control. For instance, naphthalene was reduced by 25% (w/w) below the control value, whereas the corresponding percentage reductions for the other major components were: 48% for fluorene; 30% for phenanthrene; 71% for pyrene; 24% for chrysene. Thus rate limiting factors, even within the same soil system, will affect different compounds in different ways and it is therefore the more slowly degraded components which become rate-limiting during any remedial programme. Such compounds should therefore be targeted at an early stage of treatment.

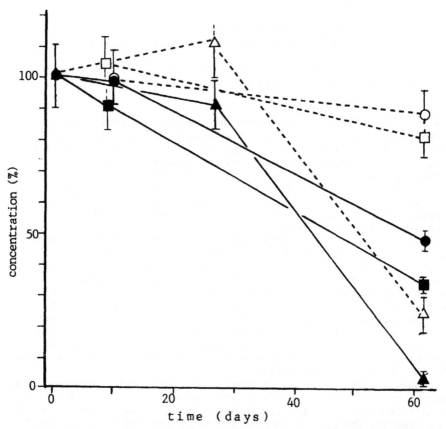

Fig.1. Treatment of various pollutants in soil pot experiments.
Soils from various contaminated sites were treated with optimised formulations of nutrients, microbes and surfactants in 1kg capacity soil pots. Figures are mean values of 5 replicate pots and error bars indicate 95% confidence limits. Open symbols represent controls (water addition only) and the treated contaminants were:● creosote (initial conc.=2858 mg kg⁻¹;incubated at 10°C, 17% moisture);■ oil hydrocarbons (initial conc.=2400 mg kg⁻¹ 15°C, 15% moisture);▲ phenols (initial conc.=211 mg kg⁻¹; 15°C, 18% moisture). Analysis was by HPLC with UV detection .

2.2   The effectiveness of adding microorganisms

Under certain circumstances stimulation of the indigenous microflora
within contaminated soil can result in adequate rates of degradation
of pollutants to achieve the specific target concentrations. Thus,
the careful adjustment of moisture content, addition of nutrients and
increased aeration, etc., may enable soil treatment without the direct
addition of particular microorganisms. The efficacy of such treatment
depends upon site specific circumstances and assumes that certain rate
limiting factors such as the lag times for growth of the appropriate
microorganisms are not inhibitory in terms of cost-effective
treatment. Thus, the aim of such treatment is to optimise the soil
conditions so that the indigenous "genetic pool" can express itself
adequately to enable effective degradation to take place.

The addition of microorganisms as a seed to enable rapid rates of
degradation to occur is established in many liquid-based industrial
processes and is examplified by the activated sludge system of sewage
treatment whereby aerobic microbial degradation of organics is
optimised by continuous recirculation of a proportion of the activated
sludge. In soil a more complex and heterogeneous system exists but
the effect of adding microorganisms can be illustrated by examining
data from a range of soil pot experiments. Figure 2 indicates the
reduction of a range of compounds in various soils using either
nutrients only or nutrients plus microorganisms. Each treatment can
be compared to the control system in which the conditions of
incubation, temperature and moisture content were indentical.
Clearly, the addition of inorganic nutrients allowed significant
reductions in all compounds when compared to the control. For
instance, after 7 weeks, 2-chlorobenzoic acid was reduced from 245.1
to 104.8 mg kg^{-1} (significant at 0.05). However, when microbes were
added (in this case a single bacterium, Pseudomonas putida at the rate
of 4×10^6 cells g^{-1} soil) a statistically significant reduction was
observed after only 4 weeks of incubation and the final concentration
achieved was much lower (2 mg kg^{-1}) when compared to the nutrient-only
treatment (105 mg kg^{-1}). Varying differences in the extent of
degradation of chlorophenols, phenol and 3-chlorobenzoic acid were
observed between the nutrients-only and nutrients plus microbe
treatments and in the case of m-chlorophenol, nutrient supplementation
was not necessary since adequate concentrations were already present
in the soil.

Thus, microbial seeding can affect the rates of degradation of some
pollutants in certain soils. However, such improvements are sometimes
only measurable during certain periods of treatment and prolonged
treatment without microbial seeding may result in degradation to
acceptable residual concentrations of the pollutants. When successful
remediation is attainable on a particular site by stimulation of the
indigenous soil microflora, careful monitoring is nonetheless still
essential since over-supplementation (e.g. of inorganic nutrients) can
result in soil or groundwater poisoning.

Thus, although microbial seeding may often have a significant
effect upon the rate of pollutant disappearance, the decision as to
its inclusion in the overall site treatment strategy will be dependent
upon an economic assessment of each situation: i.e. the cost of
microbial amendments may not be justified if the relative benefits

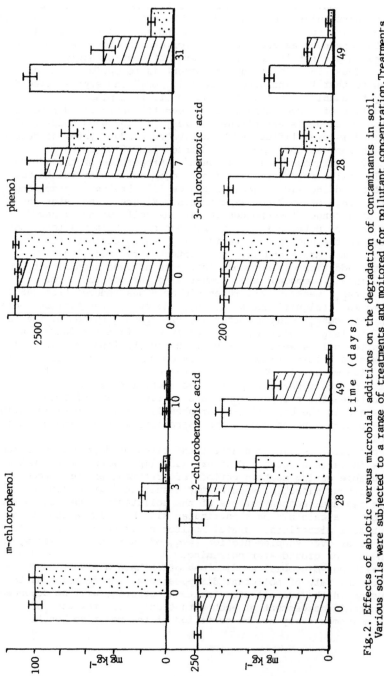

Fig.2. Effects of abiotic versus microbial additions on the degradation of contaminants in soil. Various soils were subjected to a range of treatments and monitored for pollutant concentration. Treatments were in 1kg-capacity soil pots and consisted: ☐ control (water and mixing only); ▨ abiotic (inorganic nutrient addition, water and mixing only); ⊡ nutrients + microorganisms. Incubation was at 15C, 15% moisture with regular mixing. Error bars indicate 95% confidence.

compared to the use of abiotic supplements only are marginal.

3. Field Trials

Laboratory studies are useful in providing the basic information necessary to decide a treatment strategy. However, new treatments require scaling up in order to evaluate and refine the methodology prior to its full-scale implementation. The major model for ex situ soil clean-up has been the "treatment bed" which can take many forms but generally consists of processed soil laid out in mounds of up to one metre deep. Treatment supplements such as water, nutrients, microorganisms and surfactants can be added periodically. Regular turning of the soil with an agricultural spader ensures adequate aeration.

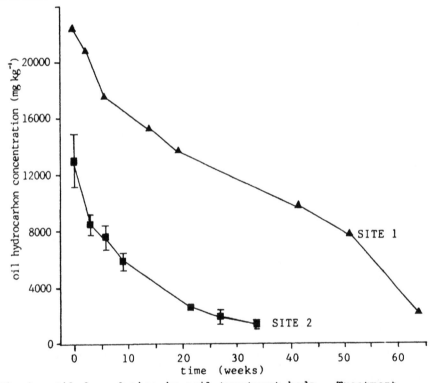

Fig.3. Oil degradation in soil treatment beds. Treatment beds consisted of approximately 200m³ of oil-contaminated soil. Treatments involved the periodic addition of nutrients, microorganisms, surfactants and water. Analysis was by infra red spectroscopy (standard German method-DIN H-18) and is reported on a dry weight basis. Site 1 shows the mean value of 10 pooled samples while for Site 2 the mean value of 99 samples per point is indicated with 95% confidence intervals.

Figure 3 illustrates data from 200m^3 capacity soil treatment beds which were monitored during remediation at two separate oil refineries. In both cases the systems were enclosed in polythene tents in order to control temperature and moisture. Oil degradation rates varied due to the differences in soil type and the nature of the oil contamination. Thus, at Site 2 the mean oil concentration was reduced from 12980 mg kg^{-1} to 1273 mg kg^{-1} in 34 weeks whereas at Site 1 a more gradual degradation rate was achieved from 22600 mg kg^{-1} to 2000 mg kg^{-1} in 63 weeks.

A further field trial study at Site 2 indicated that soil contaminated with relatively low concentrations of oil could be successfully treated in situ (Ellis, et al, 1990). Here, treatment was effected by inducing groundwater movement and recirculation through a reactor tank. Oxygen, nutrients, microorganisms and surfactant were added to the system by pumping through gravel-filled infiltration trenches around the perimeter of the treatment area. Thus, an area of 10m x 20m was treated to a depth of 8m during a 15 week period. One hundred monitoring points (20 points at 5 depths) were used in order to validate the treatment statistically and the final concentration achieved was 26 mg kg^{-1} i.e. an 86% reduction.

BioTreatment Limited has now accumulated field trial and treatability data on a wide range of contaminants including phenolics, chlorinated aromatics, polyaromatic hydrocarbons, herbicides, oil hydrocarbons and insecticides. Using such information it is possible to minimise the time necessary to evaluate treatment strategies for particular sites.

4 Full scale reclamation

Contaminated sites are usually a mixture of patches of 'clean' and polluted soil. The most economic remedial plans for such sites therefore consist of applying several technologies since no one methodology is adequate for every pollution problem. This is illustrated by the examples shown in Table 1 which refers to three contaminated gas works sites reclaimed by BioTreatment Limited. Thus, at Site A, 70% of the contaminated material was treated microbiologically whereas the remainder was encapsulated on-site or sent to landfill facilities. At Site C, approximately 1000 m^3 of contaminated liquid was treated by chemical oxidation and precipitation before being disposed to sewer. Details of the treatment regimes used have already been described (Bewley and Theile, 1988; Bewley et al, 1989) and relied significantly on the use of treatment beds in which surfactant, nutrients and microorganisms were added to the soils in order to achieve removal of the organic contaminants (polyaromatic hydrocarbons and phenols) to below concentrations set by the local authorities.

In contrast to the treatment bed system of decontamination after excavation, BioTreatment Limited is currently undertaking a programme of in situ clean up of a gasworks site in Sweden. Here, for logistic reasons, excavation of the whole site was not feasible. Therefore, following a 9 month treatability study, a biological treatment strategy based on liquid recirculation was initiated. The contaminated area (approximately 4500m^2) has been surrounded by metal

Table 1. Range of techniques used during reclamation of three gas works sites.

Treatment	Volume of soil treated (m^3)		
	Site A	Site B	Site C
Microbiological	18,200	30,000	9,200
Encapsulation (on-site)	7,400	12,000	0
Landfill Disposal	380	0	10,000
Chemical	0	0	1,000*

* 1000 m^3 effluent contaminated with phenols, cyanide, metals and sulphide.

sheetpiles in order that building work can be established during the treatment period of 3 years. A network of pipes has been laid in order to pump leachate vertically through a depth of 3m facilitating the clean up process. Contamination consists of polyaromatic hydrocarbons and phenols at concentrations exceeding 10000 mg kg^{-1}. By using microbiologically based techniques the land will be treated to a concentration of <200 mg kg^{-1} total creosote. For technical reasons a further 5000 m^3 of contaminated soil is being microbiologically treated off site in treatment beds.

5 **Conclusion**

The diverse nature of contaminated land has resulted in the evolution of a wide and expanding range of measures available to remediate these sites. Since there are often several technically feasible clean-up techniques available for each situation, a detailed investigation of the site is usually necessary if the most efficient remedial programme is to be chosen. Such decisions depend not only on the prevailing site-specific conditions, such as geophysical factors and the nature of the chemical contaminants, but also on the proposed future use of the site.

Once the efficacy of biodegradation has been demonstrated for the particular circumstances of a polluted site then the economic feasibility of using the technology within a full scale remedial programme is assessed. The various remedial techniques are often applied using conventional engineering practices and in order to achieve economically sound site clean-up, skilful project management is required. Biodegradation techniques are now a poven technology but the benefits of science and engineering can only be fully exploited once both disciplines have striven for mutual understanding of the problems involved, particularly when scaling up from laboratory to field situations.

6 References

Bewley, R.J.F. and Theile, P. (1988). Decontamination of a Coal
Gasification Site Through Application of Vanguard Microrganisms.
Contaminated Soil '88, Eds. K. Wolf, J. vanden Brink and
F.J. Colon. Kluwer Academic Publ. pp 739–743.

Bewley, R.J.F., Ellis, B., Rees, J.F.R., Theile, P., and
Viney, I. (1989) Microbial Clean–up of Contaminated Soil. Chem.
Ind. 1989, 23 December 778–783.

Ellis, B., Balba, T and Theile, P. (1990) Bioremediation of Oil
Contaminated Land. Environ. Technol. Lett. (In Press).

House of Commons Environmental Committee First Report: Contaminated
Land (1990). H.M.S.O. London.

Petts, J. (1990) Contaminated Land: Is the U.K. Cleaning Up or
Covering Up? Contaminated Land Policy, Regulation and Technology.
Proceedings of the Conference held at Cafe Royal, London (22–23
Feb.) IBC Technical Services Ltd.

Smith, M.A. (1985). In: Contaminated Land, Reclamation and Treatment.
NATO: Challenges of Modern Society 8. Ed. Smith, M.A. Plenum Press,
London, pp 1–11.

20 CLOGGING PROBLEMS IN A SUBSURFACE PRESSURE DISTRIBUTION SYSTEM FOR WASTEWATER DISPOSAL

G. CHERIER, J. LESAVRE and A. ZAIRI
A.F.B.S.N./C.R.E.A.T.E., Colombes, France

Abstract
Pressurized underground treatment of wastewater from small communities is a way of obtaining efficient results in both the purification and draining-off of effluents. Setting up a localized underground distribution network using techniques developed for micro-irrigation allows for an even distribution and dosage of the effluent over the whole infiltration area. The reliability of such systems consists essentially in overcoming or minimising the clogging problems encountered in the use of sewage water. With the aim to avoid and limit all clogging problems due to physical, biological or chemical factors, several technological solutions have been studied : selection of suitable emitters available with sewerage effluent; choice of wastewater pretreatment in relation with the selected emitters; utilization of geotextile wrapping to protect the buried pipes. These experiments carried out initially outside the soil were subsequently studied in experimental underground sewage disposals.
Keywords: Wastewater treatment, Underground sewage disposals, Effluent distribution, Clogging characterization, Emitters, Geotextile, Geophysical resistivity.

1 Introduction

Soil as a treatment reactor is a very effective treatment system if we manage to mobilize the large quantities of aerobic micro-organisms which could decompose the organic matter and nitrify the ammoniacal nitrogen in the wastewater. CATROUX (1974) estimates the quantity of micro-organisms in one hectare at 1-2 tons of dry matter. This is equivalent to 600 m^3 of activated sludge with a concentration of 3 g/l.

The large variety of micro-organisms allows the process
to adapt easily to variations in the quality and quantity
of the waste. Moreover, the endemic presence of this
microflora considerably reduces the starting-up period
compared to traditional treatment plants.

Underground sewage disposal consists of bringing
wastewater into contact with the soil by means of a
network. This network must ensure an even distribution and
dosage of effluent in order to regulate the pollution and
the oxygen transfer. This in turn will ensure satisfactory
hydraulic functioning which will give high-efficiency
treatment results.

The different soil treatment techniques used in France
for single-house wastewater treatment are described in the
law of 3rd March 1982 and in the circular of 20th August
1984. The transposition of these techniques to small
communities necessarily implies an increase in the surface
area used. The hydraulic load allowed on a soil being a few
centimetres per day, the surface area needed is therefore
very large. However, as the traditional process for single
houses uses a gravitational distribution, the simple
transposition of this technique to large surfaces cannot
ensure an even distribution (FAZIO 1987). This uneven
distribution was observed on 369 collective spreading areas
by PLEWS and DEWALLE (1984) and confirmed by WARD and
MORRISON (1984), who noted that the improvement of this
process depended on :
- an even distribution of the effluent over the whole
  infiltration area,
- the control of the dosage of the effluent.

Underground micro-irrigation meets these requirements.
However, it must be adapted to the bad quality of the
water. The main problem is clogging, either in the network,
in the soil, or in both. The solutions to these clogging
problems are the object of this study.

2 The biological clogging in this process

The pollution carried by wastewater can be classified in
two categories :
- the particular pollution which must be reduced to a
  minimum by a primary treatment, in order to avoid the
  physical clogging of the emitters,
- the dissolved pollution which will be transformed into
  a biomass during the treatment process . This biomass
  can disturb the network and the infiltration of the
  water in the soil.

The different kinds of clogging (physical, chemical,
biological) are difficult to identify. We present below the
technological solutions tested during our study where the
microbiological element is the most important factor in the
clogging (fig. 1)

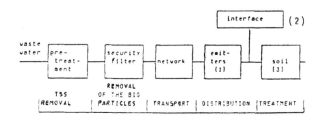

Fig 1 : Schematic diagram of the process

The study concentrates on :
- (1) the selection or conception of emitters which are
  resistant to clogging,
- (2) finding a protection for the network against
  external clogging : the study of geotextiles.
- (3) elaborating a method of investigation which allows
  for an evaluation of the soil clogging without
  disturbing the surroundings : electric resistivity.

3 Experimental procedures

3.1 Selection of emitters
3.1.1 Emitters
The test was carried out using wastewater from an urban
collector. This wastewater, treated either by decantation,
or by screening, was injected into four ramps, each fitted
with 13 different emitters selected for their resistance to
clogging.

Table 1. Characteristics of the tested emitters

	Name	N°	Type	Discharge (l/h) under 1 bar
	GANA	1	shunt connected (s.c.)	4
	PLASTIF	2	s.c., self compensating	4
	D.I.I.	3	in line, long flow path	2
Emitters	D.I.I.	4	in line, long flow path	4
	D.I.I.	5	s.c., long flow path	4
	LEGO	6	shunt connected	4
	BIP	7	shunt connected	4
	REGLAIX	8	s.c., self compensating	3.8
Mini Jets	NOVOJET	9	black base	22
	MINIJET	10	green	52
	MINIJET	11	red	63
Nozzle	BAS-RHONE	12	⌀ 1.3 mm	50
Microtube	EUROFILTRE	13	microtube ⌀ 1.1 mm L100 mm	15

There are two emitters of each type on every ramp. Two ramps undergo a pressure of 1.5 bars and the two others a pressure of 1 bar.

On two ramps (1 bar and 1.5 bars) a chlorine solution was injected twice a week for one hour (total residual chlorine 17ppm).

3.1.2 Sewage effluent quality analysis

Table 2. Average content of the effluent (mg/l)

	Settling and Digestion	Screening (120 μm)
COD	277	393
BOD₃	182	195
TSS	263	281

3.1.3 Experimental results

3.1.3.1 Clogging of emitters
The emitters can be classified in 3 categories (see Table 3) :
- those which have a high resistance to clogging (average loss in flow rate between 2 and 10%) : these are simple, with a large aperture and a high flow rate (50 l/h). Examples are the red and green MINIJETS, the BAS-RHONE nozzle and the microtube,
- those which are subject to gradual clogging : these are emitters with a long flow path and a small aperture. TSS gradually build up and eventually clog them completely unless maintenance is carried out. Examples are the GANA emitters, D.I.I. in line (2 l/h and 4 l/h) and LEGO in derivation,
- those which work irregularly : these emitters clogg and unclogg by themselves. Nevertheless, the flow rate never recovers to the initial level. Examples are D.I.I. in derivation, REGLAIX, NOVOJET, BIP.

Table 3. Results of the primary treatment and chlorination on the clogging of emitters

TREATMENT		NIL	CLOGGING GRADUAL	IRRE-GULAR
SETTLING	Without Cl	10,11,12	1,2,3,4	5,6,7,8,9
	With Cl	3,4,6,10,11,12	1	2,5,7,8,9
SCREENING	Without Cl	10,11,12,13	1,2,3,4,5,6,	7,8,9
	With Cl	2,4,5,7,10,11,12,13	1,3,6,8,9	

3.1.3.2 Chlorination influence
Chlorination only appears to have a significant effect on a few emitters. The effect also depends on the primary treatment undergone by the effluent.

3.1.3.3 Pressure influence
This influence was observed by comparing the ramps pressurised at 1 bar and 1.5 bar. Our tests did not demonstrate a significant change in the emitters' resistance to clogging when the pressure was increased.

3.1.4. Selection of emitters
Experiments have shown that the clogging- resistant emitters are simple and with a large aperture. We therefore selected the BAS-RHONE emitter (Fig. 2) to equip two experimental buried networks. Protection against clogging is assured by geotextile sheets for the first network and by a single-house drain for the the second network (Appendix B).
  We made "elastic emitters" by making 9 mm slits in nitrile flexible pipe (Fig. 3). The tests found them to be well adapted for injecting water directly in the soil without a sophisticated pretreatment, due to the elasticity of the slits.

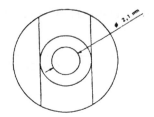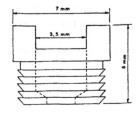

Fig. 2 : BAS RHONE emitter

slit

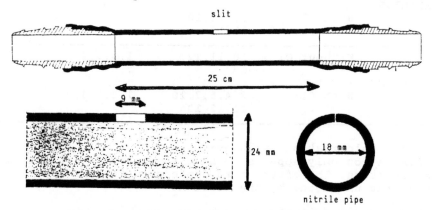

25 cm

9 mm

24 mm

18 mm

nitrile pipe

Fig. 3 : Slit flexible elastic emitter

3.2 Protection of the underground network
The use of geotextiles for protection against the clogging of underground tubes has been widely used in agricultural drainage.(CESTRE 1985, SCHNEIDER 1982). Experiments have tested the efficiency of these products as a wrapping material for the emitter-soil interface of an injection network. Seven selected geotextiles were tested for their influence on clogging.

Table 4 : Characterization of selected geotextile

CHARACTERISTICS	NAME	POLYFELT TS 600 R1	POLYFELT TS 800 R2	SOMDRAIN 175 R3	SOMDRAIN 175/17 R4	NAUE 317 RP 10 R5	GRILTEX B 536 R6	DRAINFELT R7
THICKNESS (mm)		2,1	3,3	7,9	8,28	20	7,7	12
POROMETRY (µm)		110	70	100	100	63	130	259
TRANSMISSIVITY $(10^{-4}\ m^2/s)$		2,1	2,6	3	3	4,44	0,11	6
PERMITTIVITY (s^{-1})		2,3	1,2	0,58	0,55	1,42	3,7	0,41
TEXTILE FIBERS		POLYPROPYLENE	POLYPROPYLENE	POLYPROPYLENE + POLYAMIDE	POLYPROPYLENE + POLYAMIDE	POLYESTER + POLYPROPYLENE	POLYPROPYLENE	POLYPROPYLENE

3.2.1 Experimental Conditions
- A 2 m long horizontal ramp (diameter=25mm) set up above ground and equipped with a BAS-RHONE nozzle of a diameter of 2.1 mm (Flow rate : 60 l/h).
- Ramps completely wrapped in geotextiles sewn up at the seams. For one of the seven geotextiles used, an unwrapped ramp acted as a control.
- Primary treatment : settling and then filtering at 180 microns.
- Batch feeding : 4 times a day. Each batch lasts 10 mins, and the daily flow rate is thus estimated at 40l.
- Experimental period : 330 days.

3.2.2 Experimental follow-up
Clogging of the wrapped ramps due to the build-up of TSS and bacterial development was studied by comparing the changes in the flow rate of the wrapped ramps with those of the control ramp (Appendix A).

The textiles R5, R6 and R7 were observed to have a high resistance to clogging. The ability of ramps R5 and R7 to unclogg themselves (on the 310th and 195th days respectively), constitutes an advantage for these materials.

Finally, the geotextile NAUE 317 RP10 (R5) was selected as the protection for the underground ramps fitted with BAS-RHONE nozzles (ϕ=2.1mm). Its thickness (20 mm) should allow it to act as a reservoir equivalent to the gravel ditches in which the drains are placed in single-house wastewater treatment systems.

3.3 Follow-up of 3 underground sewage disposals using electric resistivity
The clogging observed in underground sewage disposal systems is due to several factors, the importance of which are difficult to measure. However, there does appear to be a dominance of biochemical and biological factors.

So, clogging problems can be caused by sulphides obtained by biochemical reduction of sulphates or by accumulation of zoogloeal film issued from bacterial matabolism (AVNIMELECH 1984, MITCHELL 1964)

The clogging process starts slowly and then accelerates rapidly, thus provoking a high resistance to the flow rate. Electric resistivity was used to monitor the progress of the clogging on 3 underground sewage disposal systems. (Appendix B). This technique of investigation which does not disturb the environment enables us to obtain a cartographic representation of the soil moisture.

3.3.1 Principle of the measurement
Rocks and soils are very bad conductors of current. Their resistivity is very high. Ions ensure electrolytic conduction in the soil. Also, the resistivity of a soil will vary principally with its moisture level.

If between two injection current (I) electrodes A and B, we insert two electrodes M and N connected to a measuring instrument (V), we obtain the value of the resistivity (ρ) by the formula :

$$V = \frac{I\rho}{2\pi} \left[(\frac{1}{r_1} - \frac{1}{r_2}) - (\frac{1}{r_3} - \frac{1}{r_4}) \right] \quad (1)$$

I : intensity of the current
V : value of the potential difference
ρ : resistivity

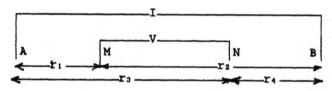

Fig. 4 : 2 current electrodes (A,B) and 2 potential electrodes (M,N) on the surface of a uniform homogeneous soil

If AM = MN = NB = a, (1) can be written : $\rho = 2\pi a V/I$

The resistivity is assigned to the mid point of the device and at a depth of AB/4. In fact, this is the apparent resistivity of a uniform and homogeneous soil which has the same value.

3.3.2 Practical applications
The electrical soil resistivity measurement was applied to an experimental underground pressurized sewage disposal. The tests were carried out on an experimental site of 500 m^2, divided into four parts, 3 of which were each equipped with two ramps injecting pressurized wastewater. The fourth part was left untouched as a climatic control (Appendix B).
    The networks were injected under pressure in order to ensure:
    - even distribution of the effluent along the ramps,
    - control of the applied doses.
    For example, the cartography obtained by measuring the resistivity at the depth of of 30 cm using surface electrodes (AB=1.2 m) shows how the moisture levels evolves after 180 days of influent injection (Appendix C and Appendix D).
    At this stage the soil is observed to be saturated (low resistivity),which provokes a horizontal draining- off of effluent. We can conclude that a rest period must be allowed for between the injection periods in order to avoid gradual obstruction of the waste disposal.

4 CONCLUSION

Subsurface distribution with a trickle irrigation type
network is one way of spreading wastewater over large
areas. Several networks were tested without clogging :
- BAS RHONE emitters protected from the soil with a house
drain or geotextiles,
- 9 mm slits in elastic pipe burried without protection.
Clogging was observed using geophysical resistivity
measurements. This non destructive investigation method
allows us to determine the rest and inflow periods.

5 REFERENCES

Avnimelech, Y. and Nevo, Z. (1964) Biological clogging of
  sand. **Soil Science** , 98, 222-226.
Cadiou, A. and Lesavre, J. (1984) Drip irrigation with
  municipal sewage : Clogging of the distrbutors in
  Proceedings of the water reuse symposium III, SAN DIEGO ,
  vol 1, 472-478
Catroux, G. Germon, J.C. and Graffin, P. (1974)
  L'utilisation du sol comme système epurateur. Ann. Agro.,
  25, 179-193
Cestre, T. (1985) L'utilisation des géotextiles comme
  produits d'enrobage de drains agricoles. Bull. d'Inf.
  C.E.M.A.G.R.E.F., 329.
Chérier, G. (1988) Suivi d'un épandage souterrain par
  mesure géophysique de la résistivité. D.E.A., Université
  PARIS VI
Fazio, A. (1987) Les filtres à sables verticaux en
  assainissement autonome : Approche pratique du
  dimensionnement. Thèse, **U.S.T.L. MONTPELLIER.**
Mitchell, R. and Nevo, Z. (1964) Effect of bacterial
  polysaccharide accumulation on infiltration of water
  through sand. **Appl. Micro.**, 12, 3, 219-233.
Plews, G.D. and Dewalle, F. (1984) Performance evalution of
  369 larger on site systems in **Proc. of the fourth
  National Symposium on Individual and Small Community
  Sewage Systems** (eds Am. Soc. of Agr. Eng.), 372-381 .
Schneider, H. and Wewerka, W. (1982) Le comportement à long
  terme des géotextiles. **GEOTEX**, 3, 82.
Ward, R.C. and Morrison, S.M. (1984) Research needs
  relating to soil absorption of wastewater in **Proc. of the
  fourth National Symposium on Individual and Small
  Community Sewage Systems** (eds Am. Soc. of Agr. Eng.), 1-
  14.
Zaïri, A. and Lesavre, J. (1988) Epuration des eaux
  résiduaires par épandage souterrain sous pression. Thèse,
  Université PARIS VI.

6 Appendix A. Discharge of ramps versus time

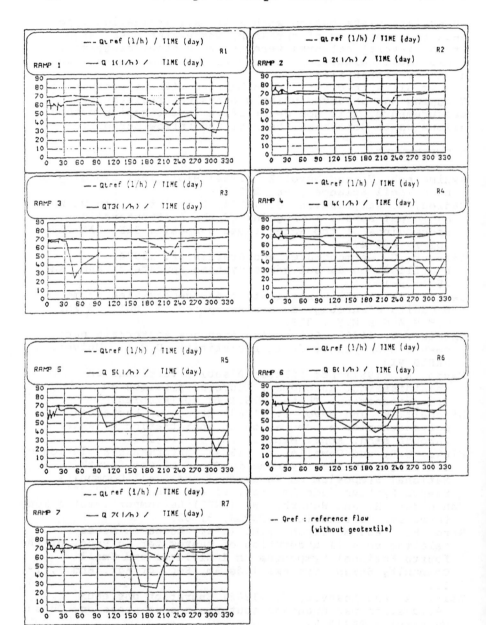

7 Appendix B. Diagram of the experimental network and of the preliminary treatment

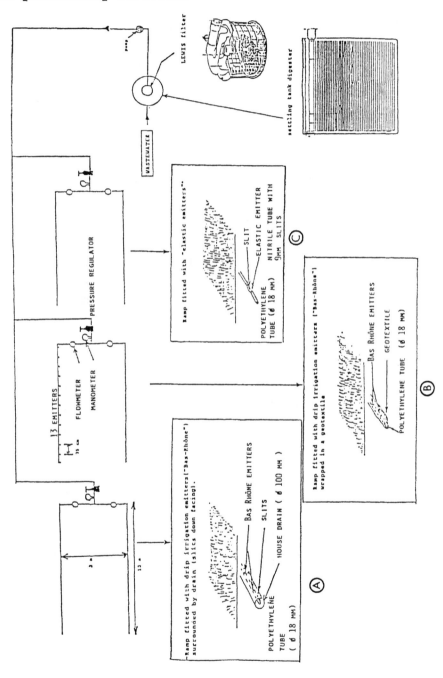

8 Appendix C. Map of isoresistivity before the first injection

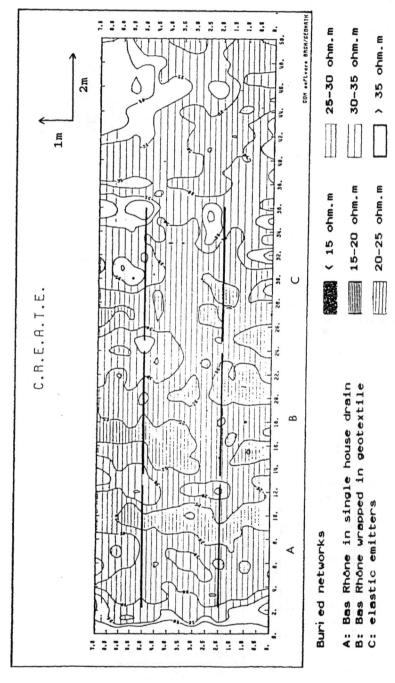

C.R.E.A.T.E.

DOM software BRGM/GEOMATH

	< 15 ohm.m	25-30 ohm.m
	15-20 ohm.m	30-35 ohm.m
	20-25 ohm.m	> 35 ohm.m

Buried networks

A: Bas Rhône in single house drain
B: Bas Rhône wrapped in geotextile
C: elastic emitters

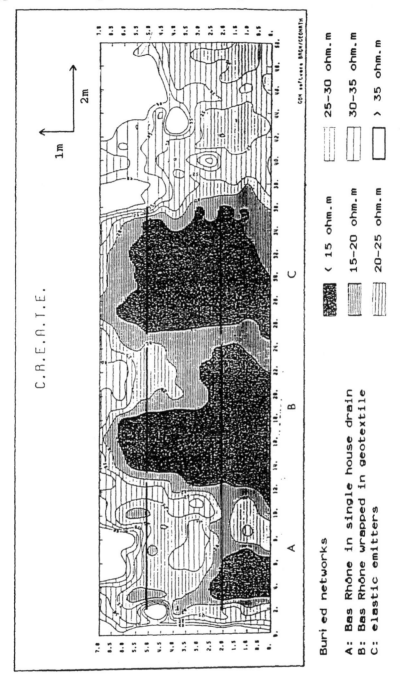

C.R.E.A.T.E.

Buried networks

A: Bas Rhône in single house drain
B: Bas Rhône wrapped in geotextile
C: elastic emitters

< 15 ohm.m
15-20 ohm.m
20-25 ohm.m
25-30 ohm.m
30-35 ohm.m
> 35 ohm.m

21 ELIMINATION OF OCHRE DEPOSITS FROM DRAINPIPE SYSTEMS

A. ABELIOVICH
The Jacob Blaustein Institute for Desert Research, Ben-Gurion
University of the Negev, Israel

ABSTRACT

The presence of ferrous ions (Fe^{2+}) in groundwater is frequently associated with clogging of drainpipe systems. Iron bacteria of the *Leptothrix-Sphaerotilus* group develop and clog the pipes. We have overcome this problem by constructing the drainage system in a way that ensures maintenance of anaerobic/microaerophilic conditions, thus creating a hostile environment for these bacteria which are obligate aerobes. At present, an area of 12 ha is being successfully drained in Ein Tamar (south of the Dead Sea) by an anaerobic drainage system designed according to this line of thought.

INTRODUCTION

Specific environmental conditions cause clogging of underground drainage systems by iron ochre produced by sheathed bacteria of the *Leptothrix-Sphaerotilus* group. The information of ochre is often associated with the presence of a high level of anaerobic groundwater containing Fe^{2+} ions, usually in peaty soils with clay.

Various measures to eliminate ochre, such as the use of cupric sulfate, $CaCO_3$, sodium bisulfate mixed with H_2SO_4, or wood bark, have been tried throughout the world, usually unsuccessfully.

In the northern Arava valley of Israel, about 10-20 km south of the Dead Sea, saline groundwater is present at 1.8 to 2.0 m below surface. Because intensive irrigation brings this saline groundwater to the surface, underground drainage pipes were installed at a depth of 1.8 to 2.0 m, with a distance of 100-300 m between the pipes. All the drainage pipes installed became clogged by iron ochre within a few months, causing a rise in the saline groundwater wherever intensive irrigation was practiced.

Although all sheath-producing iron bacteria of the genus *Leptothrix* are obligate aerobes, ochre production is always associated with the presence of Fe^{2+} in groundwater. At the pH values prevailing in this water, free Fe^{2+} can exist only under microaerophilic or anaerobic conditions. These two prerequisites (aerobic growth conditions and presence of Fe^{2+}) for the production of ochre cannot coexist at the same point in time and space, but in the drain pipe, the holes that perforate it are the points that might come closest to fulfilling these requirements, as on the inner side they are exposed to the atmosphere, while being flushed constantly with low oxygen groundwater enriched with Fe^{2+}. Hence, they are probably the sites for development of *Leptothrix* biomass causing ochre clogging of drainpipes. With this in mind, a series of field experiments was carried out to see whether it is possible to construct and operate an anaerobic drainage system, and thus avoid the whole complex of problems by changing the environmental conditions so that growth of *Leptothrix* would be prevented. The experiments were planned with the aim of achieving anaerobiosis by raising the groundwater level through installation of elevated outlets at the outlets of the drainpipe.

Several field experiments to test this approach were carried out during the years 1982-1984.

Field Work

Kaptein and Van der Zwan (1963) prevented ochre production in drainpipes by installing them below groundwater level. This was possible in the Netherlands as groundwater could be kept constantly high by adjusting the water level in adjacent canal systems. The drainpipes thus turned anaerobic, and unsuitable for *Leptothrix* growth and ochre production. As it is impossible to regulate groundwater level by this technique in the Dead Sea area, experiments was planned with the aim of achieving similar results by elevating groundwater by installing elevated outlets in manholes and at outlets of the drainpipe.

Two field experiments were carried out during 1982-1984.

1. A ten-year old drainpipe (3" diameter, 1.8-2.0 m depth), clogged by ochre deposits was used. The outlet was raised by 70 cm and, as the drainpipe was constructed at an average slope of 0.3%, the water level was affected about 200 meters upstream.

The elevated outlet was kept in place for two months, removed for one month, and then reinstalled. This procedure was repeated four times with identical results: 3-4 weeks after elevating the outlet, water was completely clear in the outlet and in the two manholes upstream, about 200 meters from the outlet. Three to four weeks after removal of the siphon, the drainpipe outlet was clogged again. Oxygen concentration in the outlet was at saturation point when the system operated without an elevated outlet (7.0-8.0 mg/liter), dropped to ≤ 0.2 mg Ob_2/l when operated with the elevated outlet in place. Presumably, the microaerophylic or anaerobic conditions that developed within and around the submerged drainpipe enabled microbial reduction and solubilization of Fe^{2+} precipitates which comprise the bulk of the gelatinous ochre mass.

2. A new drainpipe that had never before been in operation, was disconnected at 50 m from its entrance to a collecting manhole, and was separated by a plastic lining (0.2 mm polyethylene) placed 2 m deep, and 4 m on each side of the pipe so as to seal it from the rest of the field. A drip irrigation system was installed above the drainpipe, which was operated continuously for 13 months, flooding the area with about 3,000 m^3 water per month. A siphon was installed at the outlet, elevating the groundwater level in the isolated section by an average 30 cm. No ochre was seen in the water and the system operated in this way for seven months without any ochre development. Six weeks after removing the siphon, ochre appeared and completely clogged that particular drain pipe system.

Following these experiments, a drainage system for 12 ha was constructed at the same site consisting of seven drainpipes (300 m long, 60 m apart). The outlets were elevated by about 45-50 cm, and as the slope of the pipes was 0.2% (no manholes were installed), most of their volume was submerged. O_2 concentrations, measured at the outlets, dropped gradually about 0.2 mg O_2/l. This system has already operated free of ochre deposits for 4 years.

References

Hallgren, G. and Ostholm, C.O., 1957. Om utfallning av jarn i draneringsledningar. Grundforbattring, 10, 87-101.

Kaptein, L.A. and van der Zwan, L.M., 1963. Roestafzettingen bij in het veen gelegde drainbuizen. Landbouwvoorlichting, 20, 34-42.

Kuntze, H. and Scheffer, B., 1974. Organische dranfilter gegen verockerung. Zeit. fur Kultertechnik und Flurbereinigung, 15, 70-79.

Mulder, E.G. and Van Veen, W.L., 1974. The sheathed bacteria. In: Buchanan, R.E. and Gibbons, N.E. (eds.), Bergeys Manual of Determinative Bacteriology, 8th edition. The Williams and Wilkins Company, Baltimore. pp. 128-147.

Petersen, L., 1966. Ochreous deposits in drainpipes. Acta agriculturae scandinavica, 16, 120-128.

Puustjarri, V. and Juusela, T., 1952. On rust precipitates present in drainage pipes and on the means of preventing their formation. Acta agriculturae scandinavica, 2, 131-152.

Spencer, W.F. and Mackenzie, A.J., 1957. Bacterial products can be removed from tile. Progress in Soil and Water Conservation Research; Quart. Report No. 12. ARS-SWC. USDA>

Spencer, W.F., Patrick, R. and Ford, H.W., 1963. The occurrence and cause of iron oxide deposits in tile drains. Soil Science Society Proceedings, Division S-2 -soil Chemistry, 134-137.

Vaughan, D., Wheatley, R.E. and Ord, B.G., 1984. Removal of ferrous iron from field drainage water by conifer bark. Jour. of Soil Science, 35, 149-153.

Van Veen, W.L., Mulder, E.G. and Deinema, M.H., 1978. The *Sphaerotilus-Leptothrix* group of bacteria. Microbiol. Rev., 42, 329-356.

PART SEVEN

GEOTECHNICAL ENGINEERING

22 BIOLOGICAL FACTORS INFLUENCING LABORATORY AND FIELD DIFFUSION

R.M. QUIGLEY
Geotechnical Research Centre, Faculty of Engineering
Science, The University of Western Ontario, London,
Ontario, Canada
E.K. YANFUL
Centre de Technologie Noranda, Pointe Claire, Québec,
Canada
F. FERNANDEZ
Golder Associates, London, Ontario, Canada

Abstract
Microbiological activity in raw municipal solid waste leachate
initiates precipitation of calcium carbonate flocs coated with ferrous
sulphide slime. If this occurs in the reservoir of a short-term
diffusion test, desorbed Ca^{++} in the pore water of the test soil may
actually diffuse out of the soil into the reservoir distorting
anticipated diffusion controlled profiles of ion concentration.
    Observation of field liners indicates abundant microbe-induced
precipitation of carbonate and black sulphide slime in sand drainage
layers separating solid waste from underlying clay liners. This
results in diffusion curves which start at the top of the sand cushion
rather than at the top of the clay barrier. The chemical fluxes
related to diffusion are thus significantly reduced.
Keywords: Municipal Leachate, Precipitation, Clay, Diffusion,
Laboratory, Field Barriers.

1 Introduction

Municipal solid waste (MSW) leachate is a biologically active, semi-
saline, slightly organic liquid derived from rainwater percolating
downwards through solid wastes at landfill sites. Most sites are now
lined with a clayey barrier and the leachate is removed by means of
bottom gravity drains or large central collection wells.
    Migration of pollutants through the clayey barriers is by combined
advection and diffusion. In the case of high quality barriers having
either very low hydraulic conductivity ($k < 10^{-8}$ cm/s) or negligible
gradients, the migration rates are largely controlled by molecular
diffusion. Calculation of the migration velocity and corresponding
chemical fluxes requires a diffusion coefficient obtained either in
the laboratory or by back calculation from chemical profiles produced
by field diffusion.

---

[1] Centre de Technologie Noranda, 240 Boul. Hymus, Pointe Claire,
Québec, Canada, H9R 1G5.

[2] Golder Associates, 500 Nottinghill Road, London, Ontario, Canada,
N6K 3P1.

This paper presents information on the chemical stability of MSW leachate and demonstrates the influence of biologically-induced precipitation of calcite on laboratory diffusion curves. This is followed by an example of field diffusion through a clayey liner overlain by a sand drainage layer partly clogged by biologically precipitated iron sulphide.

2 Leachate Stability

The stability of municipal solid waste leachate is important in laboratory and field studies on contaminant migration since it influences the C_o values of various species subject to removal by biochemical and physicochemical processes before and during their migration. Major factors influencing leachate stability include the composition of the organic matter, the saturation state of various species, and environmental parameters such as pH, redox potential (E_h) and temperature (Apgar and Langmuir, 1971; Chian and DeWalle, 1976).

It is frequently observed that grey landfill leachate left in storage bottles at laboratory temperature (22° ± 3°C) without exposure to light or air turned black after a few days. The change in colour is caused by the formation of sulphides resulting from microbial activity which continues during bottle storage. The colour change also occurs at lower storage temperatures but requires more time. In this section, the results are presented for a study carried out to investigate some aspects of the chemical stability of waste leachate stored in the laboratory without exposure to light or air.

2.1 Materials and Methods
The MSW test leachate was obtained from the leachate collection system at the Westminster Landfill site serving the City of London, Ontario (pop. ~ 250,000). Leachate samples were equilibrated at a laboratory temperature of 22° ± 3°C in 250 mL plastic bottles wrapped with duct tape to minimize exposure to light. Special bottle caps were designed to hold a platinum redox combination electrode and a glass pH combination electrode. Concurrently, five small 30 mL polyethylene bottles, also wrapped with duct tape, were filled with the same stock leachate and tightly sealed.

E_h and pH were determined on the leachate in the 250 mL bottles during an eight-day period. Both electrodes were periodically calibrated with suitable standards. At specified times during the experiment the 30 mL bottles were sent to a specialist laboratory at the University Hospital for bacterial analysis (Agar coated plate counting of incubated cultures of leachate samples diluted by a factor of 10^4 in Ringer's solution).

At the end of the experiment, the black slimy precipitate was separated from the bulk leachate in one of the 250 mL bottles by centrifuging at 10000 rpm for 10 minutes. The slimy solids were then characterized by scanning electron microscopy (SEM), energy dispersive x-ray analysis (EDX), x-ray diffraction analysis (XRD) and atomic absorption spectrophotometry (AAS).

2.2 Results

The SEM photomicrograph presented on Fig. 1 shows the dried slimy precipitate to consist of a porous system of framboid-like flocs about 60 μm in diameter. If this floc size is similar to that which develops in the field, the flocs could migrate through gravel surrounding drainage pipes but would probably not migrate through a graded sand filter or drain.

The SEM-EDX analysis obtained on the same dried slime (Fig. 2) indicates the presence of abundant calcium, iron and sulphur, plus lesser amounts of zinc, manganese, silicon, magnesium and aluminum. X-ray powder diffraction traces (run on the same dried flocs) yielded peaks for calcite only ($CaCO_3$). Quantitative AAS cation analyses run on the precipitates dissolved in HCl yielded ~ 22% calcium, ~ 10% iron and ~ 1% magnesium. These results suggest that the slime consists mainly of calcite that is coated or intermixed with amorphous material consisting of iron sulphide.

Fig. 3 presents the results of the pH and E_h measurements and the bacterial analyses on the leachate during the 8-day experiment. As shown on Fig. 3a, the pH increased slightly from 6.9 to 7.3, while the E_h decreased from +45 mV to -240 mV over the eight-day period. The decreases in E_h are attributed to depletion of oxygen and correspond to increases in pH due to loss of hydrogen ions. The following mechanism for the formation of the slime is proposed (Stumm and Morgan, 1981):

$$Fe^{+++} + e = Fe^{++} \tag{1}$$

$$Fe(OH)_3 + 3H^+ + e = Fe^{++} + 3H_2O \tag{2}$$

$$SO_4^= + 9H^+ + 8e = HS^- + 4H_2O \tag{3}$$

During the process represented by Equations (1) and (2), iron-reducing bacteria possibly activated by exposure to room temperatures produce Fe^{++} at the initially observed E_h values. Decreases in E_h occur apparently due to depletion of the oxygen by aerobic and facultative bacteria, which derive their energy from the organic matter in the leachate (Langmuir, 1972; Baedecker and Back, 1979). As the E_h decreases, the activity of sulphate-reducing bacteria produces sulphide (Equation 3) which precipitates Fe^{++} as FeS (Chian and DeWalle, 1976). The ferrous sulphide (FeS), which imparts the black colour to the slime, does not appear on the x-ray diffraction patterns indicating that it is amorphous to x-rays. On exposure to air the slime quickly changes colour from black to reddish brown, probably resulting from re-oxidation of the amorphous FeS to $Fe(OH)_3$.

The calcite observed in the slime may have formed in the leachate as a result of HCO_3^- production due to biodegradation of organic matter. Berner (1971) and Stumm and Morgan (1981) have noted that bacterial reduction of $SO_4^=$ to either HS^- or H_2S will result in the formation of bicarbonate ion, HCO_3^-. These authors suggest that the presence of HCO_3^- may further lead to the precipitation of dissolved Ca^{++} as $CaCO_3$ (calcite).

The slight increase in the pH of the leachate (Fig. 3) may be due

Fig. 1   Scanning electron photomicrograph of flocculated precipitate
from domestic waste leachate standing at 22°C ± 3°C for eight days

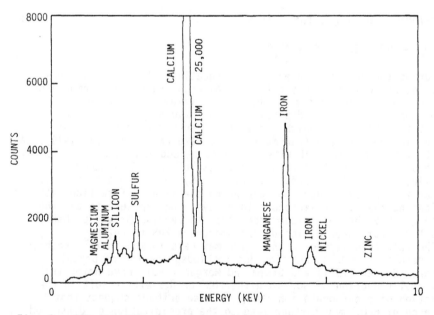

Fig. 2.   SEM-EDX (Energy dispersion analysis of x-rays) traces
obtained on dry flocs of slime

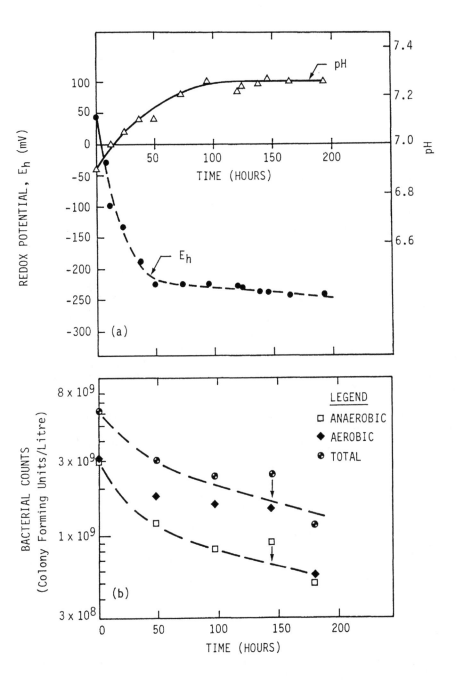

Fig 3. Changes in pH, E_h and bacterial population during 8-day experiment on domestic waste leachate stability

to the consumption of H^+ to form water during the reduction of $Fe(OH)_3$ and $SO_4^=$ (Equations 2 and 3). Apgar and Langmuir (1971), on the other hand, suggest that increases in leachate pH would result from the production of NH_4^+ and HCO_3^- during bacterial reduction of nitrate and nitrite.

The aerobic, anaerobic and hence total bacterial counts appear to have decreased slightly during the experiment indicating loss of nutrients by bacterial utilization.

3 Laboratory Diffusion

In laboratory diffusion tests it is important to use real MSW leachate as the source solution. The reason is that single salt source solutions produce lower diffusion coefficients (even for conservative chloride) due to counter-diffusion from the soil to the source solution (Barone et al, 1989). A complicating factor in the use of real leachate is calcite precipitation and the loss of Ca^{++} as already described.

The changes in cation concentration of a source solution over a Ca^{++}-rich clayey soil are illustrated on Fig. 4 for a 15-day diffusion test. The decreases in Na^+ and K^+ represent diffusion losses to the soil whereas the Ca^{++} losses represent precipitation. The slight increases in Mg^{++} represent diffusion from the soil.

These features seem confirmed by the porewater cation concentration profiles illustrated on Fig. 5. For reference, the background concentrations obtained on pore water squeezed from the samples at 25 MPa are shown by arrows at the bottom of the figure. This pressure was high enough to yield adequate water for analysis and low enough not to squeeze out double layer water. The Cl^-, Na^+ and K^+ profiles are "normal" diffusion profiles from a source containing a higher concentration than the soil. The Ca^{++} profile is complex since it represents a combination of exchange site Ca^{++} desorption by the downward diffusing Na^+ and K^+ followed by upward Ca^{++} diffusion towards the lower Ca^{++} concentration source solution. Because of reservoir precipitation this does not increase the Ca^{++} concentration in the source reservoir. On the other hand, upwards migration of desorbed Mg^{++} may be causing the source solution to increase in Mg^{++} concentration.

In addition to the "strange" looking diffusion curves, the biological precipitation of Ca^{++} in the source solution (and possibly also in the soil) will have the net result of decreasing somewhat the chloride ion diffusion coefficient due to counter-diffusion of Ca^{++}. Barone et al (1989) calculated the D_{Cl^-} to be 7.5×10^{-6} cm^2/s for the MSW leachate source illustrated on Fig. 5.

4 Field Diffusion

Two major liner exhumations have been carried out at Metro Toronto's Keele Valley Landfill as part of the clay barrier performance monitoring program. On each excavation occasion, the upper 5 to 10 cm

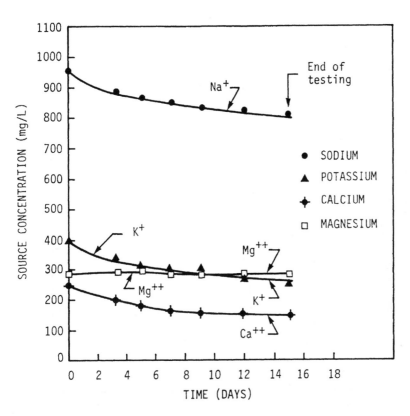

Fig. 4. Changes in source concentration with time during a laboratory diffusion test with real leachate

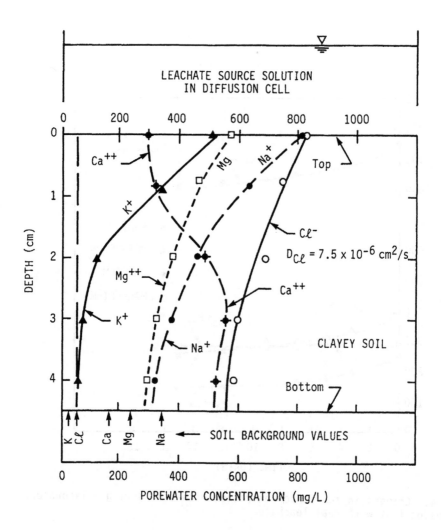

Fig. 5. Porewater ion concentration profiles in the soil at the end of a laboratory diffusion test with source solutions and legend as shown on Fig. 4 (adapted from Barone et al, 1989)

of the 30 cm thick sand drainage layer was observed to be darkened by
a black slime similar to those observed in the laboratory. In
addition, the underlying 10± cm of sand was observed to have been
reduced to a grey colour from the original light brown colour (Fig.
6).

A combination of excess leachate and drainage layer clogging
resulted in rapid leachate flow along the top of the sand layer
towards lower elevation drains (along with slower laminar flow within
the sand drainage layer). This combination of factors resulted in a
diffusion controlled profile for chloride at time = 4.25a which
started at the top of the sand layer rather than at the top of the
clay liner as clearly illustrated on Fig. 6. This resulted in a
barrier with an effective thickness of 1.5 m rather than 1.2 m.

Porewater samples from the sand were collected by high speed
centrifugation. For the clays, squeeze pressures of up to 50 kPa were
used to collect adequate water for testing without altering the cation
concentrations.

The error function solution to Fick's second law for transient
diffusion was used to calculate a diffusion coefficient for the
chloride data on Fig. 6.

$$\frac{C}{C_0} = \text{erfc} \left(\frac{x}{2\sqrt{Dt}} \right) \tag{4}$$

where C   = concentration at depth x at time t
       C_0 = constant source concentration
       D   = diffusion coefficient
       erfc = complementary error function of item in the brackets

A diffusion curve having D = 6.5 x 10^{-6} cm^2/s was fitted to the
measured chloride data. This particular calculation assumed the
velocity of advective flow was zero which was a good approximation for
the low k clay and the low gradient field situation. This field D_{Cl^-}
value is very similar to that for the lab test on Fig. 5 and to field
values calculated for the Confederation Road Landfill at t = 16a
(Quigley et al, 1987) all at similar water contents of 17 to 22%.

The corresponding cation profiles are presented on Fig. 7 for the
field case, and their similarity to the laboratory profiles on Fig. 5
is striking. The same hardness halo of desorbed Ca^{++} is visible as is
back diffusion of Ca^{++} towards the sand layer. Again, there is the
possibility of carbonate precipitation contributing to the drop in
Ca^{++} values but this was not confirmed.

5 Conclusions

Bacterial processes play a significant role in defining the short-term
stability of MSW leachate. Rapid precipitation of calcium carbonate
(calcite) and amorphous iron sulphide rapidly depletes the leachate of
calcium and iron.

In laboratory diffusion tests, chemical profiles from clayey test

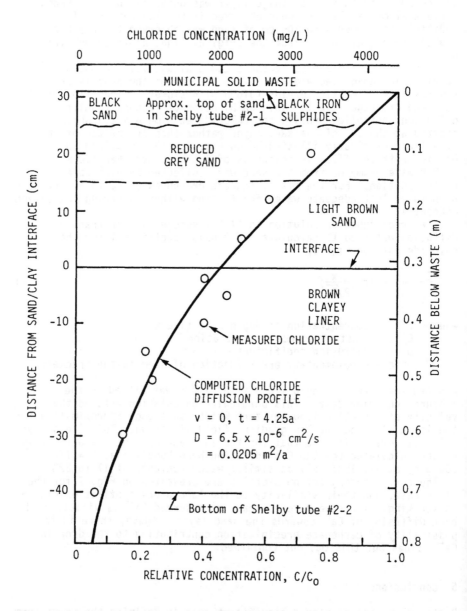

Fig. 6. Porewater chloride profile for East Pit Shelby Tubes #2-1 and 2-2. Calculated and fitted chloride diffusion profile using $D = 6.5 \times 10^{-6}$ cm^2/s and $v = 0$. Clay thickness = 1.2 m (Adapted from Reades et al, 1989)

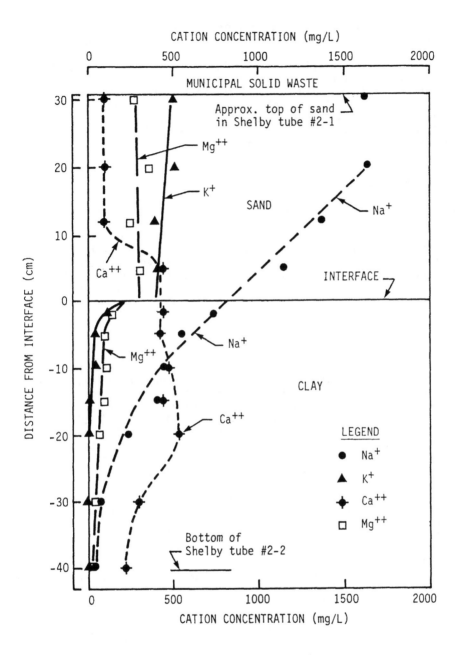

Fig. 7. Porewater cation profiles for East Pit Shelby Tubes #2-1 and 2-2.

specimens indicate Ca^{++} diffusion from the soil back into the leachate, especially if Ca^{++} is the dominant cation adsorbed on the clay.

In the field, precipitation of calcite and iron sulphide partially clogs drainage layers. This contributes to leachate flow across the top of the sand and diffusion profiles that start at the top of the sand rather than at the top of the clay liner. This adds extra "effective thickness" to the liner system. Cation profiles in the field barrier indicate complicated diffusion profiles similar to those seen in the lab using real leachate.

6 Acknowledgements

The theoretical work presented in this paper was funded by the Natural Sciences and Engineering Research Council of Canada. The bacterial analyses were performed at the University Hospital under the supervision of Dr. J. Whitby. The field work was funded by Metro Toronto through their geotechnical consultants, Golder Associates. The authors are particularly grateful to Metro Toronto and Golder Associates for permission to use some of the field data from the Keele Valley liner exhumation and monitoring program.

7 References

Apgar, M.A., and Langmuir, D. 1972. Groundwater pollution potential of a landfill above the water table. **Groundwater**, 9 (6), 76-94.

Baedecker, M.J., and Back, W. 1979. Hydrogeological processes and chemical reactions at a landfill. **Groundwater**, 17 (5), 429-437.

Barone, F.S., Yanful, E.K., Quigley, R.M., and Rowe, R.K. 1989. Effect of multiple contaminant migration on diffusion and adsorption of some domestic waste contaminants in a natural clayey soil. **Canadian Geotechnical Journal**, 26 (2), 189-198.

Berner, R.A. 1971. **Principles of Chemical Sedimentology**. McGraw-Hill, New York, 243 p.

Chian, E.S.K., and De Walle, F.B. 1976. Sanitary landfill leachates and their treatment. **Journal of the Environmental Engineering Division**, ASCE, 102, 411-431.

Langmuir, D. 1972. Controls on the amounts of pollutants in subsurface waters. **Earth and Mineral Sciences**, 42 (2), 9-13.

Quigley, R.M., Yanful, E.K. and Fernandez, F. 1987. Ion transfer by diffusion through clayey barriers. **Proc. Geotechnical Practice for Waste Disposal '87**, GT Div. ASCE, Ann Arbor, June, Geotechnical Special Publication No. 13. 137-158.

Reades, D.W., King, K.S., Benda, E., Quigley, R.M., LeSarge, K. and Heathwood, C. 1989. The results of on-going monitoring of the performance of a low permeability clay liner, Keele Valley Landfill, Maple, Ontario. **Proc. Focus Conference on Eastern Regional Ground Water Issues**, National Water Well Association, Kitchener, Ontario, October, 79-91.

Stumm, W. and Morgan, J.J. 1981. **Aquatic Chemistry: An Introduction Emphasizing Chemical Equilibria in Natural Waters,** 2nd Edition, John Wiley & Sons Inc., New York, 583 p.

23 THE EFFECTS OF HEAVY CIVIL ENGINEERING AND STOCKPILING ON THE SOIL MICROBIAL COMMUNITY

J.A. HARRIS and P. BIRCH
Environment and Industry Research Unit, Polytechnic of
East London, UK

Abstract
The soil microbial community plays a vital role in soil
with regard to both nutrient cycling and the maintenance of
a stable structure. During heavy civil engineering
operations this community can be significantly altered with
respect to it's size, composition and activity. Studies
have been made of soils from areas prior to disturbance,
during the course of stockpile construction and during the
first few months of storage. Significant changes in the
microbial community have been measured as a result of these
operations, with fungal groups being particularly adversely
affected. There is also a steady accumulation of ammonium
in the deeper parts of stockpiled soil, confirming earlier
work which suggested that anaerobic conditions prevail
within the store. Data from restored areas indicates that
considerable patchiness in soil characteristics may arise,
due to soils coming from different parts of a soil store.
The consequences for restoration are discussed, using
examples and the potential to use microbial assays as
indicators of soil quality outlined.
Keywords: Micro-organisms, civil engineering, soil
disturbance, indicators, structural stability, nutrient
cycling.

1 Introduction

The soil system that exists today has been formed over
a period of many years, thousands in the case of soils in
areas uncovered after the most recent glaciation, and
millions for some very ancient soils. The soil microbial
community has had a very active and central role in this
development, extending back to the first evolution of an
oxygen atmosphere, and subsequent development of
terrestrial communities. The most important primary
colonisers of bare rock are lichens, a mutualistic
association between algae and fungi.
In modern soils the microbial community is responsible

for mediating the cycling of many nutrients, in some cases solely, and is intimately involved in the maintenance of an open soil structure. The microbial biomass may only form a small proportion of the total of an ecosystem's biomass, but it's functional importance is critical in cycling nutrients. The majority of nitrogen (over 95%) incorporated into the soil-plant system is fixed by micro-organisms, in symbiotic association with other organisms, e.g. plants, or as free living organisms (Richards, 1987). Many of the nitrogen cycling pathways, such as nitrification, the conversion of ammonium to nitrate via nitrite, is solely mediated by micro-organisms. Molope and co-workers (1987) have demonstrated that micro-organisms have an essential role in maintaining a stable soil structure, and the importance of vesicular arbuscular mycorrhizae (VAM) in binding soil particles together in sand dunes has been reported (Clough and Sutton, 1978).

What is of central importance is that soil is a dynamic system with soil structure being the result of a constant turnover of microbial biomass and the associated cycling of nutrients, and that once a stable structure has been achieved it is by no means fixed in an engineering sense - it may very easily be destroyed by disrupting the microbial community.

This present paper aims to bring together the work carried out in recent years to demonstrate the effects of disruptions caused by civil engineering operations, and indicate some opportunities for amelioration of the damage caused.

2 Disturbances caused by civil engineering operations

The effects of heavy earthmoving equipment have been described by Ramsay (1986). There are both shear and compaction forces imparted to the soils, which are particularly pronounced when motorised scrapers are used to move soil. It has been reported that standing forces of 8 kg cm^{-2} are exerted by such equipment, and that transitory peaks during operation may be even higher.

McRae (1989) reviewed the current evidence on practice of restoration from mineral workings in the UK, and concluded that successful restoration was feasible, provided there was sufficient co-operation between mineral contractors and statutory bodies. However little mention was made of the importance of the microbial community in the restoration process.

3 Effects of moving soils

Harris (1989) reported the results of an investigation

for a soil sampled prior to moving and then on the day of
disturbance; several important effects on the microbial
community were noted. Samples were taken from 0-30cm, 90-
120cm and 180-210cm from the surface.

On the day of store construction there was no
significant effect on total bacterial numbers, as measured
by plate-counts, caused by the process of store
construction. However, there was a large increase in the
numbers of bacteria found one month after construction,
with consistently higher numbers found in the store at all
depths as compared when the control area (Fig.1, values are
expressed as a percentage of those found in an adjacent
undisturbed control area). After 3 months of storage, the
numbers had fallen to pre-disturbance levels. Fungal
propagule numbers, however, exhibited a marked decrease in
numbers at all levels in the soil store, in comparison to
the control, on the day of store construction (Fig.2).
There was a significant difference (p=0.001) between the
control area soil and the two deepest zones sampled. One
month after construction, all microbial numbers had
declined but the numbers in the 0 - 30 cm zone were now
significantly higher than those found in other samples.
After 3 months the numbers had not altered in the 0 - 30 cm
zone, had increased by more than three fold in the 90 - 120
cm zone, but neither of these values was significantly
different from the control area value. The samples from
the 180 - 210 cm zone were by now significantly lower than
those found in the control area. This is consistent with
the hypothesis that fungi are the group most susceptible to
the damage caused by earth moving.

At the same time the biomass results (ATP) remain
steady in the period leading up to store construction
(Fig.3). On the day of store construction, however, there
were significant falls in biomass in the store soils as
compared to the control. The total biomass within the
store samples declined after 3 months, when there was a
large increase in the biomass in the control area. The
biomass from the 180-210 cm level had, however, declined to
half of that of the other zones and control area, and was
significantly lower than all of them.

Incubation studies indicated that after 3 weeks
incubation there has been a greater mobilisation of organic
to inorganic nitrogen in the samples taken from deeper
within the store. This evidence suggests that after soil
has been re-spread there could be an accelerated conversion
of organic to inorganic forms. This could be lost by
volatilisation as ammonia, especially if the pH has been
increased by liming, by leaching as nitrate/nitrite or as
run-off.

There is an immediate effect, therefore, on the size of
the microbial community caused by soil movement during
construction. This appears to be almost entirely due to a

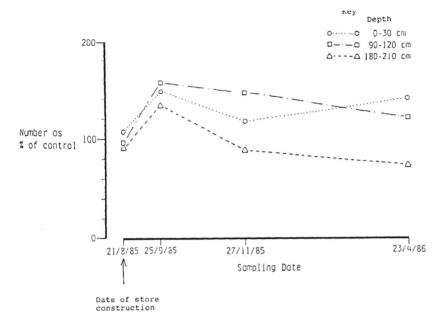

Fig.1   Bacterial numbers in three depth zones of a soil store expressed as a percentage of the undisturbed control values, plotted against time

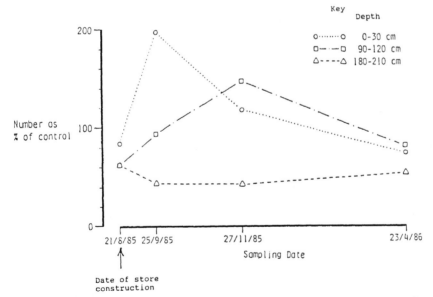

Fig.2   Fungal propagule numbers in three depth zones of a soil store expressed as a percentage of the undisturbed control values, plotted against time

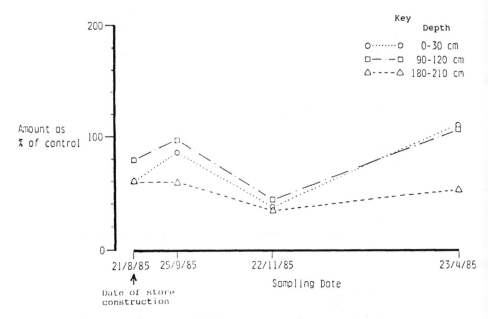

Fig.3   Microbial biomass in three depth zones of a soil store
         expressed as a percentage of the undisturbed control
         values, plotted against time

decrease in the fungal population, which declines at the
same time, although it must be noted that there is a higher
number of fungal spores recovered in the 0-30 cm zone of
the store than the control 1 month after storage. This is
accompanied by a large and significant increase in the
bacterial population, in particular in the first few months
after storage Gram negative bacteria exhibited a very large
increase in numbers in the surface of the soil store.

This response appears to be remarkably similar to that
noted by a number of workers when a soil is tilled (Molope
et al, 1987), whereby there are transitory increases in the
biomass of fungi and bacteria, accompanied by a temporary
increase in soil structural stability. However, the soils
under investigation in this study are subjected to far
greater compaction and shear forces than occurs during
ploughing (Ramsay, 1986), with a resultant fall in biomass
not found in agricultural operations.

4 Effects of storage

Abdul-Kareem & McRae (1984) reported on the results of
a survey of soil stores from a variety of civil engineering
sites, and found that there had been significant reductions
in the soil's infectivity with regard to vesicular
arbuscular mycorrhizae. Miller et al (1985) demonstrated

the importance of moisture content on the survival of VAM
propagules in stores, and this work was confirmed and
extended by Harris et al (1987) who found that soil further
than 1m from the store surface was unable to infect
bioassay plants with mycorrhizal fungi. Visser et al
(1984) found decreases in microbial carbon in stored soils
but attributed this to incorporation of subsoil material
during store construction.

Harris et al (1989) in an intensive study of several
stores on one site found profound changes in the size and
composition of the soil microbial community. There was an
overall reduction in the numbers of fungi and bacteria,
with far fewer being found in the deeper parts of an older
store. Combining the results of this and the initial
changes described above (Harris, 1989), it is possible to
see a gradual decline in the 'carrying capacity' of soils
in storage (Fig.4). Here the letters A to E refer to
stores of increasing age, followed by number denoting depth
of sample. Therefore D200 is a sample from 200cm below the
surface of store D. Store A was 3 months old, B 12 months,
C and D 48 months and E 252 months old. The letter S
refers to the store sampled in the study described above,
followed by a depth, then time after construction.
Therefore S120,3 denotes 120 cm depth sample 3 months after
construction. There are three control areas; REF an
undisturbed pasture, CONT an area adjacent to the store
monitored pre and post storage, and REFSUB an area of
subsoil which had been exposed for several months. A
cluster analysis (average method, STATGRAPHICS, 1988) has
been performed to ascribe groups to the data. The graph
shows quite clearly a decline in values from the REF area
in the top middle of the plot through the upper zones of
the younger stores (clusters 1 and 2) through middle and
upper parts of older stores (cluster 3) to cluster 4 in the
lower part of the left hand side, which represents samples
from the deeper parts of the older stores (D and E) and a
subsoil control.

The decline in numbers and biomass in the deeper parts
of the store with time are most likely due to stress,
including the lack of oxygen, preventing any increase in
the microbial biomass, and the dominance of the community
by anaerobic forms.

This work has led to the proposal of a descriptive
model of soil store, with three zones being recognised,
each with their own distinct characteristics (Harris et
al., 1989). The three zones are: an aerobic zone, which
remains anaerobic during storage; an anaerobic zone, which
remains anaerobic during the course of storage; and a zone
of transition which fluctuates between these two states.
The salient features of this model are the transitional
nature of the changes which occur in the aerobic zone such
as a flush in bacterial numbers which eventually return to

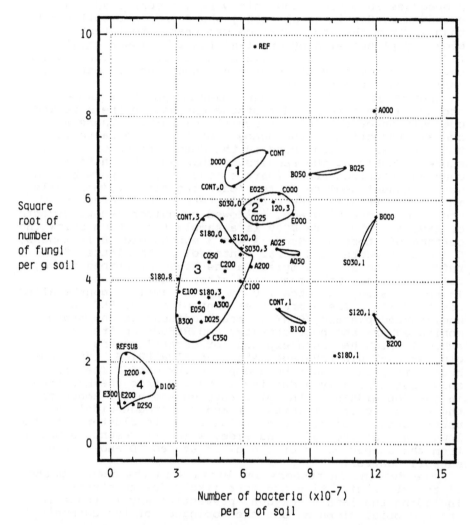

Fig 4   Square root of number of fungi (g soil) plotted against
number of bacteria ($\times 10^{-7}$/g soil) for soil store data

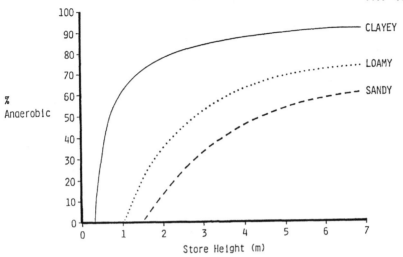

Fig.5   Percentage of stored soil which is anaerobic, dependant
         upon depth and textural class

values found in undisturbed soils, and the permanent
decreases in the populations of aerobic bacteria and fungi
coupled to an accumulation of ammonium in the anaerobic
zone.
     The obvious deleterious effects of anaerobic conditions
within stores leading to exacerbation of the damage caused
by soil movement, leads to a consideration of management of
soil stores to keep as much of the soil in an aerobic state
as possible. Evidence from a number of sources (Abdul-
Kareem, 1983; Abdul-Kareem and McRae, 1984; Harris, 1989)
suggests that the depth of the anaerobic zone is dependant
on the soil's textural classification (Fig.5).
     From this simple relationship it becomes clear that
sandy soils may be mounded higher than clay soils, without
producing anaerobic conditions. For example a 2m high
store would be 12% anaerobic if made of sandy soil, 35% if
a loamy soil, and 78% if a clayey soil. This gives rise to
the possibility to introduce guidelines, to minimise the
amount of anaerobic soil that a store more contain or,
perhaps more pragmatically, to what extent specific
remedial management will have to be applied after storage.

5 Return to agriculture/conservation

The return of soils re-instated after mineral extraction to agriculture has been well documented (Younger, 1989). The general consensus is that with timely cultivation involving conscientious attention to soil conditions, there need not be any problems with returning yields to those obtained before disturbance. However the problem of compacted, waterlogged soils remains, even where subject to the most carefully monitored management programme (King, 1988), and there remains a need to address more fully the rehabilitation of soil structure. Some significant moves in this direction have been made by encouraging the development of earthworm populations (Scullion et al, 1988) and this line of investigation holds out much promise.

There is also emerging a clearer picture with regard to the changes in the microbial community in areas previously disturbed by opencast mining. Miller (1987) has described the changes in the populations of vesicular-arbuscular mycorrhizae and there are distinct differences in soils which have been lifted and immediately replaced as compared to those which have spent some time in storage, with the consequence that sites restored using stored soils are less likely to develop a diverse plant community rich in native species. Miller (1987) has attributed this to a depletion in the buried soil seed bank, a situation also reported in a UK study (Dickie et al., 1988). This undoubtedly results from anaerobic conditions, leading to a decline in the seed and mycorrhizal populations. Further to this, any recovery which has occurred in the aerobic part of the store may be lost when the soil is handled during respreading.

Harris and Birch (1989) sampled a number of areas on opencast coal mine restorations, including a soil store and compared them with results obtained from samples of undisturbed pasture and cereal fields. Values of soil enzyme (dehydrogenase) activity ranged from 140 to 580 μg triphenyl formazan (TPF) formed g^{-1} soil $24h^{-1}$ in the undisturbed soils, whereas the range was 10 to 220 μg TPF g^{-1} soil $24h^{-1}$ in the reinstated soils, with the lowest values being recorded in the most recent restorations. Nitrogen mineralisation was found to be stimulated by storage and subsequent reinstatement, confirming the suggestion made by Harris (1989) that large losses from the nitrogen pool could result from the storage/re-instatement process. The authors also found a significant ($r = 0.945$, at $P=0.005$) and positive correlation between the nitrifying potential and the soil water holding capacity, indicating the link between improvement in soil structure and development of a competent microbial community. The authors also found that the soils were switched from a low ammonium/high nitrate nitrogen economy characteristic of arable soils (Richards, 1987) to a high ammonium/low nitrate economy more common in climax ecosystems, typically woodlands.

6 Possible use as indicators

Insam and Domsch (1988) investigated the size of the soil organic carbon pool and it's living microbial component in a chronosequence of soils reclaimed to agriculture or forestry after opencast lignite mining. They found increases in carbon and microbial biomass carbon, with increasing age after restoration, but a decrease in microbial carbon as a percentage of total carbon with increasing age. This decrease was more rapid on forest sites than agricultural ones. They concluded that these measures were superior to other parameters in determining the progression of restoration with respect to undisturbed systems, but that the sites under study had not reached equilibrium conditions even 50 years after re-instatement.

In an interesting parallel to this, Hart et al (1989) reported an investigation into areas restored to pasture 10 and 25 years previously, and had earlier data from the same sites to hand to construct a chronosequence. They confirmed the utility of the microbial-carbon/total-carbon ratio suggested by Insam and Domsch (1988), but further suggested that other parameters may be just as useful, provided that suitable control sites were available. They suggested that microbial biomass, soil respiration and net nitrogen mineralisation could all be used in this regard, as they all increased with increasing time after restoration. This contention would appear to be further supported by the work of Harris and Birch (1989), which has demonstrated an increase in soil microbial activity with increasing age after restoration.

The use of soil profiles of microbial activity has also been proposed (Harris and Birch, 1990). Soil core samples were taken from areas restored 5 and 10 years previously and compared with undisturbed areas, divided in to 5cm segments and analysed for dehydrogenase activity and carbon content. The microbial activity measurements proved to be superior to the carbon content in distinguishing between age after re-instatement and management regime, with profiles similar to the undisturbed controls developing more rapidly under cutting as opposed to grazing regimes.

7 Manipulation of the biomass to enhance restoration

There are few examples of the soil microbial community being manipulated per se to effect a speedy recovery after civil engineering operations. More usually the management programme has been aimed at the plant community directly, with the addition of fertilizers, either inorganic or organic, or the introduction of seeds or seedlings. That not all of these approaches have proved successful is not surprising, since the microbial component has only been mentioned in passing or, more often, completely ignored. Introduction of micro-organisms to a site, although beneficial, has often been accidental, such as the introduction of Rhizobium as a result of sowing leguminous species on a site. However there are one or two indicators of how progress may be made in enhancing the rate and success of re-instated areas. The effects on mycorrhizal species outlined above suggests that a programme of investigation into the use of mycorrhizal inoculation is required.

Biondini et al (1985) found an increase in the enzyme activity of reclaimed soils to be positively correlated with dominance of perennial species in the vegetative cover of a strip mine spoil. It may be possible that by introducing seed mixes dominated by perennials that a large and competent biomass may be developed rapidly, and this line of research certainly warrants further attention.

Some attempts at manipulating the microbial community directly have been made, with some degree of success but on very unusual substrates. Ripon and Wood (1975) reported on a series of investigations into the microbiological aspects of pulverised fuel ash (PFA) intended for use as a plant growth substrate. They followed an initial investigation of natural rates of colonisation with a set of amendment experiments, where either sewage sludge, chopped grass, cellulolytic fungi, phosphate solubilising bacteria, Azotobacter (a free-living nitrogen fixing bacterium), or inorganic fertilizer were added to the PFA. Barley was used as a test crop. None of the bacterial treatments had a significant effect on yield, but the Azotobacter treatment did lead to a significant increase in the total nitrogen concentration. When sewage sludge was added there was a significant increase in yield compared to the control, and there was a further highly significant increase when the cellulolytic fungus was also added. This was repeated with grass as the organic amendment, and showed significant increases with all microbiological additions. The highest yield was obtained when all three microbiological amendments were made, with both organic and inorganic fertilizers, and this was significantly higher than the yield obtained from just adding the two fertilizers. this provides clear evidence for the efficacy of microbial amendment with a suitable organic substrate, and it is perhaps unfortunate that no attempt has been made to repeat these experiments with topsoils disturbed by civil engineering.

Measurements of soil aggregate stability were made on volcanic ash from the 1980 eruption of Mount St. Helens, Washington State, USA (Lynch and Elliott 1983). They added fungi and bacteria to ash and found significant increases in the structural stability of the ash, after a period of incubation. This is consistent with the findings of Clough and Sutton (1978), who reported that fungi were major contributors to the aggregation of sand particles in dune soils. This again offers an avenue of research which should prove fruitful, the inoculation of unstable substrates with microbial species capable of binding soil particles together.

8 Conclusion

It is clear from the evidence available that the forces applied to soils during civil engineering operations have profound effects on the soil microbial community. These changes are not easily reversible, linked as they are to deterioration in soil structure, and may require many years of patient management to re-establish a stable and fertile soil system.

However, there are several pointers to the application of our knowledge of the soil microbial community to enhance restoration rates and end points. The microbial community may be used to indicate the state of the soil system, as it provides an integration of those physico-chemical characteristics that are used to describe the state of a soil. Soil micro-organisms offer a very sensitive means of monitoring changes in soil oxygen status, organic matter, plant available nutrients and similarly important soil factors contributing to good plant growth. Measurement of the size, composition and activity of the soil microbial community may be used to produce a classification of soils with respect to one another, providing 'target' restoration end points which may be achieved by manipulation of the soil microbial community of the re-instated soil.

Restoration to different end uses from those prior to disturbance are also available, with our ambitions not limited solely to agriculture. The change from a nitrate to a ammonium nitrogen 'economy' of soils provides an opportunity to return them from agricultural to conservation uses, especially woodlands.

The opportunities for useful, intelligent application of these findings are legion, which should hand in hand with further research effort. This will provide both practical benefits in the form of an enhanced environment and the extension of our knowledge of the interaction between the soil microbial community and it's physico-chemical environment.

Acknowledgements

The authors thank the Opencast Executive, British Coal for financial support during this project, and Ian Carolan, British Coal, for useful discussions.

References

Abdul-Kareem, A. (1983). The effects on topsoil of long-term storage in stockpiles. Ph.D. Thesis, University of London.

Abdul-Kareem, A. and McRae, S. (1984). The effects on topsoil of long-term storage in stockpiles. **Plant and Soil** 76, 357-363.

Biondini, M.E., Bonham, C.D., and Redente, E.F. (1985). Relationships between induced successional patterns and soil biological activity of reclaimed areas. **Reclamation and Revegetation Research** 3, 323-342.

Clough, K.S., and Sutton, J.C. (1978). Direct observation of fungal aggregates in sand soil. **Canadian Journal of Microbiology** 24, 333-335.

Dickie, J.B., Gajjar, K.H., Birch, P., and Harris, J.A. (1988). The survival of viable seeds in stored topsoil from opencast coal workings and its implications for site restoration. **Biological Conservation** 43, 257-265.

Harris, J.A. (1989). A study of the interrelationships between the microbial community, organic matter and physico-chemical conditions in stored topsoils. Ph.D. Thesis, Council for National Academic Awards.

Harris, J.A. and Birch, P. (1989). Soil microbial activity in opencast coal mine restorations. **Soil Use and Management** 5, 155-160.

Harris, J.A. and Birch, P. (1990). Application of the principles of microbial ecology to the assessment of surface mine reclamation, in **Mining and Reclamation, Proceedings of the 1990 Conference** (eds J.Skousen and J.Sencindiver) American Society for Surface Mining and Reclamation (In Press).

Harris, J.A., Birch, P., and Short, K.C. (1989). Changes in the microbial community and physico-chemical characteristics of topsoils stockpiled during opencast mining. **Soil Use and Management** 5, 161-168.

Harris, J.A., Hunter, D., Birch, P., and Short, K.C. (1987). Vesicular-arbuscular mycorrhizal populations in stored topsoils. **Transactions of the British Mycological Society** 89, 600-603.

Hart, P.B.S., August, J.A., and West, A.W. (1989). Long-term consequences of topsoil mining on select biological and physical characteristics of two New Zealand loessial soils under grazed pasture. **Land Degradation and Rehabilitation** 1, 77-88.

Insam, H., and Domsch, K.H. (1988). Relationship between soil organic carbon and microbial biomass on chronosequences of reclaimed sites. **Microbial Ecology** 15, 177-188.

King, J. (1988) Some physical features of soil after opencast mining. **Soil Use and Management** 4, 23-30.

Lynch, J.M. and Elliott, L.F. (1983). Aggregate stabilization of volcanic ash and soil during microbial degradation of straw. **Applied and Environmental Microbiology** 45, 1398-1401.

McRae, S. (1989). The restoration of mineral workings in Britain - A Review. **Soil Use and Management** 5, 135-142.

Miller, R.M., Carnes, B.A., and Moorman, T.B. (1985). Factors influencing survival of vesicular-arbuscular mycorrhiza propagules during topsoil storage. **Journal of Applied Ecology** 22, 259-266.

Miller, R.M. (1987). Mycorrhizae and Succession, in **Restoration Ecology: A synthetic approach to ecological research** (eds W.R. Jordan, M.E. Gilpin, and J.D. Aber) Cambridge University Press, Cambridge, pp. 205-219.

Molope, M.B., Grieve, I.C., and Page, E.R. (1987). Contributions by fungi and bacteria to aggregate stability of cultivated soils. **Journal of Soil Science** 38, 71-77.

Ramsay, W.J.H (1986). Bulk soil handling for quarry restoration. **Soil Use and Management** 2, 30-39.

Richards, B.N. (1987). **The Microbiology of Terrestrial Ecosystems** Longman Scientific and Technical, Harlow, Essex.

Ripon, J.E., and Wood, M.J. (1975). Microbiological aspects of pulverised fuel ash, in **The Ecology of Resource Degradation and Renewal** (eds M.J. Chadwick and G.T. Goodman), Blackwell Scientific Publications, Oxford, pp. 331-349.

Scullion, J., Mohammed, A.R.A., and Ramshaw, G.A. (1988). Changes in earthworm populations following cultivation of undisturbed and former opencast coal-mining land. **Agriculture, Ecosystems and Environment** 20, 289-302.

Statistical Graphics Corporation, STATGRAPHICS V3.0 (1988).

Visser, S., Fujikawa, J., Griffiths, C.L. and Parkinson, D. (1984). Effect of topsoil storage on microbial activity, primary production and decomposition potential. **Plant and Soil** 82, 41-50.

Younger, A. (1989). Factors affecting the cropping potential of reinstated soils. **Soil Use and Management** 5, 150-154.

24 CASE STUDIES OF FLOOR HEAVE DUE TO MICROBIOLOGICAL ACTIVITY IN PYRITIC SHALES

R.J. COLLINS
Building Research Establishment, Watford, UK

Abstract
Many shales contain iron pyrites (FeS_2) which may be oxidised to
sulphuric acid by Thiobacillus ferrooxidans. This is the origin of
acid mine drainage and of most cases of high sulphate soil both of
which can cause severe attack and disintegration of ordinary Portland
cement concrete. If there is a source of acid-soluble calcium (e.g.
calcite) within the shale, formation of gypsum ($CaSO_4.2H_2O$) can force
apart the layers and cause heaving problems. The process is normally
initiated by allowing of access air either by using the shale as a fill
material or in bedrock by the digging of foundations, services etc.
The present paper examines two cases of heave, one in bedrock and
another in fill and discusses some of the criteria to be applied in the
identification of potential for heave.

1 Introduction

It is now over 10 years since the first case of pyritic heave was
reported in the UK (Nixon 1978). Failures of ground floor slabs,
lifting of floors and internal walls and pushing out of external walls
was attributed to microbiological oxidation of finely divided pyrite
followed by growth of gypsum crystals forcing apart layers of shales
used for fill beneath the ground floor slabs. The shales were locally
quarried and were identified as Whitbian Shales (Upper Lias) of Lower
Jurassic age. X-ray diffraction analyses revealed the presence of
pyrite and calcite which in oxidised and partly oxidised shales gave
rise to gypsum, jarosite and other complex sulphates.

Since the occurrence in the Teeside area there have been two other
reports of pyritic heave in the UK, both in bedrock materials. There
is a short reference in BRE Digests (1979, 1983) to heave problems in
Glasgow which occurred in shales within the Upper Limestone Group of
the Lower Carboniferous Series. More recently Hawkins and Pinches
(1987) have published a detailed paper on heave at Llandough Hospital,
Cardiff which is built on the Westbury Formation (Rhaetic) of Upper
Triassic age.

In this paper further details are given of the Glasgow occurrence
and of a new case of heave in fill materials affecting a warehouse in
the English Midlands.

2 Bedrock Heave (Glasgow)

In 1978 BRE was called to investigate floor heave within a four storey masonry structure built about 100 years ago. The main hall (approx 12 x 25 m) on the ground floor had risen up to 125 mm above the general level and this was associated with a low wall running longitudinally along the centre of the hall with shallow foundations resting on bedrock and supporting the suspended timber floor. The basement of the building extended only under a very small part of the main hall, but in this area some arching of the concrete floor was observed. The main foundations of the building appeared to be unaffected. A trial pit dug beside the sleeper wall revealed a well-broken and partially delaminated shale with white and yellow deposits. Subsequent powder X-ray diffraction analysis confirmed the presence of gypsum, jarosite and and pyrite. These observations are typical for problems of pyritic heave. Similar observations but of lesser extent were found for material dug from beneath the basement floor slab.

Three boreholes were sunk through the bedrock and observations and analyses of the cores retrieved are detailed in figure 1. In borehole 1 beside the sleeper wall pyritic shale containing calcium carbonate as calcite and also in the mixed carbonate ankerite were recorded down to 2.3 m below the top of the borehole. Expansive reaction had taken place for about 1.7 m below current ground level. Between 2.3 and 2.7 m calcite and ankerite were absent from the pyritic shale and this stratum was correlated with shale between 0.9 and 1.2 m below the basement floor. It was considered unlikely that this particular stratum would contribute very much to the observed heave problems. Below this stratum and also in a borehole taken well away from the sleeper wall and on the opposite side of it to the basement area no significant occurrences of pyrite or calcium carbonate were observed.

Recommendations were made to found a new sleeper wall on piles bored through the potentially unstable strata and to excavate pyritic material from below the affected areas of floor in the basement area. After 10 years no further problems have been reported at this site.

3 Expansive Fill (Midlands)

In 1989 BRE was called to investigate floor heave in a 3000 m^2 warehouse. Problems had become apparent soon after construction, with cracks appearing in floors and at the top of internal walls in attached offices. Foundation walls had also been pushed outwards causing cracking in brickwork cladding. The maximum movement was about 25 mm which was recorded at one corner where the fill was thickest (about 2 m). The ground level had been reduced to about 2 m below floor level on this side of the building facilitating coring horizontally through the foundation wall into the fill material. This had been carried out by a consultant who gave the following analysis of the material:

GROUND FLOOR LEVEL

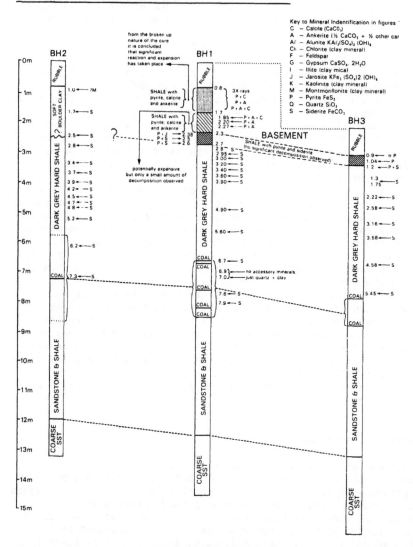

Figure 1 Bore hole data (Glasgow)

Hole no	pH	Moisture	Sulphide	Acid soluble sulphate	Pyritic sulphide
2	4.6	2.7%	17 ppm	4.1%	1.7%
3	3.6	8.0%	12 ppm	2.15%	1.8%

These are class 5 conditions for sulphate attack according to BRE Digest 250 (1981) but there was no evidence of concrete deterioration through chemical attack; either in the foundation walls or the 150 mm thick concrete floor slab; polythene protective membranes had been installed. The consultants suggested that oxidation of pyrites, which is known to cause an increase in volume, was the cause of the heave. However, all cases of heave involving pyrites, as far as is known, have involved other reactions, specifically in this context the formation of gypsum crystals which force apart layers of shale. BRE was engaged to investigate this.

Two samples of 5 kg were delivered to BRE. Both samples consisted of lumps of dark grey shale and clay fines with rust staining and yellowish deposits of jarosite. There was no evidence of inter-layer growth of gypsum crystals and pyrite oxidation products appeared to be confined to the surface of lumps and in the clay fines.

After removal of small samples of materials representative of various shale lumps and the clay fines the bulk samples were reduced in a jaw crusher and riffed to provide representative subsamples of 50 g. This was sieved and ground repeatedly to pass 150 μm. These samples were analysed for total sulphur using a Leco carbon-sulphur analyser and also for calcium oxide (HF dissolution and atomic absorption):

Hole no	Total Sulphur (%) (as S)	(as SO_3)	Total Carbon (%) (as C)	CaO (%)
1	1.6	4.0	1.1	0.36
2	1.3	3.25	1.6	0.39

The characteristic immediately of note is that the calcium oxide content is below the 0.5% threshold for pyritic heave suggested by Nixon (1978). X-ray diffraction analysis of these two samples revealed only a small quantity of gypsum, and no detectable calcite or pyrite. Jarosite was evident in trace amounts only in the sample from hole 1 and the diffractogram for this sample is illustrated in figure 2.

The main minerals in the shale were quartz and an assemblage of clay minerals typical of a wide variety of shales: kaolinite, chlorite and illite (with a peak broadened to the low angle side indicating the probable presence of mixed layer illite-smectite). Figure 2 also shows the presence of minor amounts of feldspar and alunite. The source of this shale at the time of writing is unclear as the mineralogy shows significant differences from that of the documented source material. Middlemen were involved in the supply of the material and it appears that substitution from another source may have occurred. The fill is also unlikely to be a colliery shale because of the low carbon content. Between 5 and 20% carbon should be expected for colliery shales (Collins 1976).

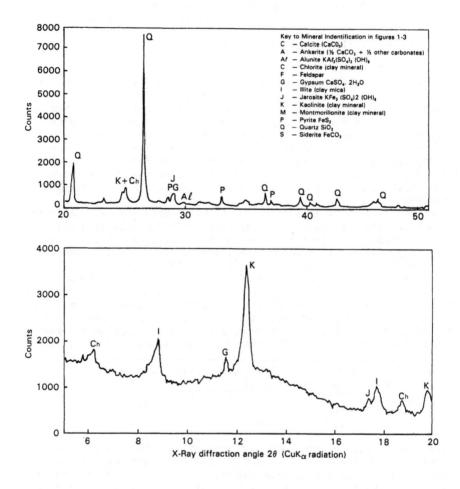

Figure 2 X-Ray diffraction traces of representative sample of fill from side hole 1

6 lumps of shale and the clay fines taken from the two 5 kg samples were also examined by X-ray diffraction. Clay fines (and to a lesser extent, one lump from hole 1) showed the presence of pyrite as well as increased quantities of gypsum and jarosite (see figure 3). During a subsequent visit of inspection to the site it was possible to select a few lumps of shale from the holes which showed some evidence of pyritic heave with crystals of gypsum between shale layers. These lumps only constituted a small proportion of the fill material. They were not observed in the samples originally delivered to BRE because extraction of the tightly-packed fill material with a club hammer and chisel broke up most of the weaker lumps. Scrapings taken from each of these lumps and from some further lumps extracted from an extra hole through the floor slab were analysed by X-ray diffraction. Large diffraction peaks were obtained for pyrite and gypsum. Jarosite was also identified in some of the samples. Less than half the samples showed any detectable traces of calcium carbonate either as calcite or as part of the complex carbonate ankerite. It was noted that calcite, ankerite and jarosite appeared to be mutually exclusive and that the samples taken from the borehole in the floor were somewhat less oxidised and contained proportionally more feldspar than the samples from the side holes.

4 Discussion

It would seem that the type of pyritic heave that has taken place under the warehouse as described above introduces some new factors for consideration. Firstly, the calcium oxide content is low and there appears to be a low proportion of expansive lumps. Expansion may thus be insufficient to cause structural distress except in large depths of fill. In the present case a linear expansion of little more than 0.1% has been sufficient to cause distress when the depth of fill is 2 m. Secondly there appears to be a shortage of calcium carbonate for the formation of gypsum and this suggests that the other calcium-containing minerals present, the feldspars, may also be dissolving in the sulphuric acid from pyrite oxidation. Feldspars would also be a ready source of alkali metals for the formation of jarosite which also may be implicated in the expansion mechanism. Evidence for this is provided by observations of the mutual exclusion of calcite, ankerite and jarosite in scrapings from pyritic layers and in the proportionally higher feldspar contents in less oxidised shales.

Coal measure shales (especially colliery spoil) have been extensively used as fill materials and no cases of pyritic heave have come to the attention of BRE apart, possibly, from one isolated case under a bungalow in S. Wales where there was insufficient evidence to provide a definite conclusion. Whilst many colliery spoils contain pyrites and also calcium oxide content above 0.5%, it is probable that the pyrites and acid soluble calcium are generally not in sufficient proximity for damaging growth of gypsum crystals. Some evidence for this mutual exclusion can be extracted from analyses of colliery spoils (Collins 1976, 1985). Other factors could be the crystal size and habit of the pyrite and the shale structure. London clay, for example, in its unoxidised state contains pyrite and acid soluble calcium but

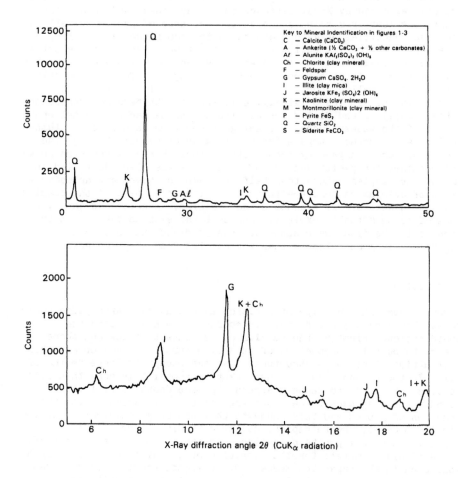

Figure 3 X-Ray diffraction traces for clay fines from side hole 2

gypsum (selenite) crystals are formed without significant heave because of the overall weakness of the rock/bedding planes.

5 Conclusions

Pyritic oxidation and heave must be assumed to take place down to at least 2 m below the surface for both bedrock and fill materials. With this thickness of material only small concentrations of expansive materials are necessary and a limiting value of 0.5% for calcium oxide may be too high.

Feldspars may be implicated in the reaction obviating the need for calcium carbonate minerals.

So far only examples of Triassic, Jurassic and Carboniferous (not Coal Measure) shales have definitely been implicated pyritic heave in the UK.

6 Acknowledgement

This paper has been written as part of the research programme of the Building Research Station of the Department of the Environment and is published by permission of the Director.

7 References

BRE (1979) Fill and hardcore. Digest 222, Building Research Establishment, Watford (withdrawn 1983).

BRE (1981) Concrete in sulphate-bearing soils and groundwaters. Digest 250, Building Research Establishment, Watford.

BRE (1983) Hardcore. Digest 276, Building Research Establishment, Watford.

Collins, R.J. (1976) A method for measuring the mineralogical variation of spoils from British collieries. Clay Minerals, 11, 31-50.

Collins, R.J. (1985) Mineralogical variation in spoil from one colliery and implications in the manufacture of synthetic aggregates. Br. Ceram. Trans. J., 84, 99-104.

Hawkins, A.B. and Pinches, G.M. (1987) Cause and significance of heave at Llandough Hospital, Cardiff - a case history of ground floor heave due to gypsum growth. Quart. J. Eng. Geol., 20, 41-57.

Nixon, P.J. (1978) Floor heave in buildings due to the use of pyritic shales as fill material. Chemistry and Industry, 4 March 1978, pp 160-164.

25 A CASE STUDY OF SULPHATE INDUCED GROUND HEAVE

A.B. HAWKINS
Department of Geology, University of Bristol, Bristol, UK

Abstract
Following heave of the ground bearing floor slab at Llandough
Hospital, Cardiff, a trial pit investigation revealed the area of
distress to be coincident with the dark mudstones of the Westbury
Formation.
  The heave was proved to be related to sulphate generation,
selenite crystals developing on the laminations within the dark
mudstones. When the ground was lowered by 3 m at the time of con-
struction the fresh material was oxidised, releasing sulphuric acid
which combined with calcium carbonate to form gypsum. The role of
Thiobacillus in the pyrite decomposition process is discussed.
Keywords: Pyrite, Sulphate, Thiobacillus, Heave

1   Introduction

An appreciation of the role of Thiobacillus in the oxidation of
sulphur as part of the process of the formation of sulphuric acid
dates back to Waksman and Joffe (1922). In 1927 Jensen recorded that
Thiobacillus thiooxidans oxidized pyrite most rapidly in acidic conditions. In 1950 Colmer, Temple and Hinkle isolated Thiobacillus
species from acid minewaters and in 1951 Temple and Colmer proved
this to be an iron oxidizing autotroph which they named Thiobacillus
ferrooxidans.
  During their investigations of two Canadian buildings which had
suffered from heave, Quigley and Vogan (1970), Quigley, Zajic,
McKeyes and Yong (1973) and Gillott, Penner and Eden (1974)
identified the presence of Thiobacilli in the dark mudstones beneath
the structures. This paper briefly discusses the problems caused by
heave of the dark mudstones beneath Llandough Hospital, Cardiff and
describes some of the work to establish the significance of the role
of bacteria.

2   Llandough Hospital, Cardiff

In April 1982 it was appreciated that the side walls of Ward 6 at the
hospital were moving outwards. As a consequence MRM Partnerhips were
appointed as structural engineers to establish the cause and significance of this and propose remedial works. The engineers immediately
appreciated the seriousness of the structural distress in the

northern part of the ward block and that part of the block was
scaffolded to ensure the stability of the external side walls
(Hawkins and Pinches, 1987a).

A level survey undertaken as part of the structural investigations indicated that the maximum heave was 81 mm compared with a
datum point in the main corridor north of the outcrop of the Westbury
Beds. At a number of locations differences in levels between the
datum point and the ground bearing floor slab exceeded 50 mm, Fig. 1
after Hawkins and Pinches, 1988. Over the years this heave had
caused continuing distress to the structure; parts of the hospital
walls having been covered with formica and the ward doors reportedly
having been periodically trimmed.

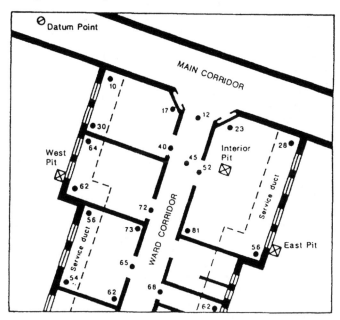

Fig. 1. Ground heave recorded in a survey of the eastern ward of
Llandough Hospital, relevant to the datum point (in millimetres)

An investigation was undertaken to establish the nature of the
ground conditions, particularly to determine whether there was any
reason why only the northern end of the ward block showed signs of
distress, the southern end being apparently unaffected. An attempt
to obtain a cable percussive borehole was unsuccessful and it became
clear that the underlying bedrock contained strong to very strong
limestone bands at about 4 m, which were effectively impenetrable
with this site investigation technique.

A decision was made to open trial pits on either side of the ward
block in the area of distress. These would have the advantage not
only of indicating the geology on which the structure was founded but
also of providing details of the actual foundation construction and
depth.

The western trial pit recorded very dark grey laminated moderately weathered claystone with some iron stained discontinuities to 1.3 m. This was followed by a 60 mm limestone horizon below which dark grey claystones were again present. At this depth, however, they were only slightly weathered; the weathering being mainly related to the iron staining on the closely spaced discontinuities. Below this was a 150 mm thick limestone horizon which overlaid fresh dark grey claystones. The foundation was a typical strip footing positioned on the lower limestone horizon.

The eastern pit again identified thinly laminated moderately weathered claystone with iron stained very closely spaced discontinuities to 1.0 m, below which the claystones were only slightly weathered until the upper limestone horizon at 1.86 m. Below this there were fresh to slightly weathered claystones with little or no iron staining on the discontinuities. When the 150 mm thick limestone horizon was exposed at 2.4 m it produced a significant seepage, implying this depth was below the ground water level. Careful examination of the exposed footings indicated that the original foundation of the building was on the limestone at 1.86 m depth but that in the 1950s this had been underpinned with a lean mix concrete below the original strip footing and a cobble plum lean mix placed adjacent to the outer wall of the ward block (see Hawkins and Pinches, 1987a).

A third trial pit was excavated beneath the ground bearing floor slab of the ward block. This interior pit showed thinly laminated moderately weathered claystone with extensively iron stained, very closely spaced discontinuities to approximately 1.75 m below the floor slab. Although no pronounced gypsum was visible above 0.95 m a considerable amount of selenite crystals was observed on the iron stained discontinuities, both the thinly laminated bedding planes and the very closely spaced fissures. Between the upper limestone band at 1.75 m and the lower band at 2.3 m the claystones appeared fresh with no iron staining or obvious gypsum.

The three trial pits, therefore, showed considerable variation in that the interior pit was both more iron stained and contained platey crystals of selenite on the bedding surfaces between 0.95 and 1.75 m. The flat selenite crystals were observed to form up to about 50% of the surface area of the bedding discontinuities which clearly had been opened to almost a millimetre by the presence of the platey monoclinic crystals of selenite.

A study of the geological map indicated that the only part of the ward block to suffer distress was that constructed over the Westbury Beds of the Penarth Group (previously referred to as Rhaetic Formation). It was clear, therefore, that a relationship must exist between the geology and the severe heave experienced by the structure.

Samples taken from the trial pits were chemically analysed; the results showed a pronounced difference in the SO_3 and pH content when tested using aliquots taken from the large samples which had been mill ground. When some of the remaining stored bag samples were re-tested in the same laboratory and using a similar test procedure in March 1984 the SO_3 percentage was found to have increased and the pH notably decreased in the intervening seventeen months.

For comparison the results from the three pits at 1.5 m below ground
level are given in Table 1.

Table 1: Changes in SO₃ and pH content of the same samples in a
seventeen month period.

			West Pit	Interior Pit	East Pit
SO₃	November	1982	0.29	2.20	0.36
SO₃	March	1984	1.20	4.15	1.60
pH	November	1982	7.10	3.75	7.10
pH	March	1984	6.25	2.65	5.30

It is important to note that the strata are dipping north east-
wards, hence although samples were taken at the same Ordnance Datum
level the material tested was actually from a slightly different
geological horizon; between the limestone bands in the west pit but
above them in the internal and east pits.

Examination of old records proved that at the time the hospital
was constructed between 1927 and 1933 up to 3 m of excavation was
required in order to establish level ground on which the northern
part of the ward block could be built. Clearly the result of this
was that the mainly geologically weathered material was removed prior
to construction, thus exposing the less weathered strata. This
grading of the land surface was not appreciated at the time of the
trial pit investigation but explained why so little highly weathered
material was encountered.

3   Comparison between Llandough Hospital and the Rideau Health Centre

Reference to the literature indicated that a similar heave situation
had occurred on the Rideau Health Centre in Ottawa where again up to
76 mm of heave was reported by Quigley and Vogan (1970) for this
building founded on Ordovician black shales of the Lorraine Formation. From its construction date they were able to determine heave
had occurred at an average rate of up to 2.5 mm/year in that portion
of the building founded on a shale plug extending above the water
table (Fig. 2). Where the shale remained saturated, little if any
heave took place.

Quigley and Vogan describe borehole samples in which flat gypsum
crystals up to 0.7 mm thick had formed on the bedding laminations and
in some other joint discontinuities. In samples which had been
obtained from the shales above the ground water table significant
oxidation had taken place and chemical analysis indicated there was
a decrease in pyrite content but an increase in sulphate content
relative to the saturated shales. Their detailed studies included
the isolation of Thiobacillus from the weathered shales and the
authors drew the conclusion that the bacteria either caused the
oxidation of the pyrite or were a major factor in exacerbating an
otherwise purely chemical reaction; similar to the conclusions drawn
by Temple and Dechamps (1953).

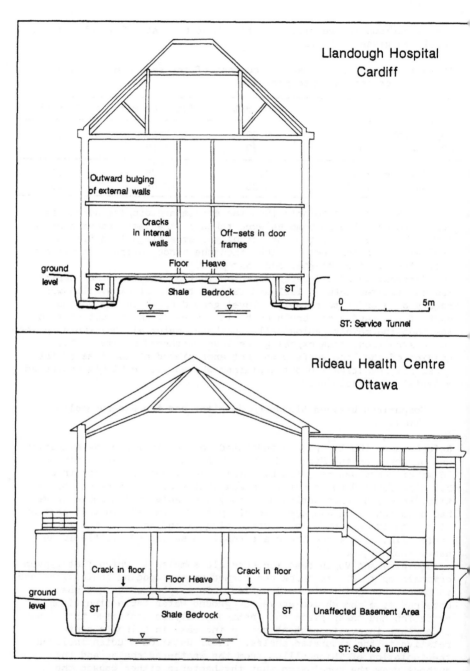

Fig. 2. Diagrammatic representation of the structures of Llandough Hospital (after Hawkins and Pinches 1987a) and the Rideau Health Centre (after Quigley and Vogan, 1970), reproduced to the same scale

Quigley and Vogan noted that the Rideau Centre had underfloor service ducts or tunnels which would undoubtedly have increased the temperature in the unsaturated shales. A similar duct system was present at Llandough Hospital (Fig. 2) and it was observed that there was a slight increase in selenite growth adjacent to the underfloor ducts. Experiments undertaken on similar Westbury Beds have been reported by Hawkins and Pinches (1987b) where samples were placed in ovens at 7.5°C, 18.5°C, 29.5°C and 41.5°C. The results show that after a period of fifteen weeks the SO_3 percentage had doubled in those samples placed in ovens at 29.5 and 41.5°C.

Samples from a pile hole (PH 9) inserted during the remedial works at Llandough Hospital demonstrated the development of oxidation when inoculated into an acidic ferrous sulphate medium. After three weeks only the samples from 1.5, 1.95 and 2.35 m had a strong to very strong visual oxidation, Fig. 3, but after six weeks all the samples above 2.0 m were in a moderately to strongly oxidized state. The sample at 2.35 m is anomalous but it is believed this sample may have been contaminated (Hawkins and Pinches, 1987a).

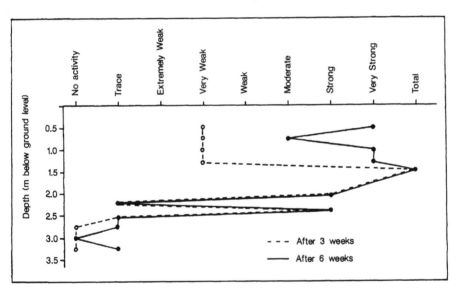

Fig. 3. Visually assessed oxidation (probably indicating varying presence of Thiobacillus ferrooxidans) from soil samples of PH 9 at Llandough Hospital after culturing for three and six weeks.

The experiment indicated that after six weeks most samples had developed a trace of oxidation yet the FeO content was below 1% above 2 m depth and above 1% below that level. It is believed that although iron is present throughout, there were fewer bacteria in the saturated lower samples compared with those from above the ground water level. Hence the visually assessed oxidation of the cultures after three and six weeks indicates the extent of the bacteria Thiobacillus ferrooxidans in the soil prior to sampling.

4   Microbiology

Kelly, Norris and Brierley (1979) suggest that the upper temperature bound for Thiobacillus species is 35°C. The work recorded by Hawkins and Pinches (1987b) implies that as the maximum sulphate generation occurs about this temperature, the bacteria is probably still very active at in excess of 40°C. Starkey (1966) supported by Kelly et al (1979) suggest that Thiobacillus is most active in acid environments where the pH is below 3 and in higher pH situations the bacteria become dormant.

The exact role of the bacteria in oxidation has not been fully explained. There are a number of authors who suggest that the important role of the bacteria is to accelerate and enhance the chemical reactions which normally take place as part of the process of geological weathering in aerated soil. Penner, Eden and Gillot (1973) considered that the biogenic oxidation was so important that if this was halted the reaction would probably stop. This view is supported by Bryner, Beck, Davis and Wilson (1954), quoted by Kuznetzov, Ivanov and Lyalikova (1963), who demonstrate the significance of bacteria on the oxidation of mine waters, measured as the amount of iron generated (Fig. 4). The sample which was sterilized (No 4) showed little change in the one hundred day experiment. In the case of Sample No 3, which was inoculated with bacteria on the 65th day and mercuric chloride added on the 70th day, the increase in total iron clearly occurred between these two periods. With Sample No 2 the inoculation was on day 28 with sterilization by autoclave on day 63; again the significant increase occurred between these two dates and ceased at sterilization. In the case of Sample No 1, inoculated with bacteria on day 31, there was a continued progressive rise throughout the experimental period.

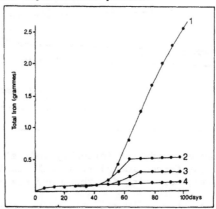

Fig. 4. Results of an experiment by Bryner, Beck, Davis and Wilson (1954) illustrating the oxidation of pyrite by Thiobacillus ferrooxidans measured as the amount of iron generated.
1:   inoculation by bacteria on 31st day after start of experiment;
2:   inoculation 28th day, sterilization by autoclave 65th day;
3:   inoculation 35th day, addition of mercuric chloride 70th day;
4:   sterile control sample.

The oxidation of pyrite is a complex process which is believed to be partly chemical and partly bacterial. Initially it is oxidized to ferrous sulphate with the release of sulphuric acid. The highly acidic environment resulting from the release of sulphuric acid may decrease the rate of the chemical reaction and it is likely that it is at this stage that the importance of the bacteria becomes significant; for details see Hawkins and Pinches (1986, 1987a).

5   Sulphate Generation

It is now established that sulphuric acid created by the decomposition of iron pyrite, either as a purely chemical process or exacerbated by the activity of bacteria, reacts with calcium carbonate within a soil mass to form the mineral gypsum and in the absence of calcium may react with illite to form jarosite. With high calcium carbonate contents the pH of the soil mass remains near neutral and hence although bacterial activity is clearly taking place the conditions are not conducive to significant profileration.

Work not yet published has demonstrated that sulphates are frequently concentrated at about the measured ground water level. Above this there is often a zone of low sulphates where clearly leaching has taken place and the soils show evidence of oxidation. The concentration at the saturated/non saturated boundary may be related to the extensive presence of bacteria in the moist aerated environment, to a natural concentration of percolating salts precipitating at the ground water table or to the lower pressures in the capillary fringe.

The localised presence of sulphate is clearly important, especially if the dark mudstones are fissile/shaley in nature. In these situations crystals of selenite can form preferentially in the slightly lower pressure areas of the natural discontinuities. Once crystals begin to grow, the opening of the discontinuities will result in the pressure along the plane being lowered, further encouraging the precipitation of the salts. In this way salts will tend to form in the area of least pressure. Clearly the load imposed by the perimeter walls of a structure are greater than those of the internal walls or the ground floor slabs. As a consequence the heave is generally experienced as a mounding of the floor which will initially cause a rise in the central area until the stresses in the concrete are such that a break occurs. Once a crack in the floor slab has opened the underlying ground may become further aerated which will favour the activity of bacteria and at the same time, by lowering the imposed loading in the area of the crack, will provide the necessary lower pressure environment in which further crystal growth can take place.

6   Conclusions

It is known that bacteria are present in dark mudstones where selenite growth has caused ground heave. The presence of the bacteria exacerbate the decomposition of pyrite; the exact relationship between the purely chemical and the bacterially induced decomposition is not yet established. Optimum conditions for bacteria

303

proliferation are an aerated environment, a low pH and a relatively high ground temperature.

The role of bacteria is clearly of considerable importance to the engineer. In many situations by removing the geologically weathered material in order to form levels on which to place the structures, fresher material is exposed on which bacterial activity can be initiated. Most developments also involve the draining of the site, resulting in a lowering of the ground water table; this also creates a new environment in which oxidation can take place. In some situations pyritic rich mudstones are used for fill. In this case the sulphuric acid may chemically modify the mudstones as well as combining with the calcium carbonate or illite to produce gypsum or jarosite. In the former situation the soil mass is likely to suffer a reduction in strength; in the latter to experience heave.

The processes involved in the oxidation of pyrite and also the speed with which the reactions can take place are rarely fully appreciated by the practising engineering. The implications of the role of bacteria in acting as a catalyst in the oxidation of iron pyrite is the subject of continuing study.

7 Acknowledgements

The author acknowledges the co-operation of the MRM Partnership and the South Glamorgan Health Authority for their help with the initial work and permission to publish the results.

8 References

Bryner, L.C., Beck, I.V., Davis, D. and Wilson, D. (1954) Microorganisms in Leaching Sulfide Minerals. **Industr. & Engng. Chem.** 46, 12.

Colmer, A.R., Temple, K.L. and Hinkle, M.E. (1950) An iron oxidizing bacterium from the acid drainage of some bituminous coal mines. **J. Bacteriology,** 59, 317-328.

Gillott, J.E., Penner, E. and Eden, W.J. (1974) Microstructure of Billings Shel and biochemical alteration products, Ottawa, Canada. **Can. Geotech. J.** 11, 482-9.

Hawkins, A.B. and Pinches, G.M. (1986) Timing and Correct Chemical Testing of Soils/Weak Rocks. **Eng. Grp. Geol. Soc. Spec. Pub.** 2, Universities Press, Belfast, 59-66.

Hawkins, A.B. and Pinches, G.M. (1987a) Cause and significance of heave at Llandough Hospital, Cardiff - a case history of ground floor heave due to gypsum growth. **Q.J.E.G.** 20, 41-57.

Hawkins, A.B. and Pinches, G.M. (1987b) Sulphate analysis on Black Mudstones. **Geotechnique,** 37, 191-196.

Hawkins, A.B. and Pinches, G.M. (1988) Expansion due to gypsum growth. **Proc. VIth Int. Conf. Expansive Soils,** New Delhi, 183-187.

Jensen, H.J. (1927) Vorkommen von Thiobacillus thiooxidans in danischen Boden. **Zentr Bakteriol Parasitenk,** Abt II, 72, 242-246.

Kelly, D.P., Norris, P.R. and Brierly, C.L. (1979) Microbiological Methods for the Extraction and Recovery of Metals. in **Micro Technology: Current State, Future Prospects.** Bull, Edwood & Rattledge (eds), Cambridge University Press.

Kuznetzov, S.I., Ivanov, M.V. and Lyalikova, N.N. (1963) **Introduction to Geological Microbiology**. translated by Broner, P.T. (ed) Oppenheimer, C.H. McGraw-Hill, New York, London.

Penner, E., Eden, W.J. and Gillott, J.E. (1973) Floor heave due to biochemical weathering of shale. Paper 26, **Proc. 8th Int. Conf. Soil Mech. & Found. Eng.** Moscow, 151-158.

Quigley, R.M. and Vogan, R.W. (1970) Black Shale heaving at Ottawa, Canada. **Canadian Geotechnical Journal,** 7, 106.

Quigley, R.M., Zajic, J.E., McKeyes, E. and Yong, R.N. (1973). Biochemical alteration and heave of black shale: detailed observations and interpretations. **Can. J. Earth Sci.,** 10, 1005-15.

Starkey, R.L. (1966) Oxidation and reduction of sulfur compounds in soils. **Soil Science,** 101, 297-306.

Temple, K.L. and Colmer, A.R. (1951) The autotrophic oxidation of iron by a new bacterium: Thiobacillus ferrooxidans. **J. Bacteriology,** 62, 605-611.

Waksman, S.A. and Joffe, J.S. (1922) Microorganisms concerned in the oxidation of sulphur in the soil. II. Thiobacillus thiooxidans, a new sulfur oxidizing organism isolated from the soil. **J. Bacteriology,** 17, 239-256.

26 BIOLOGICAL STRENGTHENING OF MARINE SEDIMENTS

D. MUIR WOOD
Department of Civil Engineering, Glasgow University, UK
P.S. MEADOWS and A. TUFAIL
Department of Zoology, Glasgow University, UK

Abstract
The presence of invertebrates and micro-organisms in marine sediments influences the strength and permeability of those sediments, particularly in the near-surface region. Formation of burrows creates permeable low strength pathways through the soil. However, many organisms secrete mucus which binds together the soil particles (and, for burrowing organisms, helps stabilise the walls of the burrow). With smaller micro-organisms this mucus secretion is the dominant effect and it enhances the overall strength of the sediment.

Although there have been a number of qualitative reports of such effects very little quantitative work has been done. Recently, however, quantitative studies have revealed modest increases in strength over depths of a few centimetres. It is known that this activity can extend to a depth of a metre or so.

Stability of many natural or man-made slope features is strongly influenced by the strength of the near-surface soil. Improvement in this strength will result in a corresponding improvement in stability. The biological influences on strength that are now being observed suggest the possibility of exercising some control over strength for civil engineering purposes.

This paper describes some preliminary studies that have been carried out, and also outlines continuing experimental work in this area.
Keywords: Marine sediments, Benthic organisms, Permeability, Strength, Laboratory tests.

1 Introduction

Rowe (1974) states, with reference to the effects of the benthic fauna on the physical properties of deep-sea sediments, that "animals can either consolidate and strengthen or disperse and weaken sediments". Though that may appear to be a somewhat imprecise statement it does serve to remind us that sediments provide a home for a vast range of organisms, and the presence of these organisms is not often taken into account in assessing the engineering properties of marine sediments. Rowe's statement also reminds us that some organisms may have a beneficial effect on the mechanical properties of the sediments, and this naturally points to the possibility of using organisms to

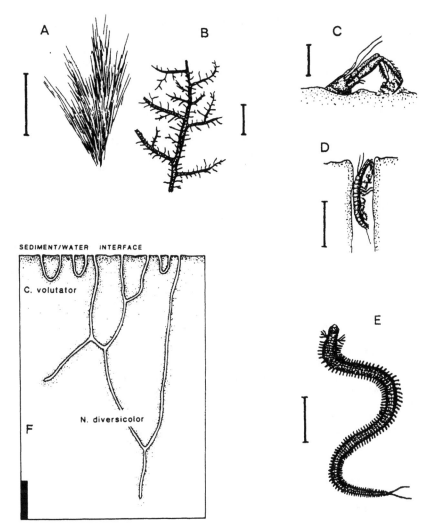

Fig. 1   A. *Enteromorpha clathrata* (Grev.) photosynthetic alga
(bar = 10mm);   B. *Enteromorpha clathrata* (bar = 0.1mm) (A
and B redrawn from Harvey, 1871);   C. *Corophium volutator*
(Pallas) amphipod, Crustacea (bar = 1mm);   D. *Corophium
volutator* (bar = 5m) (C and D from Meadows and Reid,
1966);   E. *Nereis diversicolor* (O.F. Müller) polychaeta,
Annelida (bar = 10mm) (from Green, 1968);   F. Burrow form of
Corophium volutator and *Nereis diversicolor*.   (Stippling
represents aerobic sediment at sediment surface and around
burrows in more anaerobic deeper layers of sediment.)
(bar = 50mm) (from Meadows and Tait, 1989).

exercise some control over these properties for civil engineering purposes.

Effects of organisms will be greatest in the near-surface regions of sediments, and their engineering contribution will be to situations where the condition of the near-surface soil is of primary importance. Short-term stability of slopes is dependent on the undrained shear strength of the soil that can be mobilised without volume change, usually in clayey soils of low permeability. Under drained conditions, where volume changes can occur freely, the strength of soils is predominantly frictional in origin so that the shear strength that can be mobilised increases with effective stress level and hence with depth. Near the surface of a slope the stresses, and hence frictional strengths, are low and the analysed safety of shallow slope failures is strongly dependent on the presence of even a small cohesive contribution to drained strength. Biological control of near-surface properties could be beneficial, for example, in increasing stability of seabed trenches, or in reducing the risk of erosion around pipelines, sewage outfalls and other constructions.

This paper describes in some detail the changes that occur in marine sediments in the presence of two particular burrowing organisms and one algal mat. Many organisms form submarine burrows : Weaver and Schultheiss (1983) report 5mm diameter open burrows existing to a depth of greater than 1m. The presence of such burrows may create permeable, low strength pathways through the soil. However, the stability of such burrows in weak sediments requires that the organisms should take some action to ensure stability. Many organisms including algae secrete a mucus which binds together the soil particles (Scoffin, 1970) and which may actually lead to an overall increase in strength. The presence of such a mucus lining may also help to reduce the overall permeability (Aller, 1983).

Quantitative studies of the effects of organisms on the mechanical properties of sediments have been few. A geotechnical assessment of some recent work is presented here, followed by discussion of proposals for future research studies.

2 Modification of sediment permeability and shear strength by two burrowing invertebrates

2.1 Background

The two burrowing invertebrates that have been studied (Meadows and Tait, 1989) are the amphipod *Corophium volutator* (Pallas) and a ragworm, the polychaete *Nereis diversicolor* (O.F. Müller). Drawings of these two organisms are shown in Fig 1(C,D,E), and sketches of their typical burrows are shown in Fig 1(F). *Corophium volutator*, typically about 10mm long, constructs u-shaped burrows which extend 30-40mm into the sediments. Mucus secretions are used to strengthen the wall of the burrow. *Nereis diversicolor*, which may be up to 100mm long, constructs a network of burrows which may extend 300-400mm into the sediment. Mucus secretions are pushed against the walls of the burrow : the surrounding sediment is both compacted by the pressure and bound by the mucus, so that there is a double strengthening effect.

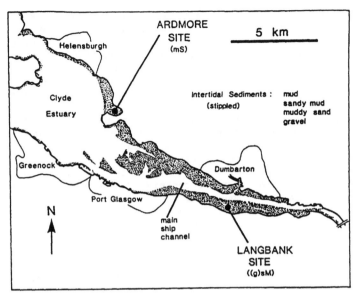

Fig. 2   Inner Clyde Estuary showing intertidal sediments (stippled),
          and sampling sites at Langbank ((g)sM gravelly sandy mud) and
          Ardmore (mS muddy sand) (from Meadows and Tufail, 1986)

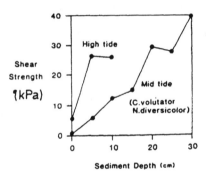

Fig. 3   Vane shear strength profiles at Langbank site (from Meadows
          and Tufail, 1986).

    Experiments have been conducted using intertidal sediment from the
shore of the inner Clyde Estuary at Langbank (Fig 2) : the sediment
here is described as (gravelly) sandy mud (Meadows and Tufail, 1986).
A profile of shear strength as measured in situ over the top 0.3m with
Pilcon vanes (19mm dia, 28mm long and 32.5mm dia, 49mm long) is shown
in Fig 3. This diagram illustrates a problem with studies at this
scale. The strength profiles vary with the state of the tide, and in
fact a large number of repetitions of such measurements are required
to produce a statistically representative profile. It may be noted

that these measurements imply a very rapid increase of strength with depth - for normally consolidated sediments a rate of strength increase with effective overburden pressure c_u/σ_v' of about 0.25 might be expected, implying a strength of only about 0.6 kPa at a depth of 0.3m. No field tests have yet been performed to assess the effects of *Corophium volutator* and *Nereis diversicolor* on in situ strength.

Recovery of soft sediments and preparation of specimens for laboratory seeding experiments can introduce various spurious effects if care is not taken. If *Corophium volutator* are to burrow into the sediment then the sediment must be attractive to them. Deans, Meadows and Anderson (1982) report that autoclaving makes sediment particularly unattractive, and sieving can also reduce the attractiveness if it leads to a reduction in the fine detrital fraction of the sediment which contains the highest organic content and greatest number of bacteria. They note that even mechanical mixing of samples can destroy in-situ structure - leading to a large decrease in strength and permeability by comparison with in-situ values.

2.2 Experimental procedure
Since one of the beneficial effects of marine organisms may be to increase resistance to erosion, it is natural to use direct observation of critical erosion velocity to monitor their contribution. Typically a box sample of sediment is placed in a flume and the critical velocities for particle movement and suspension are recorded. Girling (1984) (quoted by Meadows and Tufail, 1986) shows that the binding secretions in the casts of the lugworm *Arenicola marina* raised the critical erosion velocity from 26m/s to around 44m/s, even though the casts protruded into the flow. A similar result is reported by Rhoads, Yingst and Ullman (1978) for the polychaete *Heteromastus filiformis*.

The experiments reported by Meadows and Tait (1989) use a more direct strength measurement with a fall-cone (Hansbo, 1957). For a cone of mass m, which penetrates a distance d into a soil the undrained strength c_u of the soil is given by

$$c_u = k \; mg/d^2 \qquad\qquad\qquad (1)$$

where g is the gravitational acceleration and k ~ 0.3 for 60° cones (Wood 1985). For a soil with a strength of around 1 kPa, this expression implies a penetration of about 13mm so that the fall cone is measuring an average strength over the top 13mm of the sample.

Experiments to study the effects of *Corophium volutator* and *Nereis diversicolor* on strength of sediments were combined with experiments to study their effect on permeability. The tests were therefore conducted in falling-head permeameters of internal diameter 50mm. A depth of sediment of about 100mm was used, with a water depth of 50mm maintained above the sediment. Once individuals of the species being studied had been added to the sediment they were allowed to burrow for 24h before the experiment was begun. Falling-head permeability tests were conducted by raising the water depth to 250mm and recording the time taken for it to fall by 50mm. The strength was measured with the fall-cone at the surface of the sediment with the overlying water removed. Each experiment was performed in triplicate.

The sediment was prepared by wet sieving through a 710μm sieve to remove animals and larger particles and maintained in aerated seawater until used for the experiments, within a day of being collected. Specimens of *Corophium volutator* and *Nereis diversicolor* were obtained by sieving another sample of sediment also through a 710μm sieve. Experiments were performed with four different population densities : zero, low, medium, and high (Table 1). The medium values were chosen to match the average densities observed on the shore at Langbank.

Table 1.   Population densities used in experiments

	Individuals/m^2(/specimen)		
Species	Low	Medium	High
Corophium volutator	2500 (5)	7500 (15)	22500 (45)
Nereis diversicolor	1000 (2)	3000 (6)	9000 (18)

2.3 Permeability

The effects of each species on permeability are shown in Fig 4a.   The permeability is affected within 1 day of seeding and the effect remains over the duration of the experiment. With *Corophium volutator* the permeability is reduced.   (There is a factor of about 2 between the unseeded samples and the samples seeded at high density.)   The dominant effect here is probably the creation of somewhat impermeable burrow linings which impede the water flow over a significant proportion of the cross-sectional area. (With just one 2.5mm diameter burrow per individual and a population density of 22500 individuals/m^2 roughly 10% of the cross-sectional area would be occupied by burrows at the surface of the sediment.)

Nereis diversicolor, on the other hand, produced an increase in permeability (by a factor of about 2 at the highest population density).   It may be supposed that the longer, slightly larger diameter burrows, are more effective at making the sediment porous than is the secreted mucous at sealing it.

2.4 Strength

Effects on strength are shown in Fig 4b.   Both species produce an increase in fall-cone strength, but the effect is more marked with *Corophium volutator*, where the undrained strength has been rather more than doubled, from about 0.4 kPa to about 1 kPa, at the highest population density.   At a strength of 1 kPa the cone penetration is about 13mm, the volume of soil displaced about 770mm^3 and the area of the indentation at the surface is about 177mm^2 compared with an average area of about 44mm^2 occupied by each individual at this population density.   The fall-cone is therefore measuring something like an average surface undrained strength.   (For the tests with *Nereis diversicolor*, a strength of about 0.4 kPa implies a penetration of about 20mm, a displaced volume of about 2800mm^3, indentation

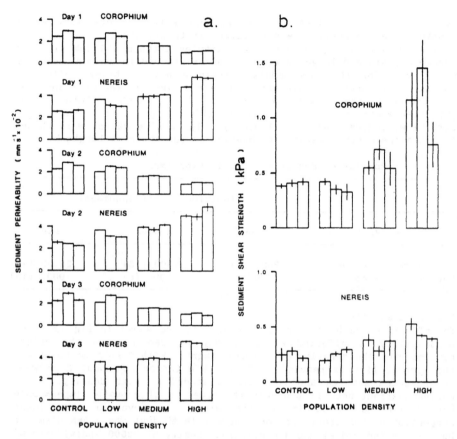

Fig. 4   *Corophium volutator* and *Nereis diversicolor*.
(a) Relationship between population density and sediment
permeability. Experimental days 1,2,3; three replicate cores
each day; (b) Relationship between population density and
sediment shear strength. Three replicate cores; measurements
taken on day 3 (bars indicate standard deviations) (from
Meadows and Tait, 1989).

area 420mm^2, compared with 111mm^2/individual at the highest density.)
    The presence of the organisms in the sediments appears to lead to a
reduction in water content.   It is a standard hypothesis of critical
state soil mechanics (Schofield and Wroth, 1968) that undrained
strength c_u and water content w should be related by an expression of
the form:

$$c_u = \exp[G_s w - A)/\lambda] \tag{2}$$

where G_s is the specific gravity of the soil particles, λ is a soil
constant – related to compressibility – and A is a reference
constant.

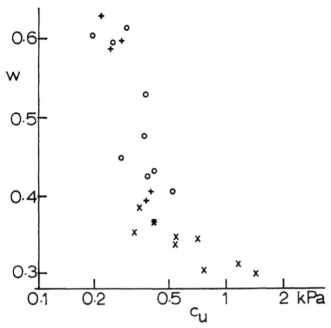

Fig. 5   Data of strength and water content from: + control samples; x
          samples seeded with *Corophium volutator*; o samples seeded
          with *Nereis diversicolor*.

Strength and water content values are plotted in Fig 5.   The data are
not sufficiently extensive to draw any precise conclusions; some
suggestions can, however, be made.
     The strength : water content relationships for the control
specimens and for the specimens treated with *Nereis diversicolor* seem
to be roughly similar indicating that the dominant effect of this
species may be one of water content reduction, perhaps by local
compaction of the sediment.   On the other hand, *Corophium volutator*
seems to produce a more rapid increase of strength with water content.
There are not sufficient control data to be able to determine whether
this is an indication merely of a variable λ in expression (2)
(compare Wood (1985))   or whether it indicates some direct
strengthening effect of the mucus secretions.

3 Modification of sediment strength in presence of algal mat

A brief series of field and laboratory experiments has been performed
to examine the effect on fall-cone penetration (and hence, from (1),
deduced undrained shear strength) of photosynthetic algae at the
surface of samples of sediment.   Samples of intertidal sediment were
obtained from Ardmore on the north shore of the Clyde (Fig 2) : the
sediment here is described as muddy sand (Meadows and Tufail, 1986).
     Samples were placed in plastic containers in a controlled light and
temperature environment in groups of two replicates.   Samples were

seeded with the alga *Enteromorpha* (Fig 1(A,B) and left for 5 and 10 days. Two sets of control samples without algae were kept in identical conditions, one in light, the other in darkness.

The experimental results, interpreted as values of undrained strength, are shown in Table 2. The presence of the algal mat appears to increase the strength by a factor of about 3. No difference was observed between measurements made at 5 and 10 days.

The results of fall-cone tests on the shore at Ardmore are also shown in Table 2. The strength increase between localities with and without the presence of the algal mat is about a factor of two, though the standard deviation of these field data is rather large.

The fall cone is most suited for determining undrained strength of clays. Clearly the presence of some real cohesion between particles of a more sandy sediment, resulting from the polysaccharide secretions of the algae, will reduce the fall-cone penetration in the same way as an increase in undrained strength, but the interpretation of fall-cone penetration in terms of the cohesive contribution to strength is not so straightforward.

Table 2. Effect of mats of the alga Enteromorpha on the undrained shear strength of muddy sand.

	undrained shear strength (kPa)				
	algal mat present		algal mat absent		increase with algal mat
	mean	s.d.	mean	s.d.	
field data	4.7	2.9	2.3	0.8	105%
laboratory	2.1	0.7	0.7	0.1	181%

4 Discussion

From an engineering point of view, although the absolute values of strength are low, a controlled doubling of strength of surface sediments would be beneficial particularly since it is the surface soil that controls the extent to which erosion will occur. In the laboratory experiments this strength increase occurred quite rapidly and was sustained. In the field one would have to be certain that the environment was receptive to the species that were to be seeded.

From an engineering point of view, however, it is necessary to discover a little more about the sources of the changes in mechanical properties that are observed so that an appropriate model can be generated. Three contributions to both permeability and strength changes can be identified.

(i) Effects of changes in water content (and void ratio) of soil around burrows can be examined in terms of conventional soil mechanics understanding (noting that the sediment is soft and the initial water content and liquidity index are high so that we are operating a little away from the range of more conventional geotechnical engineering).

(ii) Secretion of mucus stabilises the walls of burrows and also binds the soil particles together more generally and thus has a more general strengthening effect on the soil (and also tends to reduce the permeability).

(iii) The creation of burrows must lead to the presence of weak, permeable regions in the soil : the overall effect of the organisms can only be beneficial if contributions (i) and (ii) outweigh this effect.

(iv) The presence of algae and other micro-organisms can have a direct beneficial effect on strength without the weakening associated with burrowing organisms.

5 Future testing

Any suggestions for testing procedures have to take account of the fact that the soils that are being studied are extremely soft, so that handling of samples is not straightforward, and the forces required to produce failure may be lost in the resolution of conventional testing equipment. The testing techniques should also aim to separate the effects of friction and cohesion of the sediment so that the addition of a binding cohesion to a sandy or silty sediment can be identified.

The fall-cone is not ideal for this purpose because it is a rapid, dynamic test designed to discover undrained properties of clays. However, a static penetration test can give more direct information on the force: penetration relationship which can be decoded more readily in terms of basic soil parameters. The standard bearing capacity equation for a cohesive/frictional soil states that the ultimate bearing capacity q_f of a circular footing of radius a at depth D in soil of drained cohesion c' and unit weight γ is

$$q_f = 1.2c'N_c + \gamma DN_q + 0.6\gamma aN\gamma \qquad (3)$$

where N_c, N_q and N_γ are bearing capacity factors which are dependent on the angle of friction of the material. Data from tests with different radii a of penetrometers, and following the variations of bearing capacity with depth D, backed up with basic information concerning the strength of the unseeded sediments at the appropriate void ratio and stress level will provide a suitable basis for decoding. Continuous penetration will also lead to an assessment of the depth over which the properties of the soil have been improved.

Miniature bearing capacity tests with a penetrometer lead naturally to larger scale measurements of strength improvement using model pad footing tests - and those then point the way to pad footing tests in the field from which the effects of seeding certain areas of natural sediment with micro-organisms can be assessed in situ.

5 Conclusions

Experimental results have been shown here to indicate the beneficial effects on strength of marine sediments produced by various benthic organisms. The effects are significant and reveal the possibility of biological control of engineering properties of soils. It is beginning to be possible to make quantitative statements about effects which have been known qualitatively for some time. As this collaborative work continues it will be of interest to extend the ranges both of organisms and of soil environments that are studied.

6 References

Aller, R.C. (1983) The importance of the diffusive permeability of animal burrow linings in determining marine sediment chemistry. J. of Marine Research, 41, 2, 299-322.

Deans, E.A., Meadows, P.S. and Anderson, J.G. (1982) Physical, chemical and microbiological properties of intertidal sediments, and sediment selection by Corophium volutator. Int. Revue ges. Hydrobiol, 67, 2, 261-269.

Girling, A.E. (1984) Interactions between marine benthic macroinvertebrates and their sedimentary environment. PhD thesis, Glasgow University.

Green, J. (1968) The Biology of Estuarine Animals. Sidgwick & Jackson, London.

Hansbo, S. (1957) A new approach to the determination of the shear strength of clay by the fall-cone test. Royal Swedish Geotechnical Institute, Proceedings 14.

Harvey, W.H. (1871) Phycologia Britannica or A History of British Sea-Weeds. IV, Chlorospermeae or Green sea-weeds. Synopsis No. 280 to 388. Reeve, London.

Meadows, P.S. and Reid, A. (1966) The behaviour of Corophium volutator (Crustacea : Amphipoda). J. Zool. Lond., 150, 387-399.

Meadows, P.S. and Tait, J. (1989) Modification of sediment permeability and shear strength by two burrowing invertebrates. Marine Biology, 101, 75-82.

Meadows, P.S. and Tufail, A. (1986) Bioturbation, microbial activity and sediment properties in an estuarine ecosystem. Proc. Roy. Soc. of Edinburgh, 90B, 129-142.

Rhoads, D.C., Yingst, J.Y. and Ullman, W.J. (1978) Seafloor stability in central Long Island Sound. Part I: Temporal changes in erodibility of fine-grained sediment. Estuarine Interactions (ed M.L. Wiley), Academic Press, 221-244.

Rowe, G.T. (1974) The effects of the benthic fauna on the physical properties of deep-sea sediments. Deep-sea sediments (ed A.L. Inderbitzen), Plenum Press, 381-400.

Schofield, A.N. and Wroth, C.P. (1968) Critical state soil mechanics. McGraw-Hill, London pp 310.

Scoffin, T.P. (1970) The trapping and binding of subtidal carbonate sediments by marine vegetation in Bimini Lagoon, Bahamas. J. Sedimentary Petrology, 40, 249-273.

Weaver, P.P.E. and Schultheiss, P.J. (1983) Vertical open burrows in deep-sea sediments 2m in length. Nature 301 (27 Jan), 329-331.

Wood, D.M. (1985) Some fall-cone tests. Geotechnique, 35, 1, 64-68.

27 DESIGN OF PRESSURE RELIEF WELLS WITH AN INTEGRAL CLEANING SYSTEM FOR A LARGE EARTH-FILL DAM

C.M. JEWELL
Coffey Partners International Pty Ltd, Sydney, Australia

Abstract
Mardi Dam is a large earthfill embankment located near Wyong on the Central Coast of New South Wales, Australia. The dam was constructed by the NSW Public Works Department in the early 1960's; neither a foundation cut-off nor a chimney drain were incorporated in the design. Shortly after the dam was filled, artesian groundwater pressure above ground level was observed at the downstream toe of the main embankment. Although various remedial measures, including a weighting berm and a system of pressure relief wells were installed, these proved ineffective in reducing the artesian pressures. In particular the relief wells rapidly became clogged by bioaccumulations of iron. In 1987 a study was carried out to assess the situation and design new remedial works. A number of investigation bores were drilled: these showed artesian pressures more than 5m above ground level at the downstream toe. Gallionella were identified in the blocked relief wells. A new system of pressure relief wells, constructed from corrosion resistant materials and equipped with an integral cleaning system utilising hydroxyacetic acid as the main biocidal and cleaning agent, operated by gas pressure, was designed.
Keywords: Water Supply Dams, Pressure Relief Wells, Iron Bacteria.

1 Introduction

Mardi Dam is a 7,250 ML off-stream water storage, located approximately 5 km south-west of Wyong on the central coast of New South Wales, Australia, and is a key component of the Gosford-Wyong water supply system. The location is shown in Fig. 1. Mardi Dam was constructed in the early 1960's; since that time numerous developments have taken place downstream, including construction of residential and commercial property and the Sydney - Newcastle Freeway. The dam has therefore been assigned a high hazard rating, meaning that substantial property damage and loss of life would be likely in the event of failure.

The dam is founded on alluvial valley-fill containing quite extensive sand lenses. Although the potential for seepage through this alluvium was recognised during the initial design studies, the dam was constructed without either a positive foundation cut-off or any provision for internal drainage, as was normal practice at the

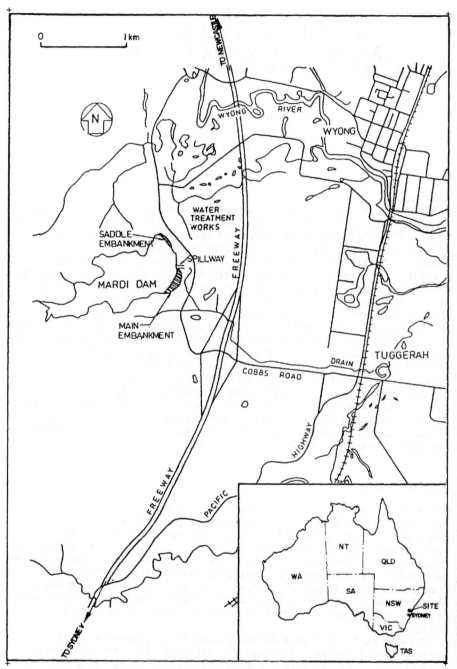

Fig.1. Location of Mardi Dam.

time. Provision for foundation drainage was, however, made in the original design and comprised:

. a horizontal filter blanket at foundation level, incorporated in the embankment base;
. 36 drainage wells constructed along the upstream side of the filter blanket (underneath the crest of the embankment) and feeding into the blanket.

Upon first filling of the dam in 1963 high seepage pressures in the foundation alluvium were recorded, indicating that the drainage wells were not performing effectively. To control these pressures additional works were carried out, comprising:

. 29 drainage wells at the toe of the embankment;
. 16 spearpoint piezometers to monitor seepage pressures;
. an earthfill weighting berm constructed at the downstream toe next to the left abutment.

Despite these measures, high seepage pressures continued to be recorded around the downstream toe of the embankment. In addition, persistent wet patches were observed on the downstream face of the dam. In view of these observations, the importance of the dam and its high hazard rating, a full safety review was commissioned in 1980.
    This review concluded that artesian foundation pressures (then believed to be 1-2m above ground surface) were not acceptable due to the risk of liquefaction under earthquake loading, and consequent failure of the embankment. Intra-plate earthquakes do occur on the east coast of Australia, indeed in December 1989 a Magnitude 5.6 quake caused extensive damage and loss of life in Newcastle, 70km north of Mardi Dam. The most direct solution to the seepage pressure problem was judged to be a system of high-efficiency relief wells, located at the downstream toe of the dam. As blockage of the existing drainage structures by deposits of iron oxy-hydroxides appeared to be a major contributor to the problem, an effective cleaning system for such wells was required.
    An investigation to assess the feasibility of this proposal and provide the data necessary for design of the relief wells was commissioned in December 1986.

2 Investigation programme

Following assessment of the substantial volume of existing geological data, an investigation programme was planned. This comprised:

    drilling three multi-tube piezometers (IB1, IB2, IB3) completed in both the alluvial deposits and the underlying sandstone;
    drilling two test wells (PB1, PB2) completed at different levels in the alluvium and underlying sandstone;
    pumping tests carried out on the test wells with observations in the piezometers;
    chemical analysis of groundwaters and microscopic examination of

deposits in the blocked drainage wells.
Locations of the investigation boreholes are shown on Fig.2.
Completion intervals for the wells and piezometers are given in Table
1.

Table 1. Completion intervals for wells and piezometers

Borehole number	Piezometer number	Completion interval (m below surface)
IB1	P1	4.9 - 5.9
	P2	8.7 - 9.2
	P3	5.4 - 5.9
IB2	P1	13.8 - 14.3
	P2	9.3 - 10.0
	P3	2.6 - 3.1
IB3	P1	16.8 - 20.3
	P2	11.4 - 13.0
	P3	5.6 - 6.6
PB1	Screen	4.5 - 10.5
PB2	Screen	16.0 - 22.0

3 Results of investigation

3.1 Geology

Fig.3. is an interpretive geological section along the downstream toe
of the embankment, based on data acquired during the investigation
programme and on information available from earlier work. From Fig.3.
it can be seen that up to 15m of Quaternary alluvium overlies Triassic
sandstones and siltstones. The alluvium comprises a scour and fill
valley infill sequence in which five main units can be identified.
These are:

Unit 1: A surficial layer of recent fill, presumably placed at the
         time of dam construction. This fill is recorded in the
         borehole logs for all 29 downstream drainage wells.
Unit 2: A sandy clay and clayey sand sequence which extends across
         the valley.
Unit 3: A denser sequence of brown to black sand and silty sand.
         This unit was intersected in 22 of the 29 downstream toe
         drainage wells, and all the boreholes drilled in this
         investigation.
Unit 4: A complex zone of boulders, slopewash, and extremely
         weathered sandstone and siltstone. Information about this
         unit is based mainly on the investigation boreholes. The
         boundaries are not readily distinguished from Unit 3 above
         and Unit 5 below. The depth of refusal of 36 original
         drainage wells beneath the crest of the embankment were
         used to approximate the upper boundary of Unit 4.
Unit 5: Highly weathered sandstone, with open, subvertical, water
         filled joints.

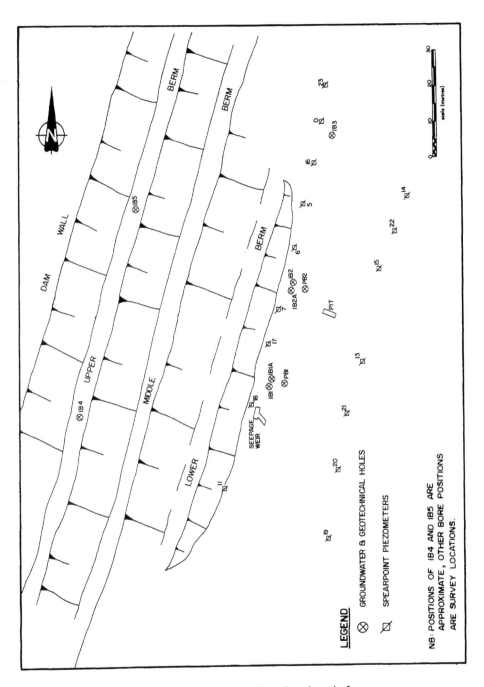

Fig.2. Locations of investigation boreholes.

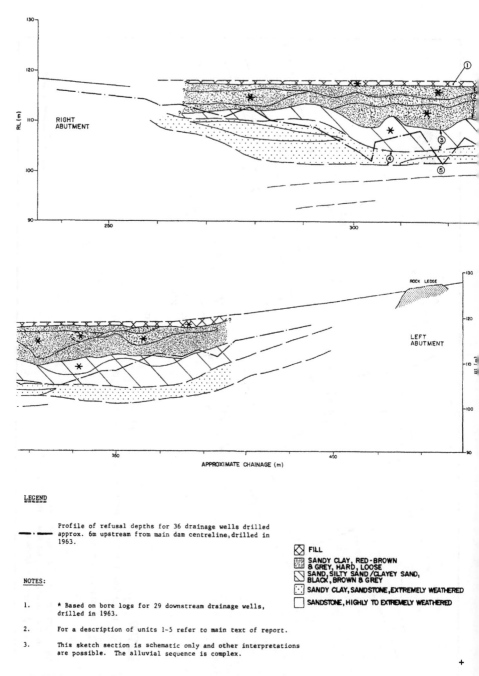

Profile of refusal depths for 36 drainage wells drilled
approx. 6m upstream from main dam centreline, drilled in
1963.

NOTES:

1.   * Based on bore logs for 29 downstream drainage wells,
drilled in 1963.

2.   For a description of units 1-5 refer to main text of report.

3.   This sketch section is schematic only and other interpretations
are possible. The alluvial sequence is complex.

FILL

SANDY CLAY, RED-BROWN
& GREY, HARD, LOOSE

SAND, SILTY SAND/CLAYEY SAND,
BLACK, BROWN & GREY

SANDY CLAY, SANDSTONE, EXTREMELY WEATHERED

SANDSTONE, HIGHLY TO EXTREMELY WEATHERED

Fig.3. Interpretive geological section along downstream toe of dam.

Additionally, a more recent channel scoured almost to the bottom of Unit 2 and infilled with two upward fining sequences of light grey sandy clay and clay, overlain by sand and sandy clay, appears to be present in the centre of the valley.

3.2 Piezometry
The investigation programme found artesian pressures even higher than anticipated. IB1-P2 encountered a measured head of 3.2m above ground level in the alluvium, which caused considerable difficulties during drilling; IB2-P1 recorded a head of 5.2m above ground level in fractured sandstone underlying the alluvium.

3.3 Hydraulic characteristics
Pumping tests of 10 and 24 hours duration were carried out on test wells PB1 and PB2 respectively. Considerable variation in the responses of the many piezometers and spearpoints measured during the tests was observed; this is indicative of the complexity of the alluvial sequence, which appears to behave as a complex leaky aquifer system. Although the response of piezometers and spearpoints was variable, significant drawdown was observed in some observation points which were separated both spatially and stratigraphically from the screened horizons.

Aquifer parameters were calculated from the response of piezometers completed in the same horizon as the test wells. The discharge rate declined continuously during both tests; results were therefore analysed using Aron-Scott's method (Aron and Scott, 1965). The interpretation is summarised in Table 2.

Table 2.   Summary of pumping test results

Borehole number	Unit tested	Depth (m)	Hydraulic conductivity (ms^{-1})	Storage coefficient
PB1	3	4.5 - 10.5	8.1×10^{-6}	6.6×10^{-4}
PB2	3,4,5	16 - 22	1.2×10^{-5}	2.0×10^{-3}

3.4 Chemical and biological characteristics
A chemical analysis was carried out on a groundwater sample obtained during the pumping test on PB1. Results are given in Table 3; a significant concentration of dissolved iron is evident.

Microbiological examination of a sub-sample from borehole PB1 showed the presence of desulphovibrio sp. (sulphate reducing bacteria); iron metabolising bacteria were not found in this sample but Gallionella sp. were detected in samples of precipitate taken from the blocked drainage wells. Much of this precipitate was, however, a floc of amorphous iron oxy-hydroxides.

Although redox potential was not directly measured, these results are indicative of quite strongly reducing, as well as acidic, conditions in the deeper part of the alluvium. Under these conditions iron is stable as the ferrous (Fe^{++}) ion. Diffusion of oxygen, or mixing with oxygenated water in the existing drainage wells allows oxidation to the insoluble ferric (Fe^{+++}) form. This oxidation is at least in part facilitated by iron oxidising bacteria (Gallionella sp). Hanert (1981) found that Gallionella are most active at low, but measurable dissolved oxygen concentrations. Thus these bacteria probably act to increase the rate of iron precipitation under marginal conditions, and thus extend the range of conditions under which iron precipitation becomes a problem.

Table 3. Chemical characteristics

Analyte	Value	
pH*	5.9	
EC* (microsiemens/cm)	835	
	(mg/l)	(meq/l)
Calcium as Ca	11.8	0.59
Magnesium as Mg	21.5	1.77
Sodium as Na	120	5.22
Potassium as K	2.05	0.05
Bicarbonate as HCO_3	170	2.79
Sulphate as SO_4	28	0.58
Chloride as Cl	145	4.09
Nitrate as NO_3	0.08	
Orthophosphate as P	0.002	
Sulphide as S#	<0.1	
Iron as Fe#	5.8	
Manganese as Mn#	0.40	

* field measurement   # filtered and preserved sample

3.5 Summary of results
The results of the pumping tests, and in particular the drawdown measured after quite short times in distant piezometers, indicate that despite the complexity of the alluvial sequence, there is drainage between the different units. Hydraulic interconnection is especially apparent between Units 3,4 and 5. Horizontal hydraulic conductivity in the more permeable units appears to be high enough to support flow to drainage wells, with these units acting as drains to less permeable underlying and overlying strata. Thus depressurisation by gravity drainage wells should be feasible.

The chemical and bacteriological results indicate that blockage of such drainage wells by both chemical and biochemical precipitates of iron is likely to be a continuing problem.

4 Design of depressurisation system

4.1 System design
A computer simulation, using an analytical model based on the Theis solution for non-steady radial flow to a well (Theis, 1935), was used to predict the sensitivity of the performance of a depressurisation

well system to changes in well spacing, and to deviation of the aquifer hydraulic characteristics from the values derived from the pumping tests. The model uses the principle of superposition to predict the effects of interference between wells, and of impermeable and recharge boundaries.

The configuration used for this simulation incorporated impermeable boundaries on either side of the valley, and a recharge boundary located 300m upstream of the line of wells.

On the basis of the model output the optimum spacing for the depressurisation wells was calculated to be 8m for the area between the right abutment and IB3, and 10m from IB3 to the left abutment.

A gravity collector drain, installed at a depth of approximately 3m below ground level, links the wells to a central outfall pipe running down the valley. Once flushed of air, the system is thus largely closed to the atmosphere.

4.2 Well design

The depressurisation well and cleaning system design is illustrated in Figure 4. The well design is based on a conventional water supply borehole specification, utilising high-efficiency wire-wrapped stainless steel screens of 100mm diameter, coupled with uPVC casing. This design was chosen in recognition of anticipated problems due to clogging by precipitated iron oxyhydroxides and iron-precipitating bacteria. The advantages of this type of construction are:

. A large screen open area to allow inflow to be maintained for a reasonable period between cleaning operations, despite the effect of iron clogging;
. The large open area allows cleaning agents access to the gravel-pack and formation outside the well;
. Both stainless steel and PVC have good anti-fouling characteristics, and are resistant to corrosion by chemical cleaning agents.

A 12m screen length was specified, this allows some flexibility of screen placement against the most appropriate alluvial or fractured sandstone horizons in individual bores.

4.3 Cleaning system design

Although the system has been designed to minimise contact between groundwater and the atmosphere, it is likely that some precipitation of iron and growth of iron bacteria will still occur. The large number of existing drainage structures, though now largely ineffective in their intended purpose, will allow air and aerated water to enter the system. Periodic cleaning will thus be required.

The well design incorporates a system whereby a cleaning agent can be introduced through the central tube to the bottom of the well screen and forced into the gravel pack and formation under gas pressure. Agitation can then be created by alternately releasing and applying gas pressure. This method, which has been shown to be effective in the redevelopment of injection wells (Olsthoorn, 1982), has the major advantage that no mechanical contact with the well screen is required. Thus the risk of damage during repeated cleaning

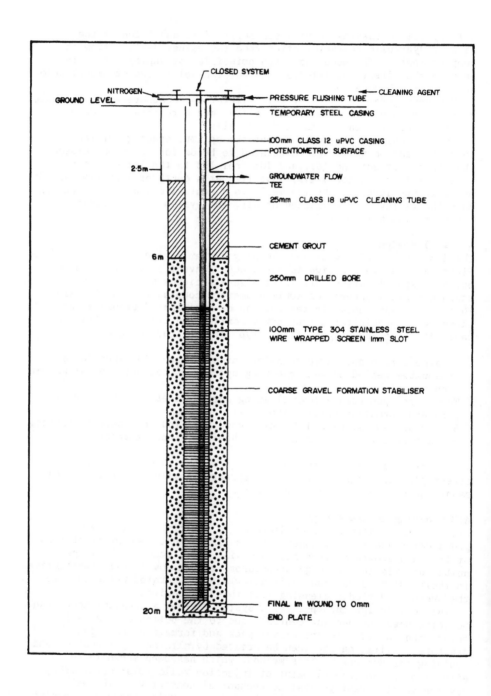

Fig.4. Design of well and cleaning system.

operations is reduced. Gas pressure is then applied to the upper part of the casing. The central tube also functions as a gas line, allowing debris to be lifted from the well in a similar manner to conventional airlift development.

4.2 Cleaning agent selection
A number of bactericides have been shown to be effective against iron bacteria (Hackett and Lehr, 1985). Of these, chlorine in various forms has been the most widely used. Bactericides on their own, however, are not effective at breaking down accumulated iron deposits around bacteria, and other agents, particularly acids and surfactants, have to be employed for this purpose. Thus a multi-stage cleaning process is required, an expensive process when many depressurisation wells are involved.

Hydroxyacetic acid (glycolic acid $HOCH_2COOH$) has been shown to be both an effective bactericide, and capable of dissolving accumulated iron deposits (Cullimore and McCann, 1977). It is also relatively non-corrosive. This liquid organic acid has therefore been recommended as the primary cleaning agent.

The depressurisation well system, as specified, is now scheduled for construction during 1990.

Acknowledgements
The author wishes to thank his colleagues in Coffey Partners International Pty Ltd, in particular Ms Simone Williams who prepared the diagrams, and Mr P.J.N. Pells and Dr L.W. Drury for their help with and criticism of the work on which this paper is based. This work was carried out for the NSW Public Works Department and the Department's permission to publish this paper is gratefully acknowledged.

References
Aron, G. and Scott, V.M. (1965) Simplified solutions for decreasing flow in wells. **Proc Am. Soc. Civ. Eng.** 91, HY5, 1-12.
Cullimore and McCann (1977) The identification, cultivation and control of iron bacteria in ground water. **Aquatic Microbiology;** (Eds F.A. Skinner and J.M. Shewan) Academic Press, New York, pp. 219-261.
Hackett and Lehr (1985) **Iron bacteria occurrence, problems and control methods in water wells.** National Water Well Association, Worthington, OH.
Hanert, H.H. (1981) The genus Gallionella; **The Procaryotes, a handbook on habitats, isolation and identification of bacteria** (Eds M.P. Starr, H. Stolp, H.G. Truper, A. Barlows and H.G. Schlegel) Springer-Verlag, Berlin, pp 509-515.
Olsthoorn, T.N. (1982) **The clogging of recharge wells.** KIWA-communications 72, Rijswijk, Netherlands.
Theis, C.V. (1935) The relation between the lowering of piezometric surface and the rate and duration of discharge of a well using groundwater storage. **Trans. Geophys. Union,** 2, 519-524.

28 A CASE STUDY OF BIOFILM FORMATION IN ASSOCIATION WITH METHANE SEEPAGE INTO AN UNDERGROUND TUNNEL

C.F.C. PEARSON
Charles Haswell & Partners, Consulting Engineers,
London, UK
M.J. BROWN
Archaeus Ltd, London, UK

Abstract
The Carsington aqueduct consists of a series of pipelines
and tunnels constructed in the early 1980's. The area has
a history of mining activites dating back to the Romans.
Early mining records reported the occurrence of firedamp
gas. Methane was detected during construction of the
Carsington tunnels, on one occasion reaching 100% of LEL.
Surveys showed methane to be present above the crown of a
substantial length of tunnel. When the aqueduct was
dewatered and ventilated naturally, the formation of a
biofilm was observed in sections of the tunnel where
methane had been detected. The biofilm developed around
points of water ingress. It consisted of a pink surface
layer coating a black/white fibrous material several cm
thick. Bubbles of gas trapped in the biofilm contained
27% methane; ingress water was indentified as the
principal route by which methane was entering the tunnel.
Methane-oxidising bacteria were indentified in ingress
water and the biofilm together with a range of
heterotrophic microorganisms. Bacteria using methane for
growth may represent the first link in the food chain,
with other microorganisms growing at their expense.
Prolific growth of biofilm in tunnels may cause local
depletion of oxygen, enrichment of carbon dioxide and
slippery, hazardous conditions for workers.
Keywords: Biofilm, Methane, Aqueduct, Tunnel,
Methane-oxidising bacteria.

1 Introduction

The Carsington aqueduct consists of a series of pipelines
and tunnels constructed during the early 1980's for
Severn Trent Water, at that time the local water
authority, as part of the Carsington Reservoir Scheme.
The scheme was promoted to provide an additional water
resource for the conurbations of Derby, Nottingham and

Leicester and the smaller towns of the North East Midlands of England.

The aqueduct comprises concrete and steel-lined tunnels, and steel and ductile iron pipelines which total 10 km in length; the finished diameter of the aqueduct varies between 1.6 and 2.4 m. The two main tunnels penetrate the areas of relatively high ground to the South of Wirksworth in the county of Derbyshire, between the valleys of the River Derwent and the River Ecclesbourne and between the valleys of the River Ecclesbourne and the River Henmore, the site of the Carsingston Reservoir. The tunnel Portals are designated A to D from west to east. Portal A is at the Carsington Reservoir, Portals B and C are either side of the Ecclesbourne Valley and Portal D is at the River Derwent. The valleys of the Derwent and Ecclesbourne are crossed by pipeline whilst the aqueduct is continued into the Ambergate pumping station by a further small section of tunnel. The maximum depth of the tunnels below ground level is approximately 135 m and 165 m in the A-B and C-D sections respectively.

2 Geology

The Carsington area lies to the south of the Peak District on the edge of the Derbyshire dome. To the north and west are the Carboniferous Limestone hills of the White Peak, whilst to the south the topography softens over the shales of the ancient basin of the Widmerpool Gulf. To the east is the exposed coalfield of East Derbyshire and West Nottinghamshire, which dips beneath the base of the Permian sandstones and limestones. The aqueduct tunnels were excavated through a series of interbedded sandstones, siltstones and mudstones of the Millstone Grit sequence of the Carboniferous period; the tunnel horizon lies at the base of the sequence within or below the Ashover Grit. The lithology and competence of the excavated rocks varies considerably from almost completely weathered mudstone clays through laminated or blocky siltstones to massive fissured sandstones.

The geological structure of the area surrounding Wirksworth is dominated by the Bolehill anticline, which plunges to the south east. To the north the anticline is terminated by the Bonsall fault with a throw of about 100 m, and to the southeast by the Gulph fault with a downthrow to the northeast of well in excess of 100 m. Both faults trend north west/south east. The faults have the effect of producing a confined graben within the Bolehill anticline. Southwest of the Gulph fault there are no discernible fold structures, the strata dipping gently to the east. The regional dip is from east to

west, with the result that the tunnels ascend the
stratigraphic sequence from west to east. At the western
end of the aqueduct the major unconformity separating the
Millstone Grit series from the Carbonferous Limestone is
estimated to be some 300 m below the tunnel horizon.

3 Preliminary investigation

Prior to the construction of the aqueduct tunnels, a site
investigation was carried out in order to indentify and
determine the properties of the rocks likely to be
encountered during tunnelling, and to identify hazards
such as water-bearing strata which might affect
tunnelling progress. Cored boreholes were drilled in the
vicinity of the portals and, initially, at 1 km intervals
along the tunnel route to a depth below that of the
tunnel invert. Later, during the construction period, a
second series of boreholes, intermediate with the first,
were installed to provide additional data. Shallow
seismic and resistivity surveys were also carried out and
packer tests were performed in each borehole at the
tunnel horizon.

In Derbyshire, rocks of the Carboniferous Limestone
have been subjected to mineralisation associated with
hydrothermal activity and many of the faults and joints
within the limestone contain minerals which have in the
past supported a thriving mining industry. In particular,
the sulphides and carbonates of lead have been mined
since the time of the Roman occupation of Britian. Later
the gangue minerals fluorite, barite and calcite were
mined and in some locations reclaimed from the old mine
waste tips.

To enable miners to pass from one mineral vein to
another, tunnels known as "gates" or "crosscuts" were
constructed, which frequently entered the mudstones and
shales associated with the mineralised limestones. Water
was always a problem for the miners and to assist in
drainage of the mines, tunnel levels, adits or "soughs"
were driven into the hillsides from the deep valleys
surrounding the orefield and these frequently penetrated
the mudstones and shales stratigraphically higher than
the mineralised limestone.

There are extensive records of the mining operations
carried out in Derbyshire during the sixteenth to
nineteenth centuries and these show that in levels and
adits in the Carsington area, firedamp (methane gas) and
blackdamp (an atmosphere containing an excess of carbon
dioxide and/or nitrogen gas) were occasionally
encountered. In one particular tunnel in the Mawstone
mine, a methane explosion resulted in eight fatalities,
including the mine manager (Naylor, 1982).

In addition to the physical site investigation carried out at the time of the promotion of the Carsington aqueduct scheme, consideration was given to the miners records of water inrush and methane occurrences in the area. This information, combined with evidence from the site investigation boreholes, led the designers of the aqueduct system to conclude that water would make tunnel driving particularly difficult and that methane gas might also be encountered. As a consequence, the specification for the construction of the tunnels stated that extensive drilling should be carried out from the tunnel face in advance of mining and that grouting should be used to reduce water inflow. In addition, the specification included requirements for ventilation, flame proofed electrical equipment, constant gas monitoring at the face during tunnel driving and regular gas monitoring throughout the length of the driven tunnels.

4 Methane incidents during tunnel construction

Tunnel excavation was carried out by roadheading machines until the spring of 1983 when drill and blast, and hand excavation methods were adopted. Probe drilling ahead of the tunnel face confirmed the presence of water and advance grouting using cement and clay/cement grouts became a regular part of the tunnel excavation cycles. During tunnel construction a number of incidents occured in which alarms on methane detection equipment were activated. Four positively confirmed methane incidents were recorded during mining of the C-D section of tunnel between 1982 and 1985. The most serious of these was during blasthole drilling when a sudden emission of methane triggered the alarms and cut off the electrical supply to the non-flameproof electrical equipment; this lead to a controlled evacuation of personnel from the tunnel. On investigation, methane levels in excess of 100% of the lower explosive limit (LEL) or 5%, v/v methane in air were recorded near to the tunnel face reducing to 25% LEL at a distance of 25 m.

As methane had been detected within the tunnels, a number of additional precautions were agreed. These included: an increased level of exhaust ventilation to create sufficient airflow in the tunnel to dilute methane emissions at the tunnel face; ventilation velocities of at least 0.5 m/s to prevent any methane seepage from forming layers of high concentration; all electrical circuitry within the ventilation ducting to be either flameproof or methane- protected; constant monitoring for methane, incorporating an automatic alarm system, to be maintained at the tunnel face at all times. In addition it was agreed that the total length of the mined workings should be monitored for methane layers at least once per shift.

Ventilation of the tunnels was by an exhaust system, which was upgraded on a number of occasions during tunnel mining. Gas detection systems were also improved from those relying on reaction to simple gas alarms to those providing automatic shutdown of non-flameproof electrical equipment in the presence of methane. These systems of detection and alarm remained in operation until the end of tunnel excavation.

Tunnel excavation was completed during the early part of 1985 without further serious incident in respect of methane gas. However, at that time drilling was being carried out for grouting behind a section of concrete lining in the C-D tunnel and methane was discovered in drilled holes in the tunnel crown. Concentrations of methane in excess of 100% LEL were found in some holes and this promoted an extensive ventilation and gas survey. Methane was found to be present above the crown of a sustantial length of the concrete-lined tunnel, although the general body of the atmosphere appeared to be gas-free when tested by methanometer. A similar examination was made in the other sectons of the C-D tunnel and in the A-B tunnel but methane was not detected at these locations.

Throughout the whole of 1986 the aqueduct was dewatered and ventilation of the tunnels was by natural means. The tunnels were open at each end and air velocities measured during routine ventilation surveys were generally between 0.5 and 1.0 m/s. On occasions the direction of the air flow reversed and at these times, for short periods, air flow velocities were zero. It was during this period that microbial deposits were first noticed. In certain sections of tunnel, particularly where water ingress was high, bright red iron staining was apparent; a red "flocky" deposit began to form in these areas. A white, calcite-like deposit appeared in areas where seepage was low. The deposit formed fine, intricate, stalactite-like fans inclined to the principal flow of ventilation. Most unusual of all, in sections of C-D tunnel a biofilm deposit, pink/grey in colour, began to form. The biofilm reached a thickness of up to 5 cm in places, and covered several m^2 of tunnel wall in total. Initial analysis of this material, carried out to investigate any toxicity, failed to identify it positively.

5 Research into the methane hazard at Carsington

Following the testing of the aqueduct tunnels in early 1987, substantial quantities of methane gas were detected in C-D tunnel after the aqueduct drain down. Severn Trent Water gave consent for a programme of research to be carried out jointly by staff from Charles Haswell and Partners Ltd and Nottingham University Mining Department, to attempt to discover the source and extent of the methane problem affecting the tunnels and attendent structures. Tunnel air and water samples were taken for subsequent analysis. Air samples of about 1 litre were collected using a hand pump to pressurise the sample tubes to approximately 10 bar; whilst water samples were collected in 2 l bottles from positions of maximum water inflow. The results of analysis for dissolved methane, carried out by the Mining Department of Nottingham University are shown in Table 1 below.

Table 1. Carsington Aqueduct: dissolved methane concentrations (8/4/87)

Distance from portal C (m)	Dissolved CH_4 (mg/l)
0	nd
870	2.16
1264	15.22
1677	nd
2139	nd
2438	3.74
3453	8.31
3969	10.48
4035	7.44
4287	1.43

The analysis confirmed that the principal means of methane ingress into the tunnel was in infiltrating groundwater in which the gas was dissolved. Although significant dissolved methane concentrations were found at numerous points throughout the C-D tunnel, there appeared to be two locations where the concentrations were extremely high. A repeat sampling and analysis was carried out (Table 2).

As the aqueduct remained dewatered, a further examination, survey and analysis was carried out in September 1987, and this confirmed the continued entrance of methane into the tunnel; general body atmosphere concentrations of methane gas were the order of 0.01% (v/v) when analysed by gas chromatography. In addition the deposit of the pink/grey biofilm appeared to be more extensive, particularly in those areas where methane-bearing ground water was entering the tunnel. Bubbles were forming on the surface of the biofilm and when these were broken, field detection equipment confirmed the release of methane at high concentration.

Table 2. Carsington Aqueduct: dissolved methane concentrations (20/5/87)

Distance from portal C (m)	Dissolved CH_4 (mg/1)
0	nd
509	nd
1180	2.47
1264	13.52
1304	15.29
2439	6.75
3291	10.02
3453	11.95
3957	0.35
4035	7.65

The gas samples collected in the tunnel during 1987 were also analysed for the presence of hydrocarbons other than methane; none of the higher alkanes (ethane, propane, pantane etc.,) were detected. In addition a sample of the methane gas was analysed for the stable carbon isotope ratio in an attempt to determine the origin and maturity of the gas. The ratio of the two stable isotopes of carbon, ^{12}C and ^{13}C, is usually expressed as the "carbon 13 delta value" defined by the equation:

$$\delta^{13}C = [(R-R_f)/R_f] \times 1000 \text{ parts per thousand (ppt)}$$

where R and R_f refer to the ratio $^{13}C/^{12}C$ in the sample and reference respectively (Craig, 1957). The $\delta^{13}C$ value for the methane from the Carsington tunnel was -65 ppt suggesting that the gas may be of a microbiological origin, a conclusion supported by the absence of the higher hydrocarbons.

An examination of local, natural groundwaters and old mine dewatering soughs in the area of Carsington tunnels was carried out to see whether methane was present. Trace amounts of methane were detected in waters from the Cromford sough and from boreholes and wells along the tunnel route. No dissolved methane was found in local streams and water from the Merebrook sough.

In February 1988 the aqueduct was filled for further testing and an air sampling regime was installed at C portal to detect methane concentration changes as the tunnel was filled. Methane concentration in the tunnel atmosphere increased dramatically as the water level rose, reaching a maximum value of 35% LEL (1.75%, v/v). When completely full the C-D tunnel was isolated and the water level was reduced so that a void formed entirely within the steel-lined section of the tunnel. After several days the atmosphere in the void was expelled and was found to contain 12% LEL methane.

6 Post testing research

Following the period of testing in 1988 the aqueduct was again dewatered. Ventilation surveys were reintroduced and monitoring for methane gas continued. As concentrations of the gas in the general body atmosphere of the tunnel remained below those detectable using conventional methanometers, samples of atmosphere were removed from the tunnel for laboratory analysis by gas chromatography. These analyses confirmed the presence of methane gas in concentrations up to 0.012% (v/v) with air velocities of between 0.5 and 0.75 m/s, gas concentration increasing with air flow and distance from the downstream tunnel portal.

Additional surveys were undertaken to detect any further build up of biofilm. In areas within the tunnel where water ingress was relatively high, deposits of biofilm formed rapidly and within six months from dewatering, significant quantities of material had formed containing large bubbles of gas. Two samples of this gas were removed and analysed by gas chromatography (Table 3).

Table 3. GC analysis of gas from biofilm bubbles (April 1989)

Gas	Sample 1 (%)	Sample 2 (%)
H_2	<0.5	<0.5
O_2	2.2	2.0
N_2	70.6	71.4
CH_4	27.0	26.5
CO_2	0.2	0.1

The analysis of this gas demonstrates that methane is being trapped by the biofilm as it enters the tunnel. Oxygen may be diffusing into the bubbles from the tunnel

atmosphere, creating an ideal environment for microbial
methane oxidation.

In the area where biofilm was abundant, analysis of
infiltrating ground water indicated dissolved methane
concentrations in the order of 11 mg/l. Furthermore, in
two areas gas appeared to be bubbling from ground water
as it entered the tunnel beneath the biofilm mat. This is
consistant with water, supersaturated with gas under
pressure, degassing on pressure relief at the tunnel
wall.

7 Microbiological analysis of ingress water

Samples of ingress water and biofilm material were
collected from C-D tunnel in January 1990 for
microbiological analysis.

Ingress water samples were spread onto agar plates
(Cornish, Tryptone Soya and Malt Extract) to investigate
the microbial population. Cornish medium (Cornish et al.,
1984) contains no carbon source; these plates were
incubated in an atmosphere containing 10% methane.
Tryptone Soya Agar is a complex nutrient medium to detect
and enumerate growth of a wide variety of microorganisms
at pH 7.3, whereas Malt Extract Agar promotes growth of
yeasts and moulds at a pH of 5.4. Since the pH of the
ingress water was 8.2, the Tryptone Soya Agar appeared
most likely to detect the microbes present.

Only one distinct colony type was visible after 7 days
on Cornish medium with methane as sole carbon source. The
colonies had a pink colouration characteristic of the
methane-oxidising bacterium Methylomonas methanica. This
organism may have been responsible for the pink
colouration of the biofilm. At least six distinct colony
types were present on the Tryptone Soya Agar, indicating
a diverse bacterial population. Only one type of colony
was present on the Malt Extract Agar.

Microorganisms detected on Tryptone Soya Agar were
present at concentrations of 10^3 cells/ml, while the
methane-oxidising bacteria and the microorganisms
detected on Malt Extract Agar were present at
10^2 cells/ml.

8 Microbiological analysis of biofilm

A layer of pink biofilm material on the tunnel wall
coated the surface of a black and white material with a
fibrous appearance. Samples of biofilm material with pink
and black colouration were suspended in ingress water and
spread onto agar plates as described above. Some ten
different colony types were apparent on Tryptone Soya

Agar, three of which had a strong orange colouration, while only white transparent colonies were visible on Cornish agar. This preliminary analysis revealed the presence of methane-oxidising bacteria and a diversity of other heterotrophic microorganisms (requiring organic compounds for growth) in the biofilm.

9 Ability of biofilm to oxidise methane

Biofilm material with pink or black colouration (5 g wet weight) was added to 5 ml of ingress water in a 150 ml Wheaton bottle. The bottles were sealed with butyl rubber bungs; 7 ml of air was withdrawn and replaced with 7 ml of methane to give a final concentration of approximately 4% methane (v/v). A control bottle contained water without biofilm material. Methane and carbon dioxide concentrations were measured at intervals during a 7 day period (Fig 1).

Methane and carbon dioxide concentrations in the control bottle remained virtually constant throughout the experimental period, indicating that no significant methane oxidation had occurred. Methane concentration in the bottles containing biofilm material decreased progressively during the 7 day test period from an initial concentration of 4% to 0.2-0.4%, demonstrating the activity of methane-oxidising bacteria. Carbon dioxide concentration increased rapidly in the bottles containing biofilm during the first 20 h, before any significant methane oxidation had occurred, indicating microbial activity within the biofilm independent of methane oxidation. The yield of carbon dioxide after 7 days was 13% (v/v) in both biofilm samples; this is well in excess of the amount of carbon dioxide that would be generated as a result of oxidising 4% methane.

Methane appears to be the only significant carbon source present in the tunnel other than the biofilm itself. Bacteria capable of oxidising methane as sole source of carbon and energy have been shown to be present in the biofilm and the ingress water (collected inside the tunnel) and are likely to represent the first link in the food chain. Other microorganisms will be able to grow at the expense of decaying methane-oxidising bacteria and hence a microbial community develops as a biofilm.

The bacteria convert methane and oxygen into carbon dioxide and water according to the following reaction:

$$CH_4 + 2O_2 = 2H_2O + CO_2$$

Hence where significant concentrations of methane are entering a tunnel dissolved in groundwater and causing prolific growth of methane-oxidising bacteria, local

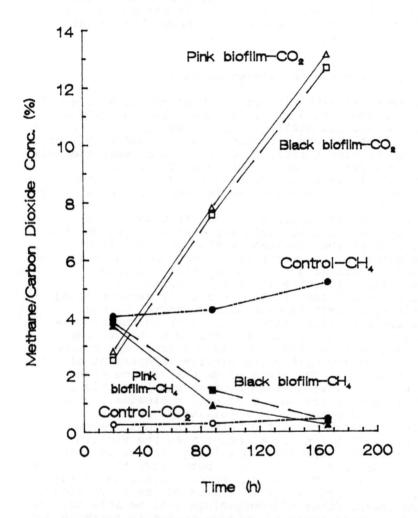

Fig. 1. The effect of incubating samples of biofilm with air containing 4% methane on methane and carbon dioxide concentrations.

oxygen depletion and carbon dioxide enrichment may become a problem. Within the Carsington aqueduct, although ventilation velocities are naturally maintained at around 0.5 m/s, on occasions changing climatic and barometric conditions cause a change of direction of air flow. This results in a short period when air flow is zero. On one occasion recently, low oxygen alarms were triggered requiring an evacuation of the tunnel system. It is possible that this was caused by an oxygen-depleted slug of air moving down the tunnel following a period of zero air flow as ventilation was reestablished.

The simplest and safest way of removing the biofilm was found to be by physical means. Great care has to be taken due to the slippery conditions on the tunnel invert and nailed sewer boots are essential. There is no evidence at this time that the formation of the biofilm is detrimental to the concrete lining of the tunnel or that it will continue to form when the aqueduct is in its normal operating condition, ie. full of water, with a probable trickle back of water toward the River Derwent.

10 Conclusions

The indentified presence of methane within the Carsington aqueduct reinforces the need to monitor atmosphere condition and quality and quantity of ventilation within tunnels at all times.

The presence of a biofilm, similar to that indentified at Carsington, in other tunnels may signify that methane gas is entering those systems in groundwaters. This should prompt a comprehensive atmosphere and ventilation study within those tunnels.

Where the presence of methane is suspected, suitable monitoring of atmosphere and ventilation are required before gaining access to any underground system.

If biofilm formation is suspected within dewatered tunnels, great care must be taken when gaining access. Safety precautions should include adequate protective clothing including suitable footware and sufficient detection equipment to safeguard personnel. Safe systems of entry and evacuation should be devised and implemented during tunnel inspection.

11 Acknowledgement

The permission of Severn Trent Water to carry out the research work and to publish this paper is greatly appreciated. The authors would also like to thank members of Charles Haswell & Partners Ltd and Archæus Ltd for the support and advice in preparing the paper, and the site staff on the Carsington project for the assistance. We would also like to express our gratitude to Mr J. Edwards of Nottingham University and Dr Durukan of Imperial College for the work in the research into the presence of methane gas in the aqueduct.

12 References

Craig, H., (1957) Isotopic Standards for carbon and oxygen and correction factors for mass spectrometric analysis of carbon dioxide. **Geochim. Cosmochim. Acta,** 12, pp133-149.

Cornish, A., Nicholls, K.M., Scott, D., Hunter, B.K., Aston, W.J., Higgins, I.J. and Saunders, J.K.M. (1984)) In vivo ^{13}C NMR investigations of methanol oxidation by the obligate methanotroph Methylosinus trichosporium OB3b. **J. Gen. Microbiol.** 130, 2565-2575.

Ford, T.D. and Rieuwerts, J.H. (1975) **The History of Mining-Part Two, Lead Mining in the Peak District.** Peak Park Joint Planning Board, Bakewell, England, pp 7-28.

Health and Safety Executive, (1985), **The Abbeystead Explosion,** HMSO, England, 22 pp.

Naylor, P.J. (1982) The Mawstone Mine Tragedy of 1932, in **Bulletin of the Peak District Mines Historical Society** Vol 8, No 3, pp 171-174.

Oakman, C.D. (1979) **The sough hydrology of the Wirksworth-Matlock-Youlgreave area, Derbyshire,** unpubl M. Phil thesis, University of Leicester, England.

Pearson, F. and Edwards, J. Methane entry into the Carsington Aqueduct System, in (1989) **Methane - Facing the Problems** Seminar 1989. Paper 4.3.

Schoell, M. (1983) Genetic Characterisation of Natural Gases. The American Association of Petroleum Geologists Bullentin, Vol 67, No. 12, pp 2225-2238.

29 BIOFOULING OF SITE DEWATERING SYSTEMS

W. POWRIE
Queen Mary & Westfield College, London, UK
T.O.L. ROBERTS
WJ Engineering Resources Limited, Watford, UK
S.A. JEFFERIS
Queen Mary & Westfield College, London, UK

Abstract
This paper describes the effects of biofouling by the iron-related
bacteria Gallionella on site dewatering systems at a number of
construction sites. The extent of the problem and the remedial
measures required in each case are compared and contrasted with
reference to factors such as the nature of the ground, the quality
of the groundwater, the rate of groundwater extraction and the type
of dewatering system employed.
Keywords: Construction Dewatering, Groundwater Control, Biofouling,
Microbiology, Gallionella.

1 Introduction

The essential component of a construction site dewatering system is
an array of wells which is pumped continuously in order to reduce
the local groundwater level so that excavations below the natural
watertable will remain dry and stable (Fig. 1). Dewatering systems

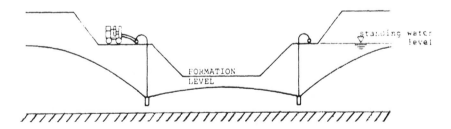

Fig.1. The control of local groundwater levels by wellpoint
dewatering

are normally temporary, being removed once construction is complete.
Dewatering systems are designed primarily on the basis of
geotechnical considerations such as the permeability of the ground
and the magnitude of the drawdown required. The quality of the
groundwater is sometimes investigated in relation to chemical

effects such as corrosion and encrustation, but where it is
necessary to operate the dewatering system for a prolonged period,
microbiological aspects such as the susceptibility of the system to
biofouling become increasingly important. Biofouling of the
wellscreen, filter pack or pipework may reduce both the capacity and
the efficiency of the dewatering system, perhaps resulting in a
gradual recovery of piezometric levels which may threaten the
stability of the excavation unless remedial action is taken.

Although construction dewatering techniques are superficially
similar to groundwater extraction systems for water supply, there
are a number of important differences which will be relevant to
biofouling:

(a) Dewatering systems are often employed in shallow unconfined
aquifers which may receive substantial surface water recharge from
precipitation, rivers, estuaries or even leaking drains. This water
may contain significant levels of dissolved oxygen and/or nutrients.

(b) Electric submersible pumps installed in wells represent just
one method of groundwater extraction. Other pumping techniques
provide different environments which may be more or less
advantageous to microbiological growth.

(c) For economic reasons dewatering systems are designed to
provide the maximum drawdown with the minimum installation cost.
Furthermore, to allow some margin for increased flows, most
dewatering systems are operated on 'snore' - that is, water levels
are drawn down to the intake level, so air is drawn into the pumping
system.

(d) Dewatering systems are often only required for a few weeks.
Occasionally for major schemes dewatering may be necessary for more
than a year. In the experience of the authors it is primarily these
longer-running schemes that may be susceptible to biofouling.

The susceptibility of a given dewatering system to biofouling,
therefore, must be investigated in relation to the ground
conditions, the groundwater quality, the extraction flowrates, the
pumping period and the method of pumping employed.

2 Mechanism of biofouling

The mechanism of biofouling by the iron-related bacteria Gallionella
is described by Howsam (1988). In addition to oxygen and ferrous
iron, certain elements - carbon, nitrogen, potassium and phosphorus
- are required for bacterial growth. These nutrients are brought to
the bacteria by moving water, which also carries away metabolic
wastes. Gallionella derive energy for metabolism from the oxidation
of soluble ferrous iron to insoluble ferric iron and gain carbon by
the fixation of carbon dioxide. The bacteria can tolerate a wide
range of conditions and will grow in waters with temperatures 5 to
30° C and pH 5.5 to 7.5 with iron concentrations as low as 0.1
mg/litre, (Howsam, 1988).

The bacteria attach themselves to a surface and the population
grows to form a biofilm. The biofilm captures organic and inorganic
material and in time builds up a clogging mass made up of soil
particles, detritus, bacteria and iron oxide, which gives the

biofilm its characteristic orange-brown colour. As the biofilm thickens it becomes possible for anaerobic micro-environments to develop beneath it, as the available oxygen is used by the Gallionella at the outer surface. This can result in the rapid corrosion of metal components to which the biofilm is attached, as described by Stott (1988). If sulphates are present in the groundwater, the anaerobic conditions can allow sulphate-reducing bacteria to develop. These bacteria generate sulphides which are very corrosive to cast iron and steel. Stainless steels may be rendered susceptible to attack by chloride ions because the passive oxide layer which normally prevents corrosion cannot re-form in the anaerobic environment beneath a biofilm of Gallionella.

Since many groundwaters appear to contain nutrients in sufficient quantity, the development of the biofilm is generally controlled by the availability of ferrous iron, oxygen and the presence of the appropriate bacteria. The difficulty of identifying a threshold concentration of soluble iron below which biofouling will not occur is illustrated by the diversity of values quoted in the literature and summarised in Table 1. This is partly because the severity of the biofouling will depend on the flowrate as well as the iron concentration, and also because the acceptable level of biofouling from a geotechnical standpoint may vary substantially for different installations.

3 Characteristics of dewatering systems with reference to their potential for biofouling

3.1 Submersible pump system
This is the standard system, in which an electric submersible pump is placed down the well. For dewatering purposes typical operating parameters are: well depth, 10 to 30m; liner diameter, 150 to 250mm; extraction flowrate, 60 to 1200 litres/min/well; number of wells in array, 2 to 30. The distinctive feature of this arrangement is that the top of the well is not sealed, so air has access to the well. In dewatering installations water levels are often drawn down close to the pump intake, resulting in the aeration of the (normally anaerobic) groundwater. These conditions are ideal for Gallionella growth if the groundwater provides sufficient nutrients, ferrous iron and initial bacteria to infect the well.

With time, a biofilm may build up both outside and inside the pump and also inside the riser pipe and the discharge main. Apart from the potential restriction in the flow capacity of the pumping system, the development of an anaerobic micro-environment beneath the biofilm can promote the corrosion of the pump, or the pump may simply sieze as the impeller becomes clogged by biomass.

3.2 Wellpoint system
In wellpoint systems a self-priming vacuum pump is used to extract water from a number of wellpoints, typically up to 100, connected to a common 150mm suction main. The use of the vacuum means that wellpoint systems are only effective for drawdowns of less than 6m

Table 1. References to soluble iron concentrations in groundwater

Author	Concentration of iron in groundwater mg/litre	Comment
Bell & Cashman (1985)	0.5	Potential for encrustation
Driscoll (1986)	1-25	Usual range of concentration
Howsam (1988)	0.1	Concentration for bacteria growth
Loughney (1986)	5	Cleaning of ejectors every six months
	> 5	Regular cleaning of ejector systems necessary
Powers (1981)	0.5	Noticeable signs of encrustation
	2-3	Encrustation may be severe
	20	Upper limit of soluble iron concentration
Raghunath (1982)	2	Encrustation may occur
Walton (1970)	0.01-10	Usual range of iron concentration
Wilkinson (1986)	2	Potential for encrustation

depth. Wellpoints are typically spaced 1 to 3m apart and the capacity of individual wellpoints is approximately 60 litres/min.

Despite involvement in more than a hundred wellpoint schemes, the authors have not come across biofouling of a wellpoint system. This must be because the environment within a wellpoint system is not conducive to bacteriological growth and/or because the necessary groundwater quality to promote biofouling is uncommon at shallow depths. With respect to the former, the high vacuum may promote anaerobic conditions which will curtail the development of an iron related biofilm. With respect to the latter, van Beek (1989) describes a case of biofouling of wellpoint filters which apparently occurred as a result of the mixing of very shallow oxygen-containing groundwater with deeper iron-containing groundwater. This confirms that appropriate conditions can exist.

3.3 Ejector system
In an ejector system groundwater is extracted by means of a water jet pump or ejector installed in each well. The ejector is driven by water at a high pressure (typically 700 kPa) which is pumped to a nozzle within the ejector casing. The supply stream is accelerated through the nozzle and emerges at a high velocity and a low pressure, entraining the groundwater, which is mixed with the recirculating water in a venturi and passes up the return pipe to ground level. In order to minimise installation costs the ejector body and return pipe may be placed concentrically in a 50mm bore well-liner, the high-pressure water being fed down the annulus. A cross-section through a concentric pipe ejector is shown in Fig. 2.

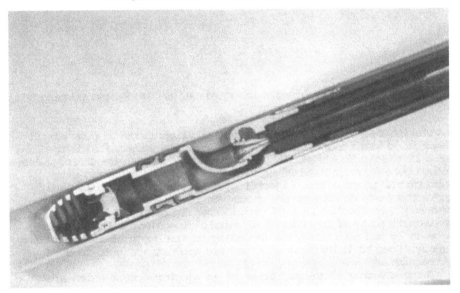

Fig.2. Cross-section through a concentric pipe ejector

Ejectors can be operated economically up to a depth of 30m and a typical system might consist of between 5 and 50 individual wells.

Normally, all of the ejectors in a system are fed by a common high-pressure supply main and discharge into a common return main which is at or near atmospheric pressure. The typical arrangement for an ejector pumping station is shown in Fig. 3. The water is continually recirculated through the system, with the excess (which should be equal in volume to the groundwater extracted during the time period under consideration) being discharged to waste.

Unlike conventional submersible pumps, an ejector can operate at near-zero rates of groundwater extraction without suffering

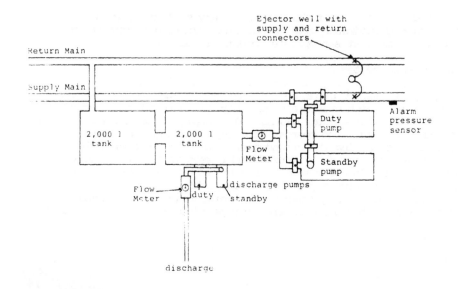

Fig.3. Schematic layout of a standard ejector system pumping station

cavitation damage. Indeed, an ejector will continue to pump air if there is no water available, or will develop a vacuum inside the well if the well is sealed. The vacuum will accelerate drainage and reduce pore water pressures further than would be achieved with submersible pumps alone. Ejector systems are most commonly used where the permeability of the ground is low (less than about 10^{-5} m/s) and the expected rate of groundwater extraction is small. The maximum capacity of an individual single-pipe ejector is typically between 20 and 60 litres/min depending on the depth of installation. Supply flows to individual ejectors are approximately 30 to 50 litres/min.

There are four characteristics of an ejector system which are particularly important in the context of biofouling:

(a) The extracted groundwater is recirculated through a tank and pump to provide the high-pressure supply flow. The recirculating water is therefore able to pick up oxygen.

(b) The continuous replenishment of the recirculating water with groundwater will result eventually in the almost complete replacement by groundwater of the water used to charge the system initially.

(c) Air as well as groundwater may be drawn into the system by the ejectors.

(d) Ejector wellscreens are normally provided with a bentonite clay seal to promote vacuum drainage. Conditions in the well filter pack will generally be anerobic.

The supply and return risers where the biofilm tends to form are comparatively small and therefore susceptible to clogging. A build-up of the biofilm in the return riser reduces the effective cross-sectional area of the pipe and increases the head loss due to friction, resulting in a reduction in the rate at which water can be pumped from the well. If the supply flow is restricted by biofilm growth in the supply riser, the efficiency of the ejector will be impaired. The headermains, supply pumps and recirculating tanks, where velocity and turbulence are generally less than in the riser pipes, tend to remain comparatively free of the biofilm.

4 Case histories

4.1 Submersible pump system

Table 2 gives case summaries of submersible pump dewatering systems at three different construction sites.

Table 2. Case summaries of submersible pump dewatering systems

Site	Soil description	Flow/well (1/min)	Groundwater iron content (mg/1) (range)	Operating period (months)	Cleaning frequency (months)
A	Alluvial sand and gravel	200	15+	16	2
B	Gravel over chalk	500	Not tested	12	Not cleaned
C	Gravel over chalk	330	0.5 (0.2-0.8)	6	Not necessary

The heaviest biofouling was experienced at Site A, where it was found necessary to lift each pump for cleaning at intervals of approximately two months. The biomass accumulated principally on the pump itself, but was easily removed by hosing with clean water.

At Site B, the dewatering system continued to perform satisfactorily without regular cleaning, although an orange biofilm was evident at the tops of the wellrisers and in the discharge lines. It was necessary to replace one pump after 9 months because the accumulation of biomass in the volute (impeller housing) was so severe that it caused the impeller to seize. On decommissioning the system, all of the pumps were found to be heavily coated with the orange biofilm. Again, the biofilm was soft and was easily removed by hosing the pumps with clean water at normal mains pressure. Considerable chloride corrosion of the stainless steel volutes

(Grade 304) had occurred in all of the pumps, with the result that the metal had become quite porous. Both sites A and B were adjacent to estuaries and the dewatering systems were pumping brackish water with a significant chloride content.

The dewatering system at Site C - which is included for comparison - is still in operation pumping fresh water: there has been no sign of iron biofouling to date.

4.2 Ejector System
Case summaries for ejector well dewatering systems at three sites are given in Table 3.

Table 3. Case summaries of ejector dewatering systems

Site	Soil description	Flow/well (1/min)	Iron content (mg/l) (range)	Operating period (months)	Cleaning frequency (months)
D (i)	Laminated silts over bedrock (Glacial lake	4	7.9 (1.8-21.0)	18	Not necessary
D (ii)	deposits over Ordovician Bedrock	20	7.9 (1.8-21.0)	18	0.5-1
E	Silty fine sand (Bracklesham Beds)	15	Not tested	4	4
F	Silty fine sand (Bracklesham Beds)	0.5	3.8 (1.9-5.6)	4	Not necessary

The high-flowrate ejector system at Site D was particularly susceptible to biofouling, with the biofilm accumulating in both the supply and return risers. As a result, pore water pressures began gradually to rise shortly after the system had been installed. It was necessary to clean the plastic well risers for the first time after about 7 weeks; thereafter, cleaning was required at intervals of 3-4 weeks on average. The biofilm remained soft, and during cleaning was simply wiped from the riser pipes with a damp soft cloth. The ease with which the biofilm could be removed was in interesting contrast to its ability to adhere to the walls of the riser pipes, which had water flowing at a velocity of up to 2.7 m/s.

In the low-flowrate ejector system at Site D, biofouling was apparently confined to the return riser pipes. Although the biofilm did not become thick enough to impair the performance of the low-flow ejector wells, it was necessary to clean the return risers of the few high-flow ejectors in this system at intervals of 4-6 weeks. Biofouling was also evident in the low-flowrate ejectors at Site E, and cleaning would have been required after four months had the system not then been decommissioned.

At site F, a very thin rust-coloured deposit was observed on the walls of the supply annulus, while the return riser seemed clear. Within a few days of the commencement of pumping, the recirculating water had become very cloudy and discoloured, but there were no adverse effects on the performance of the system. After about two months, the recirculating water was accidentally drained and had to be replaced with fresh water, which stayed comparatively clear for the remainder of the pumping period. It may be that in this case iron-related bacteria were not present, despite the high iron content of the groundwater.

5 Discussion

In all cases where the biofouling of the dewatering system was encountered, the biomass had accumulated in the pumps and pipework rather than at the wellscreen or in the filter pack. This is probably because the oxygen required by the bacteria Gallionella was introduced by the pumping device. It may be, however, that none of the dewatering systems was operated for a long enough period for clogging of the well itself to become apparent; in the cases reported by van Beek (1989) the specific capacity (ie the volumetric extraction flowrate per unit drawdown) did not begin to deteriorate until the equivalent of at least 18 months continuous pumping had taken place. (It may be noted, however, that van Beek does not mention that periodic cleaning of the pumps or pipework was necessary).

In both submersible pump and ejector dewatering systems, the biomass (if it developed) built up in areas of high turbulence and flow velocity, at or close to the borehole pumping device, where air was entrained and mixed with the extracted groundwater. In submersible pumps the biomass accumulated on the pump casing and in the volute, where in one extreme case it caused the seizure of the impeller. Since the diameter of the pipework was fairly large (75mm), the growth of the biofilm inside the discharge line was of only secondary importance.

In the ejector systems, however, the inside diameter of the return riser was small (30mm) and in some cases the development of a biofilm resulted in a serious reduction in the rate at which groundwater could be removed from the well. This effect was unimportant in terms of geotechnical performance where the groundwater extraction flowrates were low, but would have been disastrous in the high-flow system if remedial action had not been taken. Furthermore, the biomass was found to grow more quickly in the high-flow system, presumably because the supply of nutrients increases with both the flow velocity and the rate at which the system is replenished with fresh groundwater.

Although both the submersible pumps and the ejector well risers were cleaned manually, chemical treatments were considered. The addition of a biocide to the recirculating water in an ejector system would kill live bacteria, but would not remove the biofilm, which consists of dead cells, iron oxide and detritus. Thus an infested system would need to be dosed with biocide immediately after cleaning, and then at sufficiently frequent intervals to prevent the re-establishment of the biofilm. This would not have been feasible given the quantities of biomass and water involved and the difficulty of disposing of the groundwater once it had been contaminated with a biocide.

It is apparent that Gallionella thrive at an anaerobic/aerobic interface. This is because they require both ferrous iron and oxygen for respiration. Ferrous iron will be reduced to ferric iron naturally in the presence of oxygen. Thus the appropriate conditions will only occur close to an interface where anaerobic ferrous iron containing groundwater is mixed with atmospheric oxygen and/or oxygen containing groundwater. Gallionella also appear to thrive in turbulent conditions. This is perhaps because the high turbulence facilitates the transfer of oxygen and ferrous iron into the biomass.

Finally, it should be noted that it is not easy to obtain a reliable analysis of the iron content of a representative sample of groundwater, as indicated by the ranges of iron contents given in Tables 2 and 3. Where the dissolved iron content is high, a false result will be obtained if iron is allowed to precipitate out of solution between sampling and testing. Measures which should be taken to guard against this include filling completely and sealing the sampling jar; inspecting the contents for signs of undissolved iron immediately after sampling; and determining the total iron content of the sampling jar as well as the dissolved iron content of the water. A further difficulty is that the groundwater sampled prior to the installation of a dewatering system may have a different iron content from the groundwater being extracted after a prolonged period of pumping. Clearly once biofouling has occurred, the results of chemical analysis of the discharge water will be very difficult to interpret.

6 Conclusions

(1) The occurrence of biofouling in site dewatering systems is generally rare. It should be noted that iron-related bacteria are recognised as the most virulent form of biofouling of well systems. Other forms of biofouling include manganese-related bacteria, which precipitate manganese hydroxide, and sulphate-reducing bacteria, which precipitate iron sulphides. These rarer forms of biofouling are seldom encountered in dewatering systems because of the comparatively short timescale of most dewatering operations.

(2) There is evidence to suggest that the occurrence of biofouling is to some extent method-related, with systems which draw air and/or recirculate water (ie submersible pumps and ejectors) being more likely to become infested than systems which do not (ie wellpoints). The various systems are not, however, interchangeable, and geotechnical considerations will usually dictate the method which must be adopted for a given situation.

(3) It is difficult to predict whether biofouling will occur in a given dewatering system and if it does, the extent to which it will be troublesome. Factors which must be taken into consideration include the type of dewatering system; the anticipated rate of groundwater extraction; the iron content of the groundwater; and the presence of bacteria such as Gallionella. Based on the case histories outlined in this paper, however, tentative guidelines are summarized in Table 4.

Table 4. Tentative trigger levels for susceptibility to Gallionella biofouling

---

(a) Submersible pumps

Iron Concentration (mg/litre)	Frequency of cleaning
<5	6 to 12 months
5-10	0.5 to 1 month
>10	weekly (system not viable?)

(b) Ejector systems

Iron Concentration (mg/litre)

High Flow (>20 l/min/ ejector)	Low Flow (<10 l/min/ ejector)	Frequency of cleaning
<2	<5	6 to 12 months
2 to 5	5 to 10	Monthly
5 to 10	10 to 15	Weekly, system may not be viable

(c) Wellpoints

Biofouling unlikely to present difficulties under normal operating conditions and timescales of less than 12 months. Above 10mg/litre some caution should be exercised.

---

7 References

Bell, F.G. and Cashman, P.M, (1985) Groundwater control by groundwater lowering. Groundwater in Engineering Geology (Eds J.C. Cripps, F.G. Bell and M.G. Culshaw)

Driscoll, F.G. (1986) Groundwater and Wells. 2nd Edition, Johnson Division, Minnesota, USA

Howsam, P. (1988) Biofouling in wells and aquifers. Journal of the Institution of Water and Environmental Management, 2, 209-15.

Loughney, R.W. (1986) private correspondence.

Powers, J.P. (1981) Construction Dewatering. John Wiley, New York.

Raghunath, H.M. (1982) Groundwater. Wiley Eastern, India.

Stott, J.F.D. (1988) Assessment and control of microbially-induced corrosion. Journal of the Institute of Metals. April, 224-9.

van Beek, C.G.E. (1989) Rehabilitation of clogged discharge wells in the Netherlands. Quarterly Journal of Engineering Geology. 22, 75-80.

Walton, W.C. (1970) Groundwater Resources Evaluation. McGraw-Hill, New York.

Wilkinson, W.B. (1986) Design of boreholes and wells, Chapter 11 in Groundwater: Occurrence Development and Protection. (Ed T.W. Brandon), IWES London.

PART EIGHT

DIAGNOSIS, MONITORING AND CONTROL

30 THE ACHILLES EXPERT SYSTEM ON CORROSION AND PROTECTION

D.R. HOLMES
National Corrosion Service, National Physical
Laboratory, Teddington, UK
P. BALKWILL
UKAEA Technology, Harwell, Oxfordshire, England

Abstract
The paper describes the Achilles Expert System on Corrosion
and Protection, a major rule-based expert system currently
consisting of ten individual modules centred on various important corrosion and protection subjects. The topics chosen reflect the commercial interests of the major UK and
international companies sponsoring the development of the
system with additional public funding from the UK Department
of Trade and Industry. The purpose of the system is to provide expert advice for non-specialist engineers and other
technical personnel and it does this in three different
ways:

The choice of corrosion-resistant materials or protective
systems for particular environments or applications
Diagnosis of problems encountered in the operation of materials or protective systems in various corrosive environments
General education and discussion on corrosion resistant
materials and protective systems and on good design to
resist corrosion and degradation

The overall development of the system is described together
with its current contents and the facilities offered. Examples of output from it are given together with its use in
conjunction with a related fully indexed database of corrosion and protection information, ACHLIB, assembled for further consultation by Achilles users. Finally plans for the
wider development and use of the system are briefly discussed.

1 Introduction

The costs of corrosion and of providing protective systems
to resist its ravages are too well-known to warrant extensive discussion here. It is sufficient to point out that
the Hoar report, (HMSO 1971), concluded that corrosion and
degradation of materials was costing the UK economy some 3%

of the Gross National Product (GNP). Similar findings appeared later in comparable surveys carried out in Australia and the United States. Furthermore the Hoar report suggested that about one quarter of these costs could be saved by the proper application of existing knowledge and technology.
The UK Department of Trade and Industry (the DTI) accepted these conclusions and put in hand various initiatives designed to recoup for British firms and organisations some of the savings that appeared possible. One of these was the establishment at the National Physical Laboratory (NPL) of the National Corrosion Service (NCS) to provide advice and work for UK organisations requiring information on corrosion and protection topics. This was followed in 1983 by the formation at NCS-NPL of the National Corrosion Coordination Centre (NCCC). The prime purpose of the NCCC was to ascertain through its technical managers the problem areas in corrosion and protection that were causing concern and unacceptable expense to UK industry and to establish and manage collaborative research projects designed to solve them. The problems tackled have recently been discussed (Holmes et al 1989) and the results from some of the projects have been published (Holmes 1985, Holmes et al 1990, NPL 1988). This paper describes one of the multi-client collaborative NCCC projects currently in progress, the Achilles Project. This project is jointly managed by NCS-NPL and UKAEA Technology, Harwell and is supported financially by funding from the DTI and from a number of large UK and international firms and organisations listed in the acknowledgments. Their contribution, through the provision of both funding and expert advice has been vital to the development and the success of the Project.
The objective of the Achilles Project is to provide a computer based expert system that will provide expert advice for non-specialist technical personnel in three main areas of corrosion and protection:

The choice of corrosion-resistant materials or protective systems for particular environments or applications
Diagnosis of problems encountered in the operation of corrosion-resistant materials or protective systems in various environments
General educational information and advice on corrosion and protection topics

2 Historical development and funding of the project; hardware and software requirements

2.1 Historical development and funding
The initial development of the NCCC Achilles Project started in 1985 as a partnership between NCS-NPL and the Materials Technology Section of UKAEA Technology, Harwell. The latter group had been active for some time previously in developing small prototype expert systems using the Harwell-developed

SPICES expert system shell and publications on (Westcott et al 1986, Wright & Westcott 1987) and demonstrations of their early attempts had aroused industrial interest. This led the DTI, through the NCCC, to propose the formation of a major collaborative industrial project to provide a suite of expert system modules that would help relieve the load on the materials and corrosion experts of the industrial sponsors. Although it was accepted that the corrosion and protection expert could never be replaced entirely by machine based expert systems for all problem areas where expert advice and information was required, it was hoped that the systems to be developed might deal with about 90% of the more routine enquiries and would allow the expert to check his responses against those of a supposedly less fallible, but less perceptive computer expert system. On this basis the sponsoring companies and organisations listed below agreed to support the Achilles Project for an initial period of three years for the construction of various expert system modules on corrosion and protection topics chosen as the consensus view of the collaborators. This generous industrial funding was supplemented by roughly equivalent public funding from the DTI on a decreasing scale. Eventually the majority of the Project Members decided that a fourth year's participation in the Project would be beneficial for their companies and decided to fund this, again with public funding from the DTI. With supplementary seed corn and other indirect funding from both Harwell and NPL the envisaged total expenditure on the Project will be about #400K. The development of worthwhile, useful expert systems is by no means an inexpensive exercise and this sum is being exceeded many times over in other industrial nations!

As experience in the use and development of expert systems has grown during the course of the Project so have the requirements of the industrial sponsors. In the early days while the total information in the system and the facilities available were limited Members carried out consultations directly by telephone links to a host computer at Harwell, but as experience grew and the availability of PC/IBM compatible personal computers became widespread, the Project Members saw the great advantage of having the information immediately available on their own workstations. Now the various Achilles expert system modules can be rapidly consulted by Achilles Members on readily available PCs provided with the hardware and software discussed in the next section.

2.2 Hardware and software requirements
Use of the Achilles expert system is currently based on PCs or more powerful computers with the following facilities:

Hardware

IBM PC, PC/XT, PC/AT, PS/2 or 100% compatible with 640Kb RAM

10Mb (minimum) hard disk or 20Mb (minimum) if use of
ACHLIB is required
1.2Mb (preferred) or 360Kb floppy disc drive:

Software

MS-DOS operating system version 3.0 or higher
PC-SPICES inference engine
Achilles rule base

Additionally, if the user wishes to consult the supplementary free text database of corrosion and protection information, ACHLIB, the following are required:

PC-STATUS/E free text retrieval system, read-only version
(or full PC-STATUS/E if user wishes to add his own information)
ACHLIB library of relevant corrosion and protection information

3   Construction of the Achilles expert system and current
contents

3.1 Compilation of resource digests and construction of expert system modules
The methods for constructing the individual expert system
modules and the type of information in them have changed as
Members' experience and requirements have evolved during the
course of the Project. At first it was envisaged that the
modules would be largely educational and conversational,
rather like a discussion with a real life expert, when the
problem would first be put to him and after a pause for reflection a number of possibilities would be examined at
various depths and eventually the choice would home in on a
particular course of action; in fact in this type of expert
system there would be a strong emphasis on education and
training and the modules produced would resemble the teach-
ing software packages now being marketed by a number of scientific and technological institutions(Institute of Metals
1988). The Achilles Project Members soon decided that they
needed the modules to provide specific technical solutions
for the choice of materials or protective systems for particular operating environments or technical applications;
they also discovered, not surprisingly, that extensive un-
broken text appearing on the screen was unappealing and
unproductive. Instead they found that a more interactive
question and answer structured format supplemented for con-
venience with plentiful multiple choice menus was more user
friendly and allowed them to obtain the information they required more quickly and economically.
    As the Project has evolved we have developed an activity
schedule for the construction of each module, starting with

the choice of topic and the items in it to be covered and
culminating in the issue of the final expert system module
to the Achilles Members. This schedule is shown in Table 1,
where it may be seen that we have found it convenient to di-
vide the overall programme into two phases. Phase 1 is the
compilation of the Resource Digest (RD) of the information
required by a carefully chosen expert while Phase 2 covers
the construction of the expert system (ES) module from it.

Table 1. Activity schedule for construction of expert system module

Activity	Phase
1 Choice of module subject by Achilles Members	
2 Choice by Members of topics to be covered and incorporation in checklist	
3 Choice by Members and Project management of up to 3 experts to be approached	
4 Experts approached and asked to submit bids for carrying out the compilation of the RD, delivery time and checklist of topics to be included	
5 Expert chosen, differences between his check-list and that agreed by Members reconciled by discussion and contract placed	Phase 1
6 Expert submits first 25% of his RD for Project management to check that required material is being covered in succinct, abbreviated way	
7 RD received and reviewed and edited by 2 or 3 Members and Project management	
8 Editorial and review comments sent to expert for incorporation in final version	
9 Final version of RD received and approved by Members and sent to Harwell for construction of ES by knowledge engineer	
10 Ordering of RD material into rules	
11 Construction of logical tables and decision trees	
12 Construction of ES begins	
13 Issue of early version of ES to Members for comment	
14 Incorporation of Member's comments and sugges-tions for improvements and issue of final version	Phase 2
15 RD expert compiler invited for extensive test-ing of ES module to check that all relevant material has been correctly included and in-corporation of his suggestions into final form	

Particular attention is given to item 6 in the compilation
of the RD as we have found that many experts tend to treat

the task as a comprehensive scientific review of their total knowledge of the subject rather than as a highly structured succinct document designed to provide specific responses to specific queries. The latter type of document is essential for satisfactory and economical construction of the expert system module.

3.2 Contents of the Achilles expert system
By the end of the development period the Achilles expert system will have modules on 10 topics of scientific and commercial importance to the sponsors; these are listed in Table 2 with notes on their current state of development. Two of the modules, Corrosion Inspection and Corrosion Monitoring, have been combined with a common initial access path because of their close interrelation.

Table 2. Contents of the Achilles expert system

Module Title
1 Microbial corrosion
2 Cathodic protection
3 Paints and organic coatings
4 Inorganic and metallic coatings*
5 Atmospheric corrosion
6 Marine corrosion
7 Corrosion inspection
8 Corrosion monitoring
9 CO2/H2S corrosion in oil and gas production systems*
10 Aspects of stress corrosion cracking*

* Development still in progress

For module 10 it was accepted that the vast subject of stress corrosion cracking could not be covered adequately by a single expert, so the topics dealt with are mainly limited to those relevant to the Members' commercial interests.

An important adjunct to the Achilles expert system is the free text database or library of relevant information for the various modules, ACHLIB. Because all the text in ACHLIB is fully indexed, any required material can be called up to the screen by the widely-used STATUS/E software, simply by typing any words of interest (the use of specific keywords is not necessary). ACHLIB contains all the text of the RDs on which the expert system modules are based, so consulting it ensures that no relevant information discussed by the expert compiler is missed. Because all the words in the titles of the references in the RDs are also indexed further reading lists can also be provided readily if the enquirer considers it expedient to examine the source material on which the expert advice is based. Examples of the ACHLIB facilities are given in the next section.

Finally it should be pointed out that the Achilles system

contains a very useful Glossary and explanation of corrosion
and protection terms as it was accepted that many of these
words and concepts would be unfamiliar to non-specialist en-
quirers. Where possible these are based on ISO (1986)
definitions but many additional terms have been explained.
These explanations can be called up directly to an inset
window on the screen.
Currently the Achilles expert system requires 3.3Mb of
disc space and ACHLIB 6.7Mb, but both these figures are increasing rapidly as development proceeds.

4 Facilities available in the Achilles expert system and in
ACHLIB

4.1 Achilles expert system
After the expert system has been started up by moving to the
Achilles directory and initiating Achilles the main Achilles
commands available are permanently listed in a band across
the top of the screen as follows:

Back   Help   Last   Stop   Showrule   Status   ? term

All of these commands may be actuated by typing the word or
the underlined letter only.
Back returns the enquirer to the most recent decision point
(a question or multiple choice menu) to allow him to make a
different choice at that point and thus to follow another
line of enquiry from thereon but with the same input data up
to that point
Help This provides an on-screen explanation of all the commands
Last allows the Achilles text corresponding to the currently operating Achilles rule to be displayed again - useful when auxiliary information has been brought to the
screen and the last Achilles text, which may contain a question demanding a response, is required again by the enquirer
to continue the consultation
Stop stops the current consultation and presents the enquirer with the options of restarting the consultation from
the main Achilles menu or of exiting from Achilles
Showrule displays the contents of the rule which is currently being used by the Achilles expert system in its responses
Status calls up the software to allow the interrogation of
the ACHLIB library of relevant corrosion and protection
information
? term searches the Glossary to see if an explanation of
the specified term is available and if so displays it on the
screen. Words may be linked by an underscore into phrases
e.g. stainless_steel provides a brief definition of the 4
types of stainless steel. There is a useful extension of
this command; typing ? library provides an inset list of all

the terms in the Glossary which may be scrolled through to ascertain whether a particular word has been defined.

In addition to these commands which are constantly listed for the enquirer there are three important options which are offered on screen by prompts at appropriate places in the text or menus:

What if is frequently offered as a menu choice. Choosing this allows the enquirer to return to an earlier multiple choice option (menu) and to rerun the enquiry from thereon with new user inputs and the information previously input after that menu obliterated. This allows the enquirer to pursue a different line of questioning.

Why is available for use when the message "For further information to help you answer type Why" appears on the screen simultaneously with a question requiring an answer from the enquirer

More is available for use when the message "For further information type More" or "More" appears on the screen

Besides textual information Achilles also provides graphical information in all the usual forms such as graphs, histograms and charts wherever appropriate.

Finally it should be noted that Achilles automatically logs the whole consultation, including the information input by the enquirer as well as the advice provided. Hardcopy can be obtained after leaving Achilles by typing Hardcopy at the MS-DOS prompt when all the material will be recorded in an Output.txt file which can be imported directly into most word processing software and thus modified or printed as required.

4.2 ACHLIB

ACHLIB is a library of corrosion and protection information and articles related to the subject matter of the expert system modules; it is consulted by using the proprietary STATUS/E software developed by Harwell Computer Power Ltd. It contains the complete text and references of all the material in the resource digests compiled by the expert authorities.

In accordance with normal practice the articles are divided into sections with names that appear in the lefthand margin. The most important of the section names are shown in Table 3 and the whole library can be searched or the search can be restricted to one or many of the sections listed.

ACHLIB is consulted by entering one's requirements into various fields on screen display panels, with each field entry effecting different tasks. The function keys on the computer keyboard are used to change between display panels and the up and down keys are used to change between the fields on each display panel. The various functions allotted to the function keys are displayed continuously in an

Table 3   Section names in ACHLIB articles

Section name	Content
Title	Article title
Reference	Article origin
Author	Article author(s)
Abstract	Article abstract
Text	Full text of article
Conclusions	Article conclusions
Rules	Index of Achilles rules derived from article
Introduction	Article introduction
Bibliography	Article reference list
Comments	Information supplied by ACHLIB compiler
Table	Tabulated data in the article
Material	Materials considered or measured in the article
Observation	Experimental observations in the article
Design	Design matters considered in the article
Data	Numerical data given in the article
Environment	Data on the environment(s) considered in the article

inset reminder panel.

A search for information is initiated after STATUS/E has been used to call up ACHLIB by entering the subject of the search in the query field. For example to find all articles containing the word steel, steel would be entered. If we wished to restrict the search to articles containing the words stainless steel in juxtaposition then stainless steel would be entered. In a similar way entering steel + cracking would produce all articles mentioning both steel _and_ cracking, not necessarily in juxtaposition. Extending the facilities further, an entry steel, cracking will produce all articles which mention either steel or cracking or both, while articles containing the word cracking with either steel, brass or aluminium can be retrieved by entering cracking + (steel, brass, aluminium). The number of relevant articles is indicated and this can be reduced or increased as required by narrowing or widening the search area. The various entries and function keys are then used to display all or some of the titles of the articles, followed by display of the whole article chosen or any of its sections named as in Table 3 above.

5   Examples of Achilles and ACHLIB consultations

A written article such as this is a poor substitute for a

personal session on the PC consulting Achilles or ACHLIB
with a real enquiry, so we hope that its content will induce
some conference attenders to sit at the keyboard and try
out some of their problems on this PC-based expert system.
Figs 1-3 provide examples of the screens of information obtained in an enquiry to predict the possibility of occurrence of microbial corrosion in an aluminium alloy aircraft
fuel line delivering gasoline (petrol) jet fuel. Fig 4
shows a Glossary explanation of the term "cathodic protection" called up to the screen during a consultation on microbial corrosion.

Similarly Figs 5-7 show examples of screen displays obtained from an ACHLIB search to find what information is
available on the corrosion of reinforced concrete in
seawater.

6 Future development of the Achilles system

One particularly advantageous feature of the Achilles system is that it employs the SPICES expert system shell developed in house at Harwell and written in the widely-used
PROLOG programming language. Thus SPICES can be readily extended to incorporate or import from other software any
novel features needed by current or future users of Achilles
as their experience and requirements grow. We have already
capitalised frequently on this feature during the development of the Achilles system to its current level of sophis-
tication.

Finally we point out that it has always been an objective
of the companies and organisations funding the Achilles
Project that development of the system should continue with
the addition of more modules and regular review, updating
and improvement of existing modules. It is therefore our
intention that after the initial development phase described
here, marketing and selling of individual modules or of the
entire system should become the responsibility of a commercial software publisher, with part of the profits from sales
allocated to this objective. Negotiations are in progress
with suitable publishing houses for this purpose.

7 Acknowledgements

The authors wish to thank the Department of Trade and Industry and the Members of the Achilles Project listed below for their funding of the project and for their invaluable technical support during the development of the Achilles expert system.

Achilles Project Members:

British Gas
BP Research
Conoco (UK) Ltd
Department of Energy
Department of Trade & Industry (through NCS_NPL)
Esso Engineering (Europe) Ltd
Imperial Chemical Industries
Ministry of Defence
Occidental Petroleum (Caledonia) Ltd
Statoil of Norway
UKAEA Technology, Harwell Laboratory

8 References

Dewpoint Corrosion (1985), Holmes, D.R. Ed, Ellis Horwood, Chichester. ISBN 0-85312-839-1
Holmes, D.R. & Mercer, A.D (1989) Br. Corros. J., 24, 86-87.
Holmes, D.R., Strang A. & White C.H. (1990) in Diesel Engine Combustion Chamber Materials for Heavy Fuel Operation, Proceedings of the Institute of Marine Engineers, London. ISBN 0-907206-23-9
International Standards Organisation (1986) ISO 8044-1986. Terms for corrosion of metals and alloys. Also available from the British Standards Institute as BS 6918:1987
Institute of Metals (1988) The IoM Publications List 1988/89, IoM, London 1988.
NPL (1988) One day meeting on "Corrosion of Electrical Contacts: Progress in Environmental Testing, NPL, Teddington, November 29, 1988
Report of the Committee on Corrosion and Protection, (1971) HMSO, London.
Westcott, C., Williams, D.E., Bernie, J.A., Croall, I.F. and Patel, S. (1986) Corrosion '86, Paper No. 54, NACE, Houston.
Wright, E.G. & Westcott C. (1987) Paper presented at the Seminar, "Internal Corrosion Control and Monitoring in the Oil, Gas and Chemical Industries", London, 3-4 March 1987

ACHILLES V3.6: Back Help Last stoP showRule Why - sTatus - ? term
                      MICROBIAL INDUCED CORROSION

Which ONE of the following would you like to consider:

(1) Prediction of the occurrence of MIC in your chosen system
(2) Description of the remedies for MIC relevant to your system
(3) Assessing whether a failure was due to MIC
(4) MIC summary: mechanisms, phenomenology, organisms, prevention and control

(5) MIC of engineering alloys
(6) MIC: Organisms and mechanisms
(7) MIC prevention and control
(8) Detection and monitoring of MIC
(9) Prediction of MIC
(10) Documentation and other general factors associated with MIC

(11) WHAT IF: erase previous user inputs
(12) RETURN to the main ACHILLES menu

Select one option from 1-12
Achilles> 1

Fig.1. Opening menu of Achilles Microbial Corrosion expert
system module.

ACHILLES V3.6: Back Help Last stoP showRule Why - sTatus - ? term

How would you describe your environment ? Select at least one option from:

(1) Aqueous
(2) Seawater
(3) Soil
(4) Concrete
(5) Air
(6) Vapour
(7) Gaseous
(8) Organic (liquid)
(9) Steam

Select at least one option from 1-9. Separate multiple answers with a comma
Achilles> 8

Fig.2. Display during consultation on prediction of microbial corrosion of aluminium alloy jet fuel line.

I need to determine what general classes of alloy you are considering.
Please select at least one option from:

(1) Carbon steel - including high strength and low alloy steels
(2) Cast iron - including ductile and austenitic iron
(3) Stainless steel - including duplex, ferritic, austenitic etc
(4) Aluminium alloys
(5) Copper alloys
(6) Lead
(7) Other class of alloy

Select options from 1-7. Separate multiple answers with a comma.
Achilles> 4

   Is the organic phase petrol (gasoline) ? (Y/N)
Achilles> y

The main concern is the M.I.C. of ALUMINIUM aircraft FUEL tanks by
fungi and bacteria, mainly Cladosporium resinae and Pseudomonas
aeruginosa. Biofilm deposits and growth of these organisms at the
water/fuel interface, occurs resulting in pitting corrosion. The pH
of the system when monitored is in the range 3-4. Volcano-shaped
tubercles often develop. Recognition of the problem is confirmed when
a pink colour is observed. This indicates the presence and activity
of the fungus Cladosporium resinae.
Achilles>

Fig.3. Final outcome of consultation on prediction of microbial corrosion of aluminium alloy jet fuel line.

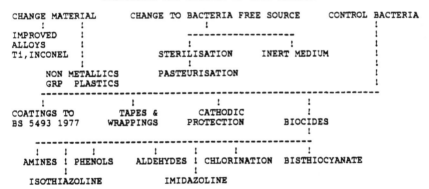

                PREVENTION AND CONTROL OF MIC: METHODS

CHANGE MATERIAL   CHANGE TO BACTERIA FREE SOURCE   CONTROL BACTERIA
    !   !   !   !
IMPROVED   !   --------------------   !
ALLOYS   !   !   !   !
Ti,INCONEL !   STERILISATION   INERT MEDIUM   !
    !   !
    NON METALLICS   PASTEURISATION   !
    GRP PLASTICS   !

    !   !   !   !
COATINGS TO   TAPES &   CATHODIC   !
BS 5493 1977   WRAPPINGS   PROTECTION   BIOCIDES
                                                   !

  !   !   !   !   !   !   !
AMINES ! PHENOLS   ALDEHYDES ! CHLORINATION BISTHIOCYANATE
  !   !
  ISOTHIAZOLINE   IMIDAZOLINE

Achilles> ? cathodic protection
The corrosion rate of components and structures can be reduced by
applying cathodic protection.Three methods are employed:
-Impressed Current: a DC generator is used to maintain a current flow
                    between the item to be protected and an inert anode
-Sacrificial Anode: the protected item is electrically connected to a
                    consumable anode
-Hybrid = impressed current + sacrificial anode
Achilles>
                                      Press <RETURN> or enter a command

Fig.4. Use of Glossary to provide information on cathodic protection during consultation on microbial corrosion.

```
Search-1 ─────────── STATUS Information Management System ──────────────
Command :

Textbase : ACHLIB

Chapters : ALL

Sections : titles

Question : reinforced concrete + corrosion





Retrieved items: 4

F1 Help F2 Info F3 End  F4     F5 Display                          F10 Set-up
```

Fig.5. Initial display panel for searching ACHLIB for information on corrosion of reinforced concrete in seawater.

```
───────────────────────STATUS Information Management System──────────────────
Titles                                 Textbase - ACHLIB            Row 1/14 Col 1

*  1.
TITLE
ASSESSMENT AND PREDICTION OF MIC.
   2.
TITLE
   INTRODUCTION TO MARINE CORROSION
   3.
TITLE
4.6  Non-Metallic Materials Behaviour in Marine Environments
   4.
TITLE
ACHILLES ATMOSPHERIC CORROSION MODULE

5   MATERIALS
                              *** End Of Text ***



F1 Help F3 Quit                                F7 Up F8 Down F9 Left F10 Right
```

Fig.6. Display of results of ACHLIB search on titles of articles on seawater corrosion of reinforced concrete.

```
-----------------------STATUS Information Management System-----------------------
Display 3/4                      Textbase - ACHLIB              Row 41/525 Col 1

4.6.2.2     Seawater Chemical Attack on Concrete

The sulphate salts present in seawater can attack the constituents of OPC,
particularly the tri-calcium aluminate (3CaO.Al2O3, C3A), with the expansive
growth of the mineral ettringite, 3CaO.Al2O3.3CaSO4.30-32H2O.  However,
structures in a marine environment normally show no evidence of the expansion
and cracking observed on land based structures, as ettringite is significantly
soluble in seawater and can be leached from the concrete.  In addition, the
presence of sulphate as magnesium sulphate in seawater breaks down the basic
cementitious structure by attacking the calcium silicate component, causing
leaching and softening rather than cracking.
The lesser occurrence of sulphate attack on marine structures is also believed
to be due to the preferential reaction of the C3A with chlorides in the
seawater.

4.6.2.3     Physical Deterioration

The most common form of physical deterioration is damage caused by freeze-thaw
cycling.  This occurs when water within the concrete expands on freezing (9%
volume increase) causing disruption.  The damage caused by cyclic freezing and

F1 Help F3 Quit          F5 Prev art F6 Next art F7 Up F8 Down F9 Left F10 Right
```

Fig.7. Information display on outcome of ACHLIB search for information on seawater corrosion of reinforced concrete.

31 CORROSION MONITORING AND INSPECTION

C.F. BRITTON
Corrosion Monitoring Consultancy-CMC, St. Neots, UK

Abstract
This paper describes the various techniques for monitoring and inspection of corrosion in process plant and structures. Also presented is guidance as to the many factors requiring consideration whenever a monitoring and/or inspection programme is undergoing design.
Keywords: Corrosion, Monitoring, Inspection, Process Plant, Structures.

1 Introduction

Corrosion is a major problem in many industries - the cost of corrosion has been estimated for a number of countries at around 3-4% of the GDP, representing a considerable drain on the economy and a waste of scarce natural resources.

The industries most severely affected by corrosion are those that lose production due to unscheduled shutdown. For oil and gas production, oil refining and chemical manufacture the economic penalties for unscheduled shutdown are particularly severe. It is these industries therefore, that have provided the major stimulus for development of corrosion monitoring techniques and associated hardware and are, therefore the main customer-base for those engaged in the manufacture of monitoring equipment.

Many other industries can also be subject to major corrosion problems - some of these may affect structural stability with consequent loss of life, injury and economic penalty to the local community. Transport and civil engineering are particular examples, therefore, methods for monitoring and inspection of corrosion are of particular interest to these industries as well, although methods and applications may be more specific depending on the actual situation.

Methods for routine corrosion monitoring in process plants include coupons, appropriate on-line electronic sensors or multi-sensors and a method of non-destructive testing such as ultrasonics. Other techniques can be called upon to supplement the above methods as required - examples are radiography, eddy current and acoustic emission. There are a number of questions to be answered whenever corrosion monitoring is under consideration eg. the objective of the programme, time-response required, kind or form of corrosion expected, (as there are many different forms of corrosion),the technology culture

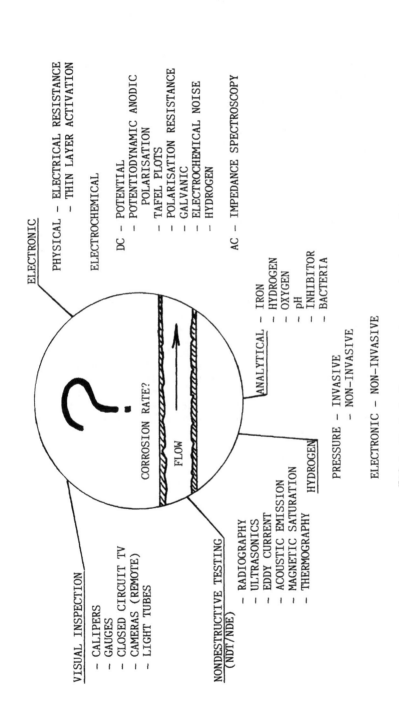

FIG.1. Methods for corrosion monitoring and inspection.

of personnel involved and application of chosen methods.   The
techniques available for corrosion monitoring and inspection are
shown in Fig. 1.   Corrosion sensing systems which produce real-time
electronic signals are available, allowing process conditions to be
correlated with 'corrosivity' signals.   Process conditions which are
'upset' resulting in an increase in corrosivity can be detected and
corrective action can be taken without delay.   This approach is
relevant to process streams, pipelines or piping systems carrying
fluids (industrial cooling water systems).   This approach is not
amenable to structural monitoring other than to provide a warning
that corrosion is taking place, and hopefully, the rate of corrosion
over a prescribed time period.
    The individual methods will now be described in more detail.
Only brief descriptions of the various techniques are given here –
additional information can be obtained from publications listed in the
bibliography.

2 Coupons

The use of metallic coupons provide base-line for any corrosion
inspection programme.   The coupons used are specimens of the metal of
interest (generally 76.2 x 25.4 x 3.2 mm), drilled with a supporting
hole, surface treated and mounted on a suitable holder for exposure to
the process stream.   Before exposure, the coupons are weighed and the
surface-area recorded.   After exposure the coupons are chemically
cleaned (involving either an acid or electrolytic bath treatment) and
re-weighed.   The weight loss is converted into a penetration rate
expressed as either mils (1 mil = 0.001 in) or mm y^{-1}.   Various
procedures and recommendations for coupon preparation, cleaning and
reporting of results are published by the National Association of
Corrosion Engineers (NACE) in the US and by ASTM.   The accuracy of
the results depends on good management regarding the laboratory treatment of the coupons and eventual data presentation.   Note the data
obtained will be an average corrosion rate over the period of the
coupon exposure.   Coupons provide the base-line data for comparison
with any realtime methods used.

3 Electrical resistance (ER)

A metal loop, similar to a hair pin, made of the metal of interest is
mounted in a holder called a probe and exposed to the process stream.
As corrosion occurs, the cross-sectional area of the loop is reduced
and its electrical resistance is consequently increased.   The change
in resistance can be monitored on a portable instrument, or the probe
can be permanently connected to a control room instrument or data
handling equipment;   equipment is available that can handle 30–40
probes.   Other elements can be used in place of the loop-strips,
tubes or flush strips – depending on the application, the sensitivity
required and the duration of the test.   An automatic instrument and
associated sensor (probe) are shown in Fig. 2.   A portable
instrument is shown in Fig. 3.
    This system can be used in any liquid or gaseous environment and

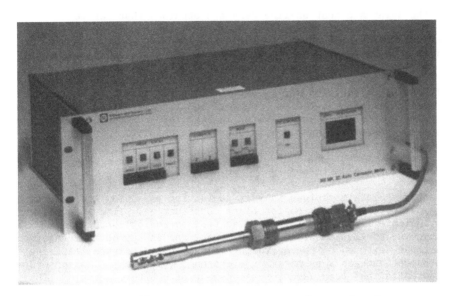

Figure 2. ER automatic instrument and associated
sensor (probe) (courtesy SSL Ltd.)

Figure 3. ER portable instrument (courtesy Cormon Ltd.)

has the advantage that on-line data are obtained that can be correlated with process conditions or any methods used to mitigate the corrosion.   The method averages the corrosion rate in a similar way to coupon exposure.   The increase in electrical resistance can be related to the reduction in diameter of the element due to corrosion which can be expressed as a penetration rate in the same units as given for coupons.   Special ER probes have been designed for sheet piling, car bodies, concrete and for aircraft.   This method has also been used to assess water ingress into concrete harbour installations and to assess atmospheric corrosivity.   Industrial instrumentation is available.

4 Electrochemical methods

Potential - This technique is very useful for assessing the conditions of a number of metals used for plant construction.   The information can indicate if the metal is passive (non-corroding) or active (corroding).   Equipment is simple, requiring an appropriate reference electrode (exposed to the corroding medium) and a voltage measuring instrument such as a pH meter or electronic voltmeter.   Measurement of potential has been used for assessment of concrete deterioration (reinforcement corrosion), and is called 'potential mapping'.
Polarisation resistance (LPR) - This method is of much interest as it can respond instantaneously to a change in corrosivity.   A sensor or probe represents (on a macro scale) the actual corrosion cell on the metal surface of interest, and by measuring the potential and the current an estimate of the actual corrosion current can be made.   The method is a simplification of laboratory electrochemical techniques used for studying corrosion reactions and is confined to liquids that meet certain conductivity requirements.   DC voltages are utilised. Industrial instrumentation is available.
Galvanic corrosion - This measurement can assess the possibility of corrosion between two different metals.   A probe containing two electrodes, where each electrode is the metal in question, is exposed to the corroding liquid.   A zero resistance ammeter is used to measure the current produced by the galvanic cell.   The magnitude of the current indicates the tendency for galvanic corrosion to occur if the two metals are coupled together and are exposed to a corroding liquid.
AC Impedance Spectroscopy (ACIS) - The use of LPR can be subject to limitations and errors.   The latter can arise because the 'resistance' of the electrode reaction is frequency dependent with capacitive components.   The impedance arising has the following components -

        corrosion
        ohmic resistance
        capacities - diffusion processes
                    - films on electrode

Measurements at a single frequency are unable to separate all the components outlined above - hence development of ACIS made possible by recent rapid developments in electronic technology and instrumentation

374

A 'test' electrode is analysed over a wide frequency spectrum enabling values to be obtained for the reaction impedance Rp, solution ohmic resistance Rs, diffusion impedance and capacitance. Measurements using ACIS have been made in concrete and other 'resistive' corrodents.
Electrochemical Noise (ECN) - This is a non-perturbation method in which the natural random current/voltage fluctuations between two identical electrodes (or test electrode and reference electrode), are measured. The voltages are characteristically of the order of a few millivolts. Statistical analysis (FFT or MEM), can provide quantitative data as to pitting and general (uniform), corrosion. For passive (non-corroding), situations the potential remains essentially constant where any fluctuations are slow and have a long-term nature. The initiation of localised corrosion is shown by sharp changes in potential. With aggresive conditions these events (potential changes), become more frequent. This method has been successfully utilised in detecting upset operating conditions in power station equipment.

5 Thin Layer Activation (TLA)

Coupons or actual structural components can be bombarded with protons or deuterons in an accelerator. Loss of material can be directly monitored by detecting the loss of activity from the small area treated (few square mm). A necessary condition for this technique is loss of corrosion product. The reduction in activity can be directly related to metal loss. Activity levels are low, and surface properties of the treated component are not affected (only one atom in 10^{10} is converted to ^{57}Co). The method is not dependent on the corrosive environment, is non-invasive and is cost competitive with other methods for corrosion monitoring.
    TLA treated coupons have been utilised in industry and can now be obtained commercially. The method has been applied in a subsea oil installation (Fig. 4), and in various process plant.

6 Nondestructive testing (NDT)

NDT comprises a number of methods of which ultrasonics, eddy currents, radiography, acoustic emission and thermography are the most important as far as corrosion is concerned.
    The most common method is ultrasonics and is frequently used to detect metal loss in industrial processes. A beam of ultrasound is transmitted through the pipewall. The beam is reflected at the inner surface and is picked up by the receiver part of the probe. The 'time of flight' can be related to metal thickness using the appropriate factor for the metal in question. The thickness measurement can be displayed in digital form. Problems associated with this technique are (i) coupling of the probe to the metal surface (as ultrasonic waves are reflected by a metal/air or liquid interface) if oxide layers (rust) are present and (ii) high temperatures can destroy the piezoelectric properties of the crystal used in the probe. Results can also be displayed visually, a technique used for more

Figure 4.   TLA monitoring system for subsea
application (courtesy Cormon Ltd.)

specialised requirements such as the detection of pits or other forms
of localised corrosion. A number of specialised ultrasonic systems
designed for specific applications of corrosion measurement are now
available. A device for inspecting crude oil storage tanks is shown in
Figure 5. NDT methods are called upon for special situations as they
occur, e.g. eddy currents are used to detect corrosion inside con-
denser tubes (offstream), in non-ferrous materials (ultrasonics can be
used for this as well) and radiography for corrosion detection in
pipes and process vessels.

Special instrumentated devices called 'pigs' are passed through
pipe lines for corrosion inspection - these are available using magnetic saturation, ultrasonics or eddy currents. NDT techniques are
widely utilised in a wide spectrum of industry.

7 Microbiology

Severe corrosion damage can occur due to the presence of bacteria.
Separate papers are included on this subject at this conference and

Figure 5. Crawler 'Beetle' inspecting crude-
oil storage tank. (courtesy RTD BV).

will not be covered therefore in this paper, other than to emphasise the work carried out by the Corrosion Engineering Association Work Group which has produced a document relating to current practice.*

8 Miscellaneous devices

There are a number of specific developments for corrosion monitoring/inspection that have been developed for specific purposes. Examples include the 'washer-stack' which can measure the expansion due to production of oxide films which is used to detect atmospheric corrosion. Other examples include the 'laser punch' developed for measuring corrosion in nuclear reactors.

9 Bibliography

Britton, C.F. (1976) Corrosion. 2nd. Edit. Chapter 20.3
    'Corrosion Monitoring and Chemical Plant' (ed L. Shreir),
    Newnes-Butterworth, London.
ibid (1990) 3rd. Edit. Chapter, Corrosion Monitoring and Inspection.
Britton, C.F. and Tofield, B.C. (1988) Effective Corrosion
    Monitoring. Materials Performance, 27, 41-44.
Wanklyn, J. (Ed) (1982) Proc. Conference 'Corrosion Monitoring in
    the Oil, Petrochemical and Process Industries', Oyez Scientific
    and Tech. Services Ltd., London.
Industrial Corrosion Monitoring, (1978) Dept. of Industry,
    HMSO, London.
Proc. Conference, (1984)'Corrosion Monitoring and Inspection in the
    Oil, Petrochemical and Process Industries' Oyez Scientific and
    Tech. Services Ltd., London.
Proc. Symposium, (1984) 'Corrosion Monitoring in Industrial Plants
    Using Nondestructive Testing and Electrochemical Methods,
    Sponsored by ASTM Committee E-7 and G-1, ASTM Special Technical
    Publication 908.

---

*Task Group E1/1 Review of Current Practices for Monitoring
Bacteria Growth in Oilfield Systems (1987).

INDEX

This index has been compiled using the keywords provided by t
authors of the individual papers, with some editing and addit
to ensure consistency.   The numbers refer to the first page o
the relevant paper.